The Dictionary of
CELL
BIOLOGY

The Dictionary of

CELL
BIOLOGY

EDITED BY

J. M. Lackie AND

J. A. T. Dow

AUTHORS

C. J. Brett

A. S. G. Curtis

J. A. T. Dow

J. G. Edwards

J. M. Lackie

A. J. Lawrence

G. R. Moores

*At the Departments of Cell Biology and Botany,
School of Biology, University of Glasgow,
Glasgow, G12 8QQ, UK.*

ACADEMIC PRESS

Harcourt Brace Jovanovich, Publishers

London   San Diego   New York   Berkeley
Boston   Sydney   Tokyo   Toronto

ACADEMIC PRESS LIMITED
24/28 Oval Road
London NW1 7DX

United States Edition published by
ACADEMIC PRESS, INC.
San Diego, CA92101

British Library Cataloguing in Publication Data
The dictionary of cell biology.
 1. Organisms. Cells
 I. Lackie, J.M. (John Michael)
 II. Dow, J.A.T.
 574.87

 ISBN 0-12-432560-2
  0-12-432561-0 pbk.

This book is printed on acid-free paper. ∞
Typeset in Northern Ireland by Textflow Services Ltd, Belfast
Printed in Great Britain by Mackays of Chatham plc, Chatham, Kent

Preface

The stimulus to write this dictionary came originally from our teaching of a two-year Cell Biology Honours course to undergraduates in the University of Glasgow. All too often students did not seem to know the meanings of terms we felt were commonplace in cell biology, or were unable, for example, to find out what compounds in general use were supposed to do. But before long it became obvious that although we all considered ourselves to be cell biologists, individually we were similarly ignorant in areas only slightly removed from our own – though collectively the knowledge was there. It was also clear that many of the things we considered relevant were not easy to find, and that an extensive reference library was needed. In that we have found the exercise of preparing the Dictionary informative ourselves, we feel that it may serve a useful purpose.

An obvious problem was to decide upon the boundaries of the subject. We have not solved this problem; modern biology is a continuum and any attempt to subdivide it is bound to fail. "Cell Biology" implies different things to zoologists, to biochemists, and indeed to each of the other sub-species of biologists. There is no sensible way to set limits, nor would we wish to see our subject crammed into a well-defined niche. Inevitably, therefore the contents are somewhat idiosyncratic, reflecting our current teaching, reading, prejudices, and fancies.

It may be of some interest to explain how we set about preparing the Dictionary. The list of entry words was compiled largely from the index pages of several textbooks, and by scanning the subject indexes of cell-biological journals. To this were added entries for words we cross-referenced. The task of writing the basic entries was then divided amongst us roughly according to interests and expertise. We all wrote subsets of entries which were then compiled and alphabetized before being edited by one of us. Marked copies were then sent out to a panel of colleagues who scrutinised entries in their own fields. All entries were looked at by one or more of this panel, and then the annotated entries were re-edited, corrections made on disc, and the files copy-edited for consistency of style. A very substantial amount of the handling of the compiled text and the preparation of the final discs was done by Dr A M Lackie who also acted as copy-editor.

Glasgow is a major centre for Life Sciences, and we are fortunate in having many colleagues to whom we could turn for help. We are very grateful to them for the work which they put in and for the speed with which they checked the entries that we sent. Although we have tried hard to avoid errors and ambiguities, and to include everything that will be useful, we apologise at this stage for the mistakes and omissions, and emphasise that the blame lies with the authors and not with our panel (though they have saved us from many embarassments).

Since there is no doubt that Cell Biology is developing rapidly as a field, it is inevitable that usages will change, that new terms will become commonplace, that new proteins will be christened on gels, and that the dictionary will soon have omissions. Were the subject static this dictionary would not be worth compiling – and we cannot anticipate new words.

Because the text is on disc, it will be relatively easy to update; please let us have your comments, suggestions for entries (preferably with a definition), and (perhaps) your neologisms. A sheet is included at the back of the dictionary for this purpose.

A note regarding entries

The main entry word is followed by synonyms in brackets. Words in bold in the definition are cross-references to other entries which might contribute usefully to the entry being consulted, although other words within the definition may well have entries.

Generally speaking compounds or substances which have a greek letter prefix have been alphabetized ignoring the greek prefix; where the prefix is spelled out, this is considered to be the more general usage. Numbers are ignored for purposes of alphabetization.

John Lackie

Tables

Tables

A

A23187 A monocarboxylic acid extracted from *Streptomyces chartreusensis* that acts as a mobile-carrier calcium *ionophore*.

A cells (α cells) Cells of the endocrine pancreas (*Islets of Langerhans*) that form approximately 20% of the population; their opaque spherical granules may contain *glucagon*. See *B cells*, *D cells*.

A9 cells Cells of an established line of heteroploid mouse fibroblasts which are deficient in *HGPRT*.

A-band That portion of the *sarcomere* in which the thick myosin filaments are located. It is anisotropic ("A") in polarised light.

A-DNA Right-handed double-helical DNA with approximately 11 residues per turn. Planes of base-pairs in the helix are tilted 20° away from perpendicular to the axis of the helix. Formed from *B-DNA* by dehydration.

ABLV The Abelson murine *leukaemia* virus, a species of mammalian *Retroviridae*. Its transforming gene, *abl*, encodes a protein with tyrosine-specific *protein kinase* activity closely related to the **src gene** product.

AB toxin Multi-subunit toxin in which there are two major components, an active (A) portion and a portion which is involved in binding (B) to the target cell. The A portion can be effective in the absence of the B subunit(s) if introduced directly into the cytoplasm. In the well-known examples, the A subunit has *ADP-ribosylating* activity. See *cholera toxin, diphtheria toxin, pertussis toxin*.

abortive infection Viral infection of a cell in which the virus fails to replicate fully, or produces defective progeny. Since part of the viral replicative cycle occurs, its effect on the host can still be cytopathogenic.

abortive transformation Temporary transformation of a cell by a virus which fails to integrate into the host DNA.

abrin Toxic *lectin* from seeds of *Abrus precatorius* that has a binding site for galactose and related residues in carbohydrates but, because it is monovalent, is not an *agglutinin* for *erythrocytes*.

abscess A cavity within a tissue occupied by pus (chiefly composed of degenerating inflammatory cells), generally caused by bacteria which resist killing by phagocytes.

abscisic acid Mediator of plant cell activity found in vascular plants (a *plant growth substance*). Originally isolated from cotton bolls and identical to dormin. Although at first thought to be important in abscission (leaf fall) now known to be involved in a number of growth and developmental processes in plants including, in some circumstances, growth promotion.

Acanthamoeba Soil amoeba 20–30μm in diameter which can be grown *axenically* and has been extensively used in biochemical studies of cell motility. Some have been isolated from cultures of monkey kidney cells, and are pathogenic when injected into mice or monkeys.

acanthosome 1. Spinous membranous *organelle* found in skin *fibroblasts* from *nude mice* as a result of chronic ultraviolet irradiation. 2. Sometimes used as a synonym for *coated vesicle* (should be avoided).

accessory cells Cells which interact, usually by physical contact, with T-lymphocytes and which are necessary for induction of an immune response. Include antigen presenting cells, antigen processing cells etc. They are usually MHC Class II positive (see *histocompatibility antigens*). Monocytes, macrophages, dendritic cells, Langerhans cells, B-lymphocytes *inter alia* may all act as accessory cells.

accessory chromosome See *B-chromosome*.

accessory pigments In *photosynthesis*, pigments which collect light at different wavelengths and transfer the energy to the primary system.

ACE See *angiotensin*.

acellular slime moulds Protozoa of the Order Eumycetozoida (also termed true slime moulds), which have a multinucleate plasmodial phase in the life cycle and exhibit shuttle-flow (tidal) *cytoplasmic streaming*.

acentric Descriptive of pieces of *chromosome* which lack a *centromere*.

Acetabularia Giant single-celled *alga* of the Order Dasycycladaceae. The plant is 3–5cm long when mature and consists of *rhizoids* at the base of a stalk, at the other end of which is a cap that has a shape characteristic of each species. The giant cell has a single nucleus, located at the tip of one of its rhizoids; the nucleus can easily be removed by cutting off that rhizoid. Nuclei can also be transplanted from one cell to another.

acetylation Addition, either chemically or enzymically, of acetyl groups.

acetylcholine (ACh) Acetyl ester of choline. Perhaps the best characterised *neurotransmitter* released by vertebrate motoneurons, pre-ganglionic sympathetic and parasympathetic neurons. ACh can be either excitatory or inhibitory, and its receptors are classified as *nicotinic* or *muscarinic*, according to their pharmacology. In *chemical synapses* ACh is rapidly broken down by *acetylcholine esterase*, thereby ensuring the transience of the signal.

acetylcholine esterase An enzyme, found in the *synaptic clefts* of *cholinergic neurons*, which cleaves the *neurotransmitter acetylcholine* into its constituents, acetate and choline, thus limiting the size and duration of the *postsynaptic potential*. Many nerve gases and insecticides are potent acetylcholine esterase inhibitors, and thus prolong the timecourse of postsynaptic potentials.

acetyl CoA The acetylated form of coenzyme A, which is a carrier for acyl groups, particularly in the *tricarboxylic acid cycle*.

N-acetyl glucosamine (2-acetamido glucose) A sugar unit found in glycoproteins and various polysaccharides such as *chitin*, bacterial *peptidoglycan* and *hyaluronic acid*.

N-acetyl muramic acid Sugar unit of bacterial *peptidoglycan*, consisting of N-acetyl glucosamine bearing a lactyl residue in ether linkage to carbon 3. Repeating unit of the cell wall polysaccharide is N-acetyl muramic acid linked to N-acetyl glucosamine *via* a β(1–4)-glycosidic bond, which can be cleaved by the enzyme *lysozyme*.

N-acetyl neuraminic acid See *neuraminic acid*.

achondroplasia Failure of endochondral ossification responsible for a form of dwarfism; caused by an *autosomal* dominant mutation. Relatively high incidence (1:20,000 live births), mostly (90%) new mutations. Also known as chondrodystrophia fetalis.

acid growth theory Theory explaining the growth-promoting effect of *auxins* in higher plants. Auxin is thought to activate, probably indirectly, a proton pump in the plasma membrane, leading to acidification of the cell wall. This causes cell-wall weakening, leading to cell-wall stretching under the influence of *turgor* pressure, and hence to cell growth (enlargement).

acid hydrolases Hydrolytic enzymes which have a low pH optimum. The name usually refers to the *phosphatases*, *glycosidases*, *nucleases* and *lipases* found in the *lysosomal* compartment. They are secreted during *phagocytosis*, but are considered to operate as intracellular digestive enzymes.

acid phosphatase (EC 3.1.3.2) Enzyme with acidic pH optimum, which catalyzes cleavage of inorganic phosphate from a variety of substrates. Found particularly in *lysosomes* and *secretory vesicles*. Can be localised histochemically using various forms of the *Gomori procedure*.

acid protease Proteolytic enzyme with an acid pH optimum, characteristically found in *lysosomes*. See *proteases*.

acid secreting cells Large specialised cells of the epithelial lining of the stomach (parietal or oxyntic cells) which secrete 0.1N hydrochloric acid, by means of K^+/H^+ *antiport* ATPases on the luminal cell surface.

acidophilia Having an affinity for acidic dyes, particularly eosin; may be applied either to tissues or bacteria.

acidophils One class of cells found in the pars distalis of the adenohypophysis.

acinar cells Epithelial *secretory cells* arranged as a ball of cells around the lumen of a gland (as in the pancreas).

acinus Small sac or cavity surrounded by *secretory cells*.

acquired immune deficiency syndrome See *AIDS*.

acquired immunity Classically, the reaction of an organism to a new antigenic challenge and the retention of a memory of this, as opposed to innate immunity. In modern terms, the clonal expansion of a population of immune cells in response to a specific antigenic stimulus and the persistence of this clone.

acrasiales See *Acrasidae*.

Acrasidae The cellular slime moulds. They normally exist as free-living phagocytic soil amoebae, but when bacterial prey become scarce, they aggregate to form a pseudoplasmodium (cf true *plasmodium* of acellular slime moulds, Eumycetozoida), which is capable of directed motion. The grex, or slug, migrates until stimulated by environmental conditions to form a fruiting body or sorocarp. The slug cells differentiate into elongated stalk cells and spores, where the cells are surrounded by a cellulose capsule. The spores are released from the sporangium at the tip of the stalk and, in favourable conditions, an *amoeba* emerges from the capsule, feeds, divides and so establishes a new population. They can be cultured in the laboratory and are widely used in studies of cell–cell adhesion, cellular differentiation, *chemotaxis* and *pattern formation*. The commonest species studied are *Dictyostelium discoideum*, *D.minutum* and *Polysphondylium violaceum*.

acrasin Name originally given to the *chemotactic* factor produced by cellular slime moulds (*Acrasidae*): now known to be *cyclic AMP* for *Dictyostelium discoideum*.

acridine orange A fluorescent vital dye, which intercalates into nucleic acids. The nuclei of stained cells fluoresce green; cytoplasmic RNA fluoresces orange. May be carcinogenic. Acridine orange also stains acid mucopolysaccharides, and is widely used as a pH-sensitive dye in studies of acid secretion.

acromegaly Enlargement of the extremities of the body as a result of the overproduction of *growth hormone* (somatotropin), eg. by a pituitary tumour.

acrosin Serine protease stored in the *acrosome* of a sperm as an inactive precursor.

acrosomal process A long process actively protruded from the acrosomal region of the spermatozoon following contact with the egg and which assists penetration of the gelatinous capsule.

acrosome The lysosomally-derived vesicle at the extreme anterior end of the spermatozoon.

ACTH See *adrenocorticotrophin*.

actin A protein (42kD) that is very abundant in eukaryotic cells (8–14% total cell protein) and one of the major components of the *actomyosin* motor and the cortical microfilament meshwork. First isolated from *striated muscle* and often referred to as one of the muscle proteins. G-actin is the globular monomeric form of actin (6.7×4.0nm); it polymerises to form filamentous F-actin.

actin binding proteins A diverse group of proteins which bind to *actin* and which may stabilise F-actin filaments, nucleate filament formation, cross-link filaments, lead to bundle formation etc. See Table A1.

actin meshwork *Microfilaments* inserted proximally into the plasma membrane and cross-linked by *actin binding proteins* to form a mechanically resistive network which may support protrusions such as *pseudopods* (sometimes referred to as the cortical meshwork).

α-actinin A protein (100kD) normally found as a dimer and which may link actin filaments end-to-end with opposite polarity. Originally described in the *Z-disc*, now known to occur in *stress-fibres* and at *focal adhesions*.

β-actinin A protein (35kD, normally dimeric) which is thought to bind to the end of the *thin filament* furthest from the *Z-disc* serving to block disassembly. Might be homologous to *acumentin*.

actinogelin Protein (115kD) from *Ehrlich ascites* cells which gelates and bundles *microfilaments*.

actinomycin D *Antibiotic* from **Streptomyces** spp. which binds to DNA and thus blocks the movement of *RNA polymerase* and prevents RNA synthesis in both prokaryotes and eukaryotes.

actinomycins A family of chromopeptide antibiotics that differ solely in the peptide portion of the molecule. Produced by species of **Streptomyces**.

actinotrichia Aligned collagen fibres (ca 2μm diameter) which provide a guidance cue for mesenchymal cells in the developing fin of teleost fish.

action potential Brief, regenerative, all-or-nothing potential that passes along the membranes of *excitable cells*, such as *neurons*, *muscle cells*, fertilised eggs and certain plant cells. The precise shape of action potentials varies, but action potentials

Table A1. Actin binding proteins

Table A1. Actin binding proteins

(i) Monomer sequestering (Bind G actin)

Protein	MW (kD)	Source
Profilin	12–15	Various
Vit D binding protein	57	Plasma
DNAase I	31	Pancreas
Depactin	19	Starfish oocytes
19kDa protein	19	Pig brain

(ii) End-blocking and nucleating

Protein	MW (kD)	End[a]	Calcium sensitivity[b]	Source
Gelsolin	90	+	+	Mammalian cells; same as brevin and ADF
Villin	95	+	+	Amphibian eggs, avian and mammalian epithelia
Fragmin/severin	40–45	+	+	*Physarum, Dictyostelium*, sea urchin eggs
Capping protein	31, 28	+	−	*Acanthamoeba*
Acumentin	65	−	−	Mammalian leucocytes
β-actinin	37, 35	−	−	Kidney and striated muscle

(iii) Cross-linking

Isotropic gelation

Protein	Subunits (kD)	Source
Actin-binding protein	2 × 270	Macrophages, platelets, *Xenopus* eggs
Filamin	2 × 250	Smooth muscle
Spectrin	2 × 240, 2 × 220	Erythrocytes
Fodrin	2 × 260, 2 × 240	Brain
TW 260/240	2 × 260, 2 × 240	Intestinal epithelium

Anisotropic bundling

Protein	Subunits (kD)	Source
α-actinin	2 × 95	Various
Actinogelin	2 × 115	Ehrlich ascites tumour cells
Fascin	53–57	Pig brain, echinoderm gametes
Fimbrin	68	Intestinal epithelium
Villin	95	Intestinal epithelium (see ii)

(iv) Miscellaneous

Protein	MW (kD)	Source
Gelactins	23–38	4 types; from *Acanthamoeba*
MAP2	300	Brain, microtubule associated
tau	50–68	Microtubule associated
Calpactins	35, 36	Various

[a]Microfilament end to which protein binds.
[b]At high calcium sever F-actin, at low calcium nucleate microfilaments.

always involve a large *depolarisation* of the cell membrane, from its normal *resting potential* of −50 to −90mV. In a neuron, action potentials can reach +30mV, and last 1ms. In muscles, action potentials can be much slower, lasting up to 1s.

action spectrum The relationship between the frequency (wavelength) of a form of radiation, and its effectiveness in inducing a specific chemical or biological effect.

activated macrophage A *macrophage* (mononuclear phagocyte) which has been stimulated by *lymphokines* (especially γ-*interferon*) and which has greatly enhanced cytotoxic and bactericidal potential.

activation (of egg) Normally brought about by contact between spermatozoon and egg membrane. Activation is the first stage in development and occurs independently of nuclear fusion. The first observable change is usually the cortical reaction which may involve elevation of the fertilisation membrane; the net result is a block to further fusion and thus to polyspermy. In addition to the morphological changes, there are rapid changes in metabolic rate and an increase in protein synthesis from maternal mRNA.

activation energy The energy required to bring a system from the ground state to the level at which a reaction will proceed.

active site The region of a *protein* that binds to substrate molecule(s) and facilitates a specific chemical conversion. Produced by juxtaposition of amino acid residues as a consequence of the protein's *tertiary structure*.

active transport Often defined as transport up an electrochemical gradient. More precisely defined as unidirectional or vectorial transport produced within a membrane-bound protein complex by coupling an energy-yielding process to a transport process. In primary active transport systems the transport step is normally coupled to *ATP* hydrolysis within a single protein "complex". In secondary active transport the movement of one species is coupled to the movement of another species down an electrochemical gradient established by primary active transport.

active zone Special region of the *presynaptic cell* membrane which has projections of dense material on the cytoplasmic face. The area in which fusion of *synaptic vesicles* is most likely to occur.

activin *FSH*-releasing protein: dimer of two *inhibin* β-chains.

actomyosin Generally: a motor system which is thought to be based on *actin* and *myosin*. The essence of the motor system is that myosin makes transient contact with the actin filaments and undergoes a conformational change before releasing contact. The hydrolysis of ATP is coupled to movement, through the requirement for ATP to restore the configuration of myosin prior to repeating the cycle. More specifically: a viscous solution formed when actin and myosin solutions are mixed at high salt concentrations. The viscosity diminishes if ATP is supplied and rises as the ATP is hydrolysed. Extruded threads of actomyosin will contract in response to ATP.

acumentin Protein (65kD) of analogous function to β-actinin, isolated from vertebrate *macrophages*.

acute 1. Sharp or pointed. 2. Of diseases; coming rapidly to a crisis — not persistent.

acute inflammation Response of vertebrate body to insult or infection; characterised by redness (rubor), heat (calor), swelling (tumor), pain (dolor), and sometimes loss of function. Changes occur in local blood flow, and *leucocytes* (particularly *neutrophils*) adhere to the walls of post-capillary venules (margination) and then move through the *endothelium* (diapedesis) towards the damaged tissue. Although usually an acute inflammation is relatively short-term, there are situations in which persistent acute-type inflammation, with neutrophils and macrophages as the dominant cell types (unlike *chronic inflammation*) occurs.

acute phase proteins Proteins found in increased quantities in the serum of animals showing *acute inflammation*. In particular *C-reactive protein* and *serum amyloid* A protein.

acute lymphoblastic leukaemia See *leukaemia*.

acyclovir Antiviral agent that is an analogue of *guanosine* and inhibits *DNA replication* of viruses. Particularly successful against herpes simplex infections.

adaptation A change in sensory or excitable cells upon repeated stimulation, which reduces their sensitivity to continued stimulation. Those cells which show rapid adaptation are known as phasic; those which adapt slowly are known as tonic.

ADCC (antibody-dependent cell-mediated cytotoxicity) A phenomenon in which IgG-coated target cells are killed by non-sensitised effector cells (neutrophils, monocytes, NK cells) by a non-phagocytic mechanism which is independent of complement.

Addison's disease Chronic insufficiency of the adrenal cortex classically as a result of tuberculosis or, more interestingly, specific *autoimmune* destruction of the *adrenocorticotrophin*-secreting cells.

addressins Molecules expressed in an organ- or tissue-selective manner by cells or extracellular elements which are found in other locations (where they may express functionally-related counterparts). Their role seems to be to signal position or mark the address for the purpose of directing cell–cell interactions. A subset, the vascular addressins, are expressed in a tissue-specific manner by endothelium and are important in lymphocyte recirculation.

adducin *Calmodulin*-binding protein associated with the membrane skeleton of erythrocytes. A substrate for *protein kinase* C, it binds to *spectrin–actin* complexes (but only weakly to either alone) and promotes the assembly of spectrin onto spectrin–actin complexes unless micromolar calcium is present. Has subunits of 102 and 97kD and is distinguishable from Band 4.1.

adenine (6-aminopurine) One of the bases found in *nucleic acids* and *nucleotides*. In DNA, it pairs with *thymine*.

adeno- Prefix indicating association with, or similarity to, glandular tissue.

adenocarcinoma Malignant neoplasia of a glandular epithelium, or *carcinoma* showing gland-like organisation of cells.

adenohypophysis Anterior lobe of the pituitary gland; responsible for secreting a number of hormones and containing a comparable number of cell types.

adenoma *Benign tumour* of glandular epithelium.

adenosine (9-β-D-ribofuranosyladenine) The *nucleoside* formed by linking *adenine* to *ribose*.

adenosine diphosphate See *ADP*.

adenosine monophosphate See *AMP*, *cyclic AMP*.

adenosine triphosphate See *ATP*.

S-adenosyl methionine (S-(5'-deoxyadenosine-5')-methionine) An activated derivative of *methionine*, which functions as a methyl group donor, in (for example) nucleic acid or phospholipid methylation and bacterial *chemotaxis*.

Adenoviridae Large group of viruses first isolated from cultures of adenoid tissue. The *capsid* is an icosahedron of 240 hexons and 12 pentons in the form of a base and a fibre; the genome consists of a single, linear molecule of double-stranded DNA. They cause various respiratory and gastrointestinal infections in humans. Some of the avian, bovine, human and simian adenoviruses cause tumours in newborn rodents, generally hamsters. They can be classified into highly, weakly and non-*oncogenic viruses* from their ability to induce tumours *in vivo* though all of these groups will transform cultured cells. The viruses are named after their host species and subdivided into many serological types.

adenylate cyclase Enzyme which produces *cyclic AMP* from *ATP*. The best known example is the adenylate cyclase which produces a *second messenger* (cAMP) in response to external signals. Receptors are coupled to the adenylate cyclase by stimulatory or inhibitory *GTP-binding proteins*.

adhaerens junctions (USA adherens junctions) Specialised cell–cell junctions into which are inserted *microfilaments* (also known as "*zonulae adhaerentes*") or *intermediate filaments* ("maculae adhaerentes" or "spot *desmosomes*").

adhesins General term for molecules involved in adhesion, but its use is restricted in Microbiology where it refers to bacterial surface components.

adhesion Attachment of two surfaces, mediated by molecular interactions. See *cell adhesion*.

adhesion plaque Another term for a *focal adhesion*, a discrete area of close contact between a cell and a non-cellular substratum, with cytoplasmic insertion of *microfilaments* and considerable electron-density adjacent to the contact area.

adipocyte *Connective tissue* cell specialised for the storage of fat. There may be distinct types in white and brown fat.

adipofibroblasts *Adipocytes* from subcutaneous fat will lose fat globules and develop a fibroblastic appearance when grown in culture. Unlike skin fibroblasts they will take up fat from serum taken from obese donors, and probably retain a distinct differentiated state.

adipose tissue Fibrous connective tissue with large numbers of fat-storing cells, *adipocytes*.

adjuvant Additional components added to a system to affect action of its main component, typically to increase the *immune response* to an *antigen*. See *Freund's adjuvant*.

ADP (adenosine diphosphate) Unless otherwise specified is the nucleotide 5'ADP, *adenosine* bearing a diphosphate (pyrophosphate) group in ribose-O-phosphate ester linkage at position 5 of the ribose moiety. Adenosine 2'5' and 3'5'diphosphates also exist, the former as part of *NADP* and the latter in *coenzyme A* and mRNA.

ADP-ribosylation A form of *post-translational modification* of protein structure involving the transfer to protein of the ADP-ribosyl moiety of *NAD*. Believed to play a part in normal cellular regulation as well as in the mode of action of several bacterial toxins.

adrenal Endocrine gland adjacent to the kidney. Distinct regions, the cortex and medulla, produce different ranges of hormones.

adrenaline (epinephrine) A *hormone* secreted by the medulla of the adrenal gland in response to stress, and by some *neurons*. The effects are those of the classic "fight or flight" response, including increased heart function, elevation in blood sugar levels, cutaneous vasoconstriction making the skin pale, and raising of hairs on the neck.

adrenergic neuron A neuron is adrenergic if it secretes *adrenaline* or *noradrenaline* at its terminals. Many neurons of the *sympathetic nervous system* are adrenergic.

adrenocorticotrophin (ACTH) A peptide hormone produced by the pituitary gland in response to stress (mediated by corticotrophin releasing factor, a 41 residue peptide, from the hypothalamus). Stimulates the release of adrenal cortical hormones, mostly *glucocorticoids*. Derived from a larger precursor, *pro-opiomelanocortin*, by the action of an endopeptidase, which also releases β-*lipotropin*.

adventitia Outer coat of the wall of *vein* or *artery*, composed of loose *connective tissue* which is vascularised. Generally means outer covering of an organ.

aequorin Protein (30kD) extracted from jellyfish (*Aequorea aequorea*) which emits light in proportion to the concentration of calcium ions. Used to measure calcium concentrations, but has to be microinjected into cells. See also *bioluminescence*.

aerenchyma Form of *parenchyma* tissue containing particularly large intercellular air spaces, the cells being in contact at only a few points on their surfaces. Found chiefly in submerged roots and stems of plants growing in aquatic or marshy environments, permitting aeration of the tissues.

aerobes Organisms which rely on oxygen.

aerotaxis *Taxis* (tactic response) to oxygen (air).

affinity An expression of the strength of interaction between two entities, eg. between receptor and ligand or between enzyme and substrate. The affinity is usually characterised by the equilibrium constant (*association* or *dissociation constant*) for the binding, this being the concentration at which half the receptors are occupied.

affinity chromatography *Chromatography* in which the immobile phase (bed material) has a specific biological affinity for the substance to be separated or isolated, such as the affinity of an *antibody* for its *antigen*, or an enzyme for a substrate analogue.

aflatoxins A group of highly toxic substances produced by the fungus *Aspergillus flavus*, and other species of *Aspergillus*, in stored grain or mouldy peanuts. They cause enlargement and death of liver cells if ingested, and may be carcinogenic.

agar A polysaccharide complex extracted from seaweed (Rhodophyceae) and used as an inert support for the growth of cells, particularly bacteria and some cancer cell lines. Gels have the unusual property of melting at high temperature (ca 100°C) but not solidifying until the temperature is less than about 40°C.

agarose A galactan polymer purified from *agar* which forms a rigid gel with high free water content. Primarily used as an electrophoretic support for separation of macromolecules. Stabilised derivatives are used as "macroporous" supports in *affinity chromatography*. See *Sepharose*.

agglutination The formation of adhesions by particles or cells to build up multicomponent aggregates, otherwise termed agglutinates or flocs. Distinguished from *aggregation* by the fact that agglutination phenomena are usually very rapid. Usually caused by agents such as *antibodies*, *lectins* or other bi- or poly-valent reagents, and it is useful to reserve the term for situations in which an extrinsic agent is added to the system, in contrast to *aggregation*.

agglutinins Agents causing *agglutination*, eg. *antibodies*, *lectins*, *polylysine*.

aggregation The process of forming adhesions between particles such as cells. Aggregation is usually distinguished from *agglutination* by the slow nature of the process; not every encounter between the cells is effective in forming an adhesion. Useful to reserve the term for situations in which no cross-linking agglutinin is added to the system.

agonist 1. In neurobiology, of a *neuron* or *muscle*; one which aids the action of another. If the two effects oppose each other, then they are known as antagonistic. 2. In pharmacology, a compound which acts on a receptor to elicit a response. 3. In ethology, "agonistic behaviour" means aggressive behaviour towards a conspecific animal.

agorins Major structural proteins of the membrane matrix, constituting approximately 15% of total plasma membrane proteins of P815 mastocytoma cells. They form large detergent-insoluble structures when the membranes are extracted with Triton X-100 and EGTA. Agorin I, 20kD; Agorin II, 40kD.

AGP (arabinoglycan-protein) A class of extracellular *proteoglycan*, found in many higher-plant tissues, and secreted by many suspension-cultured plant cells. Contains 90–98% *arabinogalactan* and 2–10% protein. Related to arabinogalactan II of the cell wall.

agranular vesicles Synaptic vesicles which do not have a granular appearance in the electron microscope; 40–50nm in diameter, with membrane only 4–5nm thick. Characteristic of peripheral *cholinergic* synapses; some are located very close to the *presynaptic cell* membrane.

agranulocytosis Severe deficiency of *granulocytes* in blood.

agrin A protein isolated from the electric organ of the electric ray, *Torpedo californica*, that induces *myotubes* to form specialisations similar to those at the neuromuscular junction.

Agrobacterium tumefaciens A Gram negative, rod-shaped flagellated bacterium responsible for *crown gall* tumour in plants. Following infection the T1 *plasmid* from the bacterium becomes integrated into the host plant's DNA and the presence of the bacterium is no longer necessary for the continued growth of the tumour.

AIDS (Acquired Immune Deficiency Syndrome) Disease caused by infection with *HIV* (human immunodeficiency virus), causing a deficiency of *T-helper cells* with resulting *immunosuppression* phenomena; there is thus increased susceptibility to other infectious diseases and to certain types of tumour, particularly *Kaposi's sarcoma*.

alanine (Ala; A; MW 89) Normally refers to L-α-alanine, the aliphatic amino acid found in proteins. The isomer β-alanine is a component of the vitamin *pantothenic acid* and thus also of *coenzyme A*. See Table A2.

albinism Condition in which no *melanin* is synthesised.

albumin The term normally refers to serum albumins, the major protein components of the serum of vertebrates. They have a single polypeptide chain, with multidomain structure containing multiple binding sites for many lipophilic metabolites notably fatty acids and bile pigments. In the embryo their functions are fulfilled by α-*foetoproteins*. The viability of analbuminaemic individuals (those deficient in albumin) suggests albumin is not indispensible.

alcian blue Water-soluble copper phthalocyanin stain used to demonstrate acid *mucopolysaccharides*. By varying the ionic strength some differentiation of various types is possible.

aldosterone A steroid hormone (mineralo-corticoid), produced in the outermost of the three zones of the adrenal medulla, which controls salt and water balance in the kidney. Release is controlled by *angiotensin* II; excessive secretion occurs in Cushing's syndrome, decreased release in Addison's disease.

aleurone grain (aleurone body) Membrane-bounded *storage granule* within plant cells that usually contains protein. May be an *aleuroplast* or just a specialised *vacuole*.

aleuroplast A semi-autonomous organelle (*plastid*) within a plant cell, which stores protein.

Aleutian disease of mink A disease caused by a *slow virus* of the *Parvoviridae* family producing *autoimmune* symptoms, glomerulonephritis and immune complexes.

algae A non-taxonomic term used to group several phyla of the lower plants, including the *Rhodophyta* (red algae), *Chlorophyta* (green algae), *Phaeophyta* (brown algae) and *Chrysophyta* (diatoms). Many algae are unicellular or consist of simple undifferentiated colonies, but red and brown algae are complex multicellular organisms, familiar to most people as seaweeds. Blue-green algae (Cyanophyta) are a totally separate group of *prokaryotes*, more correctly known as *Cyanobacteria*.

alkaline phosphatase (EC 3.1.3.1) Enzyme catalyzing cleavage of inorganic phosphate non-specifically from a wide variety of phosphate esters, and having a high (8+) pH-optimum. Found in bacteria, fungi and animals but not in higher plants.

alkaloid A nitrogenous base. Usually refers to biologically active (toxic) molecules, produced as allelochemicals by plants to deter grazing. Examples: *ouabain*, *digitalis*.

alkaptonuria In humans, the congenital absence of homogentisic acid oxidase, an enzyme which breaks down tyrosine and phenylalanine. Accumulation of homogentisic acid in homozygotes causes brown pigmentation of skin and eyes and damage to joints; urine blackens on standing.

alleles Different forms or variants of a *gene* found at the same place, or *locus*, on a *chromosome*. Assumed to arise by *mutation*.

allelic exclusion The process whereby one or more loci on one of the *chromosome* sets in

a *diploid cell* is inactivated (or destroyed) so that the locus or loci is (are) not expressed in that cell or a clone founded by it. For example in mammals one of the *X chromosome* pairs of females is inactivated early in development (see *Lyon hypothesis*) so that individual cells express only one allelic form of the product of that locus. Since the choice of chromosome to be inactivated is random, different cells express one or other of the X chromosome products, resulting in mosaicism. The process is also known to occur in *immunoglobulin* genes so that a clone expresses only one of the two possible allelic forms of immunoglobulin.

allelochemical A little-used term, referring to substances effecting allelopathic reactions. See *allelopathy*.

allelopathy The deleterious interaction between two organisms or cell types which are *allogeneic* to each other (the term is often applied loosely to interactions between *xenogeneic* organisms). Allelopathy is seen between different species of plant, between various individual sponges, and between sponges and gorgonians.

allergic encephalitis See *experimental allergic encephalomyelitis*.

allergy An inexact term, usually referring to immediate (Type I) *hypersensitivity*.

allogeneic Two or more individuals (or strains) are stated to be allogeneic to one another when the genes at one or more loci are not identical in sequence in each organism. Allogenicity is usually specified with reference to the locus or loci involved.

allograft Graft between two or more individuals allogeneic at one or more loci (usually with reference to *histocompatibility* loci); cf *autograft* and *xenograft*.

allopolyploidy *Polyploid* condition in which the contributing genomes are dissimilar. When the genomes are doubled fertility is restored and the organism is an amphidiploid. Common in plants but not animals.

allopurinol A *xanthine oxidase* inhibitor used in the treatment of gout.

allosomes One or more chromosomes which can be distinguished from *autosomes* by their morphology and behaviour. Synonyms: accessory chromosomes, heterochromosomes, sex chromosomes.

allosteric Of a binding site in a protein, usually an enzyme. The catalytic function of an enzyme may be modified by interaction with small molecules, not only at the *active site*, but also at a spatially distinct (allosteric) site of different specificity. Of a protein, a protein possessing such a site. An allosteric effector is a molecule bound at such a site which increases or decreases the activity of the enzyme.

allotope (allotypic determinant) The structural region of an *antigen* which distinguishes it from another *allotype* of that antigen.

allotype Product of one or more *alleles* that can be detected as an inherited variant of a particular molecule. Generally the usage is restricted to those *immunoglobulins* that can be separately detected antigenically. See also *idiotype*. In humans light chain allotypes are known as Km (Inv) allotypes and heavy chain allotypes as Gm allotypes.

alloxan Used to produce *diabetes mellitus* in experimental animals. Destroys pancreatic *B cells* by a mechanism involving *superoxide* production.

allozyme Variant of an enzyme coded by a different allele. See *isoenzyme*.

α_1-antitrypsin Better named α_1-antiprotease (α_1-*protease inhibitor*). A major protein of blood *plasma* (3mg/ml in human), part of the α-globulin fraction, and able to inhibit a wide spectrum of *serine proteases*.

α-cell See *A cells* of endocrine pancreas. α-*acidophils* are cells from the adenohypophysis.

α-foetoprotein Protein from the serum of vertebrate embryos, which probably fulfils the function performed by *albumin* in the mature organism. Found in both glycosylated and nonglycosylated forms. Presence in the fluid of the *amniotic sac* is diagnostic of spina bifida in the human foetus.

α-helix (alpha helix) A particular helical folding of the polypeptide backbone in protein molecules (both fibrous and globular), in which the carbonyl oxygens are all hydrogen-bonded to amide nitrogen atoms four residues along the chain. The translation of amino acid residues along the long axis is 0.15nm, and the rotation per residue, 100°, which gives 3.6 residues/turn.

α_1-protease inhibitor (α_1-PI) Plasma protein which inhibits elastase and other serine proteases (a *serpin*). Susceptible to inactivation by oxidation or by protease attack: chronic inflammation in the lung may lead to local inactivation of α_1-PI, potentiating elastase degradation of connective tissue thus contributing to the development of emphysema. Major component of what was once called α_1-antitrypsin.

altered self hypothesis The hypothesis that the *T-cell* receptor in MHC-mediated phenomena recognises a *syngeneic* MHC Class I or Class II molecule in association with antigen. See *MHC restriction*, *histocompatibility antigen*.

alternative oxidase pathway (alternative terminal oxidase) See *alternative pathway 2*.

alternative pathway 1. See *complement*. 2. Pathway of mitochondrial electron transport in higher plants, particularly fruits and seeds, which does not involve *cytochrome oxidase* and hence is resistant to cyanide.

Alu Type II *restriction endonuclease*, isolated from *Arthrobacter luteus*. The recognition sequence is 5'-AG/CT-3'. *Alu* sequences are highly repetitive sequences found in large numbers (100–500,000) in the human genome, and which are cleaved more than once within each sequence by the *Alu* endonuclease. The *Alu* sequences look like DNA copies of mRNA because they have a 3'*poly-A tail* and flanking repeats.

alveolar cell Cell of the air sac of the lung.

alveolar macrophage *Macrophage* found in pulmonary alveoli and which can be obtained by lung lavage; responsible for clearance of inhaled particles and lung *surfactant*. Metabolism slightly different (more oxidative metabolism) from peritoneal macrophages, often has *multivesicular bodies* which may represent residual undigested lung surfactant.

Alzheimer's disease A presenile dementia characterised by the appearance of unusual helical protein filaments in nerve cells, and by degeneration in cortical regions of brain, especially frontal and temporal lobes. May be associated with *slow virus* or selective loss or dysfunction of *cholinergic neurons*.

amacrine cell A class of *neuron* of the middle layer of the *retina*, with processes parallel to the plane of the retina; involved in image processing.

Table A2. Amino acids. L-amino acids specified by the biological code for proteins.

Name	Abbreviation	Single letter	Side chain	pK_a* (Stryer)	MW (D) (rounded)	Hydropathy index** (Kyte & Doolittle)	Codons
Alanine	ala	A	$-CH_3$		89.1	1.8	GC(X)
Arginine	arg	R	$-CH_2\,CH_2\,CH_2\,NH\,(C\overset{+}{N}H_2)\,NH_2$	12	174.2	-4.5	CG(X) AGA AGG
Aspartic acid	asp	D	$-CH_2\,COO^-$	4.4	133.1	-3.5	GAU GAC
Asparagine	asn	N	$-CH_2\,CONH_2$		132.2	-3.5	AAU AAC
Cysteine	cys	C	$-CH_2\,SH$	8.5	121.2	2.5	UGU UGC
Glutamic acid	glu	E	$-CH_2\,CH_2\,COO^-$	4.4	147.2	-3.5	GAA GAG
Glutamine	gln	Q	$-CH_2\,CH_2\,CONH_2$		146.2	-3.5	CAA CAG
Glycine	gly	G	$-H$		75.1	-0.4	GG(X)
Histidine	his	H	$-CH_2$ (imidazole)	6.5	155.2	-3.2	CAU CAC
Iso-leucine	ile	I	$-CH\,(CH_3)\,CH_2\,CH_3$		131.2	4.5	AUU AUC AUA
Leucine	leu	L	$-CH_2\,CH\,(CH_3)_2$		131.2	3.8	CU(X) UUA UUG
Lysine	lys	K	$-CH_2\,CH_2\,CH_2\,CH_2\,NH_3^+$	10	146.2	-3.9	AAA AAG
Methionine	met	M	$-CH_2\,CH_2\,SCH_3$		149.2	1.9	AUG
Phenylalanine	phe	F	$-CH_2$ (phenyl)		165.2	2.8	UUU UUC
Proline	pro	P	$NH_2^+{-}CH{-}COO^-$, CH_2, CH_2, CH_2 (whole molecule)		115.1	-1.6	CC(X)
Serine	ser	S	$-CH_2\,OH$		105.1	-0.8	UC(X)
Threonine	thr	T	$-CH\,(OH)\,CH_3$		119.1	-0.7	AC(X)
Tryptophan	trp	W	$-CH_2$ (indole)		204.2	-0.9	UGG (UGA mitochondria)
Tyrosine	tyr	Y	$-CH_2$ (phenol, $-OH$)	10	181.2	-1.3	UAU UAC
Valine	val	V	$-CH\,(CH_3)_2$		117.2	4.2	GU(X)

* The value for side chain ionisation when the amino acid residue is present in a polypeptide.
** A measure of the tendency for the residue to be buried within the interior of a folded protein.

amanitins (α-, β-, γ-amanitin). Cyclic peptides, the most toxic components of *Amanita phalloides* (Death cap toadstool). Specific inhibitors of *RNA polymerase* II in eukaryotes thus inhibiting protein synthesis by blocking the production of *messenger RNA*.

amantadine Used as an antiviral agent (especially against influenza virus). Produces some symptomatic relief in *Parkinsonism*.

amber codon One of the three *termination codons*. Its sequence is UAG. See also *ochre codon*, *opal codon*.

amber suppressor Mutation which alleviates the effect of an amber mutation; the mutant tRNA of the suppressor reads the *termination codon* UAG (whereas the *ochre suppressor* reads UAA and UAG) and inserts an amino acid (Y, W, L, Q or S; for amino acid single-letter code, see Table A2).

Ambystoma mexicanum (*Amblystoma* — incorrect spelling) An amphibian, the Mexican axolotl. A salamander which shows *neoteny*; in this form the adult is aquatic and possesses large external gills. A terrestrial form without gills is also found in the wild; related species, *A.opacum* and *A.tigrinum*, do not have neotenous forms. The two forms of *A. mexicanum* were originally described as separate species until it was demonstrated that the neotenous, aquatic axolotl would metamorphose into the terrestrial form if injected with thyroid or pituitary gland extract. Neoteny thought to be due to environmental deficiency in iodine, necessary for the synthesis of thyroid hormones. Axolotls are used in studies of *metamorphosis*, early embryonic development and limb *regeneration*.

ameloblasts Columnar epithelial cells which secrete the enamel layer of teeth in mammals. Their apical surfaces are tapering (Tomes processes) and are embedded within the enamel matrix.

Ames test A procedure, used to test a substance for its likely ability to cause cancer, which combines the use of animal tissue to generate active metabolites of the substance with a test for mutagenicity in bacteria.

amiloride Drug which blocks the sodium/proton *antiport*; used clinically as a potassium-sparing diuretic.

amino acids Organic acids carrying amino groups. The L-forms of about 20 common amino acids are the components from which proteins are made. See Table A2, and Table C3 for the *codon* assignment.

amino sugar Monosaccharide in which an OH-group is replaced with an amino group; often acetylated. Common examples are D-galactosamine, D-glucosamine, neuraminic acid, muramic acid. Amino sugars are important constituents of bacterial cell walls, some antibiotics, blood group substances, milk oligosaccharides, and chitin.

amino transferases (transaminase) (EC 2.6.1.x) A family of enzymes which transfer an amino group from an amino acid to an α-ketoacid, as in the transfer from glutamic acid to oxaloacetic acid, to form aspartic acid and α-ketoglutarate. One reactant is often glutamic acid, and the reactions employ pyridoxal phosphate as coenzyme.

aminophylline An inhibitor of cyclic AMP *phosphodiesterase*.

aminopterin A *folic acid* analogue and inhibitor of *dihydrofolate reductase*. A potent cytotoxic agent used in the treatment of acute *leukaemia*, and a component of *HAT medium*.

amitosis An unusual form of nuclear division, in which the partitioning of daughter genomes is haphazard. Observed in some polygenomic Protozoa.

amniocentesis Sampling of the fluid in the *amniotic sac*. In humans this is carried out, between the 12th and 16th week of pregnancy, by inserting a needle through the abdominal wall into the uterus. By *karyotyping* the cells and determining the proteins present, it is possible to determine the sex of the foetus and whether it is suffering from certain congenital diseases such as *Down's syndrome* or spina bifida.

amnion Terrestrial vertebrates have embryos which develop in fluid-filled sacs formed by the outgrowth of the extra-embryonic *ectoderm* and *mesoderm* as projecting folds. These folds fuse to form two epithelia separated by mesoderm and body cavity (coelom). The inner layer is the amnion and encloses the amniotic sac in which the embryo is suspended. The outer layer is the *chorion*.

amniotic sac Sac, enclosing the embryo of amniote vertebrates, which provides a fluid environment to prevent dehydration during development of land-based animals. See *amnion*.

amoeba Genus of protozoa, but also an imprecise name given to several types of free-living unicellular phagocytic organism. Giant forms (eg. *Amoeba proteus*) may be up to 2mm long, and crawl over surfaces by protruding *pseudopods* (*amoeboid movement*). Amoebae exhibit great plasticity of form and conspicuous *cytoplasmic streaming*.

amoeboid movement Crawling movement of a cell brought about by the protrusion of *pseudopods* at the front of the cell (one or more may be seen in monopodial or polypodial amoebae, respectively). The pseudopods form distal anchorages with the surface.

amoebocytes Phagocytic cells found circulating in the body cavity of coelomates (particularly annelids and molluscs), or crawling through the interstitial tissues of sponges. A fairly non-committal classification.

AMP (adenylate) Unless otherwise specified, 5′AMP, the nucleotide bearing a phosphate in ribose-O-phosphate ester linkage at position 5 of the ribose moiety. Both 2′ and 3′ derivatives also exist. See also *cyclic AMP*.

amphibolic Descriptive of a pathway which functions not only in *catabolism*, but also to provide precursors for *anabolic* pathways.

amphipathic Of a molecule, having both hydrophobic and hydrophilic regions. Can apply equally to small molecules, such as phospholipids, and macromolecules such as proteins.

ampholyte Substance with *amphoteric* properties. Most commonly encountered as descriptive of the substances used in setting up electrofocusing columns or gels.

amphoteric Having both acidic and basic characteristics. This is true of proteins since they have both acidic and basic side groups (the charges of which balance at the isoelectric point).

ampicillin *Penicillin* derivative with broad spectrum activity; ampicillin resistance is often used as a marker for *plasmid* transfer in genetic engineering (eg. pBR322 is ampicillin resistant).

α-amylase An endo-amylase enzyme which rapidly breaks down starch to dextrins.

β-amylase A terminal amylase which cleaves starch to maltose units from the end of starch chains.

amyloid *Glycoprotein* deposited extracellularly in tissues in *amyloidosis*. The glycoprotein may either derive from the *L-chain* of immunoglobulin (AIO — amyloid of immune origin: 5–18kD glycoprotein, product of a single clone of *plasma cells*, the N-terminal part of the λ or κ L-chain) or from unknown plasma protein (AUO — amyloid of unknown origin: usually homogeneous in an individual case, does not cross-react with immunoglobulin). The polypeptides are organised as a *beta-pleated sheet* making the material rather inert and insoluble. Minor protein components are also found. See also *serum amyloid*.

amyloidosis Deposition of *amyloid*. A common complication of several diseases (leprosy, tuberculosis); often associated with perturbation of the immune system, although there may be immunosuppression or enhancement.

amylopectin Branched, water-insoluble, polysaccharide component of *starch* in which the glucose chain is α(1–4)-linked, with α(1–6)-linkages at branch points. The main chain has side chains of 15–25 glucose units attached every 8th or 9th glucose. Forms violet or red-violet inclusion compounds with iodine. Size indeterminate, from 500–1000kD.

amyloplast A plant *plastid* involved in the synthesis and storage of starch. Found in many cell types, but particularly storage tissues. Characteristically has starch grains in the plastid *stroma*.

amylose A linear water-soluble polysaccharide formed from 100–300 α(1–4)-linked glucopyranosyl units. Found both in starch (starch amylose) and glycogen (glycogen amylose). Smaller than *amylopectin* (10–50kD); forms blue inclusion compounds with iodine.

amytal (amylobarbitone) A barbiturate that inhibits respiration.

Anabaena A genus of *Cyanobacteria* which forms filamentous colonies with specialised cells (*heterocysts*), capable of nitrogen fixation. Ecologically important in wet tropical soils and forms symbiotic associations with the fern *Azolla*.

anabolic Of a process, route or reaction. Metabolic pathways are classically divided into anabolic and catabolic types. The former are synthetic processes, frequently requiring expenditure of phosphorylating ability of *ATP*, and reductive steps; the latter, degradative processes, often oxidative, with attendant regeneration of ATP.

anabolism Synthesis; opposite of *catabolism*.

anaemia (USA anemia) Reduced level of *haemoglobin* in blood for any of a variety of reasons including abnormalities of mature red cells (*sickle cell anaemia*, *spherocytosis*), iron deficiency, haemolysis of erythrocytes, reduced *erythropoiesis*, or haemorrhage (to name the most common). In aplastic anaemia, most or all of the haemopoietic bone marrow is lost, all cell types being equally diminished in number.

anaerobic The absence of air (specifically of free oxygen). Used to describe a biological habitat, or an organism that has very low tolerance for oxygen.

analgesia A state of insensitivity to pain, even though the subject is fully conscious.

anamnestic response Archaic term now replaced by such terms as *secondary immune response*, immune memory.

anaphase The stage of *mitosis* or *meiosis* beginning with the separation of sister *chromatids* (or homologous *chromosomes*) followed by their movement towards the poles of the *spindle*.

anaphylatoxin Originally used of an antigen which reacted with an *IgE* antibody thus precipitating reactions of *anaphylaxis*. Now restricted to defining a property of complement fragments *C3*a and *C5*a, both of which bind to the surfaces of *mast cells* and *basophils* and cause the release of inflammatory mediators.

anaphylaxis As opposed to *prophylaxis*. A system or treatment which leads to damaging effects on the organism. Now reserved for those inflammatory reactions resulting from combination of a soluble *antigen* with *IgE* bound to a *mast cell* which leads to degranulation of the mast cell and release of *histamine* and histamine-like substances,

causing localised or general immune reponses. (See *hypersensitivity*).

anaplasia Lack of differentiation, characteristic of some tumours.

anaplerotic Reaction which replenishes *tricarboxylic acid cycle* intermediates and allows respiration to continue; for example, carboxylation of *phosphoenolpyruvate* in plants.

anchorage Attachment, not necessarily adhesive in character; because the mechanism is not assumed the term ought to be more widely used.

anchorage dependence The necessity for attachment (and spreading) in order that a cell will grow and divide in culture. Loss of anchorage dependence seems to be associated with greater independence from external growth control and is probably one of the best correlates of *tumourigenic* ability *in vivo*. Anchorage independence is usually detected by *cloning* cells in soft-agarose; only anchorage independent cells will grow and divide (as they will in suspension).

androgen General term for any male sex hormone in vertebrates.

anemia See *anaemia*.

aneuploid Having a *chromosome* complement which is not an exact multiple of the haploid number. Chromosomes may be present in multiple copies (eg. trisomy) or one of a homologous pair may be missing in a diploid cell.

aneurysm Balloon-like swelling in the wall of an *artery*.

ANF See *atrial natriuretic peptide*.

angiogenesis The process of *vascularisation* of a tissue involving the development of new capillary blood vessels.

angiogenin Polypeptide (14kD) which induces the proliferation of endothelial cells; one of the components of *tumour angiogenesis factor*. It has sequence homology with pancreatic ribonuclease, and has ribonucleolytic activity, although the biological relevance of this is unclear.

angioma A knot of distended blood vessels atypically and irregularly arranged. Most angiomas are not tumours but *hamartomas*.

angiotensin A peptide *hormone*. Angiotensinogen (renin substrate) is a 60kD polypeptide released from the liver and cleaved in the circulation by *renin* to form the biologically inactive decapeptide angiotensin I. This is in turn cleaved to form active angiotensin II by Angiotensin Converting Enzyme (ACE). Angiotensin II causes contraction of vascular smooth muscle, and thus raises blood pressure, and stimulates *aldosterone* release from the adrenal glands. Angiotensin is finally broken down by angiotensinases.

angiotensin converting enzyme (ACE) See *angiotensin*.

angiotensinase See *angiotensin*.

angiotensinogen See *angiotensin*.

animal pole In most animal *oocytes* the nucleus is not centrally placed and its position can be used to define two poles. That nearest to the nucleus is the animal pole, and the other is the *vegetal pole*, with the animal-vegetal axis between the poles passing through the nucleus. During *meiosis* of the oocyte the *polar bodies* are expelled at the animal pole. In many eggs there is also a graded distribution of substances along this axis, with pigment granules often concentrated in the animal half and yolk, where present, mainly in the vegetal half.

animalised cells The early blastula (8–16 cells) of sea urchins has animal and vegetal poles; by manipulating the environmental conditions it is possible to shift more cells from vegetal to animal in their characteristics.

anionic detergents Detergents in which the hydrophilic function is fulfilled by an anionic grouping. *Fatty acids* are the best known natural products in this class, but it is doubtful if they have a specific detergent function in any biological system. The important synthetic species are aliphatic sulphate esters, eg. sodium dodecyl sulphate (*SDS* or SLS).

anisotropic Not the same in all directions.

ankylosing spondylitis Poly*arthritis* involving spine, which may become more-or-less rigid, and which shows up radiologically as "bamboo spine". Interestingly the disease seems to be associated with HLA-B27; those with this *histocompatibility antigen* are 300 times more likely to get the disease, 90% of sufferers have HLA-B2.

ankylosis Fusion of bones across a joint. This may be done deliberately by surgery, or may be a complication of *chronic inflammation*. See *ankylosing spondylitis*.

ankyrin Globular protein (200kD) which links *spectrin* and an integral membrane protein (*band III*) in the erythrocyte plasma membrane. Isoforms exist in other cell types.

annealing 1. Toughening upon slow cooling. 2. Used in the context of DNA renaturation after temperature dissociation of the two strands. Rate of annealing is a function of complementarity. 3. Fusion of microtubules or microfilaments end-to-end.

annexins Family of calcium- and phospholipid-binding proteins which includes *lipocortin, endonexin* I and II, *calpactin, p70, calelectrin*. May be involved in the control of exocytosis.

annulate lamellae Perforated membrane-bound *cisternae* described in the oocytes of several animal species. Associated with the nuclear envelope; may be associated with tubulin synthesis from mRNA accumulated in these organelles.

annulus Ring-like structure, *adj.* annulate.

anomers The α- and β-forms of hexoses. Interconversion (mutarotation) is anomerisation and is promoted by mutarotases (aldose epimerases).

anoxia Lack of oxygen.

ANP See *atrial natriuretic peptide*.

antagonist Compound which inhibits the effect of a hormone or drug; the opposite of *agonist*.

antenna complex Light-harvesting array of pigments in photosynthetic bacteria.

antennapedia *Homeotic mutant* of *Drosophila*, in which a leg is formed instead of an antenna.

anterograde transport Movement of material from the cell body of a *neuron* into axons and dendrites (retrograde axoplasmic transport also occurs).

anthocyanins Water-soluble vacuolar pigments of plant cells. Colours range from red to blue depending on the amount and type of substitution and glycosylation. Often developed in response to stress, eg. nitrogen deficiency, low temperature or high insolation.

Table A3. Mode of action of various antibiotics

Table A3. Mode of action of various antibiotics

Antibiotic	Source organism	Mode of action
Penicillins		
Ampicillin Benzylpenicillin Phenoxymethyl- penicillin	Semi-synthetic *Penicillium notatum* *Penicillium notatum*	Inhibitors of cell wall synthesis. Penicillins kill growing bacteria probably by binding to active site of the enzyme that cross-links the peptidoglycan wall. Most effective against Gram positive bacteria. Resistant strains of bacteria produce penicillinases (β-lactamase) that cleave the 4-membered β-lactam ring
Cephalosporins		
Cephaloridine Cephalothin	*Cephalosporium acremonium* (Fungi) *Cephalosporium* species	Inhibitors of cell wall synthesis. Similar mode of action to penicillins
Aminoglycosides		
Gentamicin Kanamycin Neomycin Streptomycin	*Micromonospora purpurea* *Streptomyces kanamyceticus* *Streptomyces fradiae* *Streptomyces griseus*	Inhibitors of bacterial protein synthesis. They bind to the 30S subunit of the 70S (bacterial) ribosome, though at a number of different sites. They prevent the transition from an initiating complex to a chain-elongating ribosome
Macrolides		
Erythromycin Oleandomycin	*Streptomyces erythraeus* *Streptomyces antibioticus*	Inhibit bacterial protein synthesis by binding to 50S subunit preventing the translocation step
Cycloheximide	*Streptomyces noursei*	Inhibits eukaryote, but not prokaryote, protein synthesis by preventing the peptidyl transferase reaction
Peptides		
Bacitracin	*Bacillus subtilis*	Inhibitor of cell wall synthesis
Gramicidin	*Bacillus brevis*	Ionophore. Forms "pore" in cell membrane, causing loss of K^+
Polymyxin	*Bacillus polymyxa*	Acts on cell membrane causing leakage of small molecules
Polyenes		
Amphotericin B Nystatin	*Streptomyces nodosus* *Streptomyces noursei*	Form complex with cholesterol in the plasma membrane. These complexes form a ring in each half bilayer, producing a "pore". In the case of amphotericin B, eight complexes give a pore of diameter 8Å, large enough to leak glucose. Only act on membranes containing cholesterol, so have no effect on bacteria, but kill eukaryotes including the fungi
Tetracyclines		
Chlortetracycline Tetracycline	*Streptomyces aureofaciens* *Streptomyces aureofaciens* (mutant)	Inhibit bacterial protein synthesis by preventing aminoacyl tRNA binding to the A-site of the 30S ribosomal subunit

Table A3. (Continued)

Antibiotic	Source organism	Mode of action
Others		
Actinomycin D	*Streptomyces parvullus*	Inhibits RNA synthesis by binding to DNA blocking movement of RNA polymerase
Chloramphenicol	*Streptomyces venezuelae*	Inhibits prokaryote, but not eukaryote, protein synthesis by preventing the peptidyl transferase reaction
Puromycin	*Streptomyces albo-niger*	Inhibits protein synthesis. Analogue of 3' end of aminoacyl tRNA. Is added to growing end of peptide chain but as it has no aminoacyl group causes premature chain termination
Rifamycin	*Streptomyces mediterranei*	Inhibits bacterial RNA synthesis. Binds to RNA polymerase and prevents initiation of transcription. Effective against acid-fast as well as Gram positive bacteria
Valinomycin	*Streptomyces fulvissimus*	Ionophore. Causes leakage of K^+

anthocyanidin The aglycone of *anthocyanin*.

antiauxin A growth-inhibitory substance which antagonises the action of *auxins* in higher plants.

antibiotic Substance produced by one microorganism which selectively inhibits the growth of another. Wholly synthetic antibiotics have been produced. See Table A3.

antibody General term for an *immunoglobulin*.

antibody-producing cell A *lymphocyte* of the B series synthesising and releasing *immunoglobulin*. Equivalent to plasmacyte and *plasma cell*.

anticoagulant Substance which inhibits the clotting of blood. The most commonly used are EDTA and citrate (both of which chelate calcium) and *heparin* (which interferes with *thrombin*, probably by potentiating *antithrombins*). Other compounds such as warfarin and dicoumarol act as anticoagulants *in vivo* by interfering with clotting *Factors*.

anticodon Nucleotide triplet on *transfer RNA* which is complementary to the *codon* of the *messenger RNA*.

antidiuretic hormone (ADH) See *vasopressin*.

antigen A substance inducing and reacting in an *immune response*; usually macromolecules of MW more than 1000 (but see *hapten*). The *antigenic determinant* group is termed an *epitope* and the association of this with a carrier molecule (which may be part of the same molecule) makes it active as an antigen. Thus dinitrophenol-modified human serum *albumin* is antigenic to humans, dinitrophenol being the hapten. Usually antigens are foreign to the animal in which they produce immune reactions.

antigen-antibody complex (immune complex) The product of the reaction of *antigen* and *immunoglobulin*. If the antigen is polyvalent the complex may be insoluble; see also *glomerulonephritis*, *Arthus reaction*, Type III *hypersensitivity* reactions. Immune complexes activate *complement* through the classical pathway.

antigen presentation See *antigen presenting cell*.

antigen presenting cell A cell which carries on its surface antigen bound to MHC Class I or II molecules, and presents the antigen in this "context" to T-cells. Includes macrophages, endothelium, dendritic cells and Langerhans cells of the skin. See also *MHC restriction, histocompatibility antigens*.

antigen processing Modification of an

17

antigen by *accessory cells*. This usually involves endocytosis of the antigen and either minimal cleavage or unfolding. The processed antigen is then presented in modified form by the accessory cell.

antigenic determinant Also known as an *epitope*; that part of an antigenic molecule against which a particular immune response is directed. For instance a tetra- to pentapeptide sequence in a protein, a tri- to pentaglycoside sequence in a polysaccharide; see also *hapten*. In the animal most antigens will present several or even many antigenic determinants simultaneously.

antigenic variation The phenomenon of changes in surface *antigens* in populations of **Trypanosoma** and **Plasmodium** (and some other parasitic Protozoa) in order to escape immunological defence mechanisms. At least 100 different surface proteins have been found to appear and disappear during antigenic variation in a clone of trypanosomes. Each antigen (variant-specific glycoprotein) is encoded in a separate gene. Antigenic variation is also known to occur in free-living Protozoa and certain bacteria.

anti-gibberellin A compound which competes with *gibberellin* at the site of action. Used by some researchers of growth retardants, eg. cycocell (CCC), but this is improper since these compounds interfere with gibberellin biosynthesis.

antilymphocyte serum (ALS) *Immunoglobulins* raised *xenogeneically* against *lymphocyte* populations; the term refers particularly to *antisera* recognising one or more *antigenic determinants* on *T-cell* populations. ALS is used in experimental *immunosuppression*.

antimitotic drugs Drugs which block mitosis; the term is often used of those which cause metaphase-arrest such as *colchicine* and the *vinca alkaloids*. Many anti-tumour drugs are antimitotic, blocking proliferation rather than being cytotoxic.

antimycin Inhibitor of *QH_2-cytochrome c reductase*.

antiparallel Having the opposite *polarity* (eg. the two strands of a DNA molecule).

antiplectic Pattern of *metachronal* coordination of ciliary beat in which the waves pass in the opposite direction to that of the active stroke.

antipodal cells Three cells of the embryo sac in angiosperms, found at the end of the embryo away from the point of entry of the pollen tube.

antiport A membrane protein which transports two different ions or molecules, in opposite directions, across a lipid bilayer. Energy may be required, as in the *sodium pump*; or it may not, as in the Na^+/H^+ antiport.

antiproteases (antiproteinases) Substances which inhibit proteolytic enzymes.

antiserum Serum containing *immunoglobulins* against specified *antigens*.

antithrombins Plasma *glycoproteins* of the α_2-globulin class which inhibit the proteolytic activity of *thrombin* and serve to regulate the process of blood clotting.

antitoxin An *antibody* reacting with a toxin, eg. anti-*cholera toxin* antibody.

anucleate Having no nucleus.

anucleolate Literally, having no nucleoli. An anucleolate mutant of *Xenopus* (viable when *heterozygous*) is used in nuclear transplantation experiments because nuclei are of identifiable origin.

apamin A small (MW 2027) basic peptide present in the venom of the honey bee (*Apis mellifera*). Blocks calcium-activated potassium channels and has an inhibitory action in the central nervous sytem.

apical dominance Growth-inhibiting effect exerted by actively-growing apical bud of higher-plant shoots, preventing the growth of buds further down the shoot. Thought to be mediated by the basipetal movement of *auxin* from the apical bud.

apical meristem The meristem at the tips of stems and roots. Composed of undifferentiated cells, many of which divide to add to the plant body but the central mass (the quiescent centre) remains inert and only becomes active if the meristem is damaged. Also known as eumeristem.

apical plasma membrane The term used for the cell membrane on the apical (inner or upper) surface of transporting epithelial cells. This region of the cell membrane is separated, in vertebrates, from the basal membrane by a ring of tight junctions (*zonulae occludentes*) which prevents free mixing of membrane proteins from these two domains.

aplastic anaemia See *anaemia*.

Aplysia (sea hare) Opisthobranch mollusc with reduced shell; favourite source of ganglia for neurophysiological study.

apocrine Form of secretion in which the apical portion of the cell is shed, as in the secretion of fat by cells of the mammary gland. The fat droplet is surrounded by apical plasma membrane, and this has been used experimentally as a source of plasma membrane.

apoenzyme An enzyme without its *cofactor*.

apoplast Since the *protoplasts* of cells in a plant are connected through *plasmodesmata*, plants may be described as having two major compartments: the apoplast, which is external to the plasma membrane and includes cell walls, xylem vessels etc. through which water and solutes pass freely, and the *symplast*, the total cytoplasmic compartment.

apoprotein When a protein can exist as a complex between polypeptide and a second moiety of non-polypeptide nature, the term apoprotein is sometimes used to refer to the molecule divested of the latter. For example, *ferritin* lacking its ferric hydroxide core may be referred to as apoferritin.

apoptosis Form of cell death in which DNA is fragmented into pieces (200 base-pairs or multiples thereof) but there is no lysis of the cell. Seems to be associated with T-cell killing.

APUD cells Acronym for Amine-Precursor Uptake and Decarboxylation cells: *paracrine* cells of which *argentaffin cells* are an example. Usage neither helpful nor memorable.

arabinogalactans Plant cell-wall polysaccharides containing predominantly arabinose and galactose. Two main types are recognised: arabinogalactan I, found in the pectin portion of angiosperms and containing $\beta(1–4)$-linked galactan with α-arabinose side-chains; arabinogalactan II, a highly-branched polymer containing $\beta(1–3)$- and $\beta(1–6)$-linked galactose and peripheral α-arabinose residues. Arabinogalactan II is found in large amounts on some gymnosperms, especially larches, and is related to *AGP*.

arabinose A *pentose* monosaccharide which occurs in both D- and L-configurations. D-arabinose is the 2-epimer of D-*ribose*, ie. differs from D-ribose by having the opposite configuration at carbon 2. D-arabinose occurs *inter alia* in the polysaccharide *arabinogalactan*, a neutral *pectin* of the cell wall of plants, and in the antimetabolites cytosine and adenine arabinoside.

arabinoxylan Polysaccharide with a backbone of $\beta(1–4)$-linked *xylose* with side chains of *arabinose* ($\alpha(1–3)$-linked): constituent of *hemicellulose* of angiosperm cell wall.

arachidonic acid (5,8,11,14 eicosatetraenoic acid) An essential dietary component for mammals. The free acid is the precursor for biosynthesis of *prostaglandins*, *thromboxanes*, and hydroxyeicosatetraenoic acid derivatives (*HETEs*) including *leucotrienes* and is thus of great biological significance. Within cells the acid is found in the esterified form as a major acyl component of membrane *phospholipids* (especially *phosphatidyl inositol* and *phosphatidyl choline*) and its release from phospholipids is thought to be the limiting step in the formation of its active metabolites.

arboviruses Diverse group of single-stranded RNA viruses which have an envelope surrounding the *capsid* Anthropod borne, hence the name, and multiply in both invertebrate and vertebrate host. In arthropods the viruses multiply in the gut and are present in the salivary glands; transmission occurs when the the vertebrate host is bitten. Apparently only pathogenic in the vertebrate host (though this may be anthropocentric prejudice), and in humans cause, for example, yellow fever and encephalitis. The group is very heterogeneous and taxonomically rather inappropriate since the relationships are tenuous; three major families are recognised: *Togaviridae*, *Bunyaviridae*, and *Arenaviridae*.

Archaebacteria One of two major subdivisions of the prokaryotes. There are three main Orders, extreme *halophiles*, *methanogens* (see **Methanobacterium**), and sulphur-dependent extreme *thermophiles*. Archaebacteria differ from *Eubacteria* in ribosomal structure, the possession (in some cases) of *introns*, and in a number of other features including membrane composition.

archaeocyte An amoeboid cell type from sponges (Porifera).

Arenaviridae Family of enveloped single-stranded RNA viruses including Lassa virus, lymphocytic choriomeningitis virus, and the Tacaribe group of viruses. Virions are spherical or pleomorphic (ca 120nm) and in addition to the ribonucleoprotein may contain several host ribosomes. Can replicate in many mammalian cells in culture. Not all require arthropods for transmission, despite their inclusion in the *arbovirus* group, and Lassa fever is highly contagious.

argentaffin cells So-called because they form cytoplasmic deposits of metallic silver from silver salts. Their characteristic histochemical behaviour arises from *serotonin*, which they secrete. Found chiefly in the epithelium of the gastrointestinal tract (though possibly of neural crest origin) their function is rather obscure, although there is a widely distributed family of such *paracrine* (local endocrine) cells (*APUD cells*).

arginine (Arg; R; MW 174) An essential amino acid; a major component of proteins and contains the guanido group which has a pK_a of greater than 12, so that it carries a permanent positive charge at physiological pH. See Table A2.

arrestin (S-antigen) Protein (48kD) from retinal rod cells; competes with *transducin* for binding to photorhodopsin.

Arrhenius plot A plot of the logarithm of reaction rate against the reciprocal of absolute temperature. For a single stage reaction this gives a straight line from which the activation energy and the frequency factor can be determined. Often applied to data from complex biological systems when the form observed is frequently a series of linear portions with sudden changes of slope. Great caution must be observed in interpreting such slopes in terms of activation energies for single processes.

arrowheads Fanciful description given to the pattern of *myosin* molecules attached to a filament of F-*actin*. Easier to see if tannic acid is added to the fixative. The arrowheads indicate the polarity of the filament.

arteriosclerosis Imprecise term for various disorders of arteries; often used as a synonym for *atherosclerosis*.

artery Blood vessel carrying blood away from the heart; walls have smooth muscle and are innervated by the *sympathetic nervous system*.

arthritis General term for inflammation of one or more joints. There may be many causes of arthritis, although in most cases the basis of the inflammation is not understood. This is particularly true of *rheumatoid arthritis*, though knowledge of other forms is not much better.

arthropathy Any disease affecting a joint — care should be taken not to confuse arthrosclerosis (stiffness of joints) with atherosclerosis.

arthropods Animals of the largest phylum of the animal kingdom, containing several million species. Arthropods are characterised by a rigid external skeleton, paired and jointed legs, and a haemocoel. The phylum Arthropoda includes the major classes Insecta, Crustacea, Myriapoda and Arachnida.

Arthus reaction Localised *inflammation* due to injection of *antigen* into an animal which has a high level of circulating *antibody* against it. A haemorrhagic reaction with oedema occurs due to the destruction of small blood vessels by thrombi. It may occur, as in "Farmer's Lung", as a reaction to natural exposure to antigen.

aryl sulphatase (EC 3.1.6.1) Aryl sulphatases A, B and C comprise a group of enzymes originally assayed by their ability to hydrolyze O-sulphate esters of aromatic substrates. Aryl sulphatase A, substrate cerebroside 3-sulphate, is deficient in metachromatic leukodystrophy. Aryl sulphatase B, substrate acetylhexosamine 4-sulphate in glycosaminoglycans, is deficient in Maroteaux-Lamy syndrome. Aryl sulphatase C hydrolyses oestrogen sulphates. All three are deficient in multiple sulphatase deficiency.

asbestosis *Fibrosis* of the lung as a result of the chronic inhalation of asbestos fibres. The needle-like asbestos fibres are phagocytosed by *alveolar macrophages* but burst the phagosome and kill the macrophage and the cycle is repeated. *Mesothelioma*, a rare tumour of the *mesothelial* lining of the pleura, is associated with intense chronic exposure to asbestos dust, particularly that of crocidolite asbestos.

Aschoff bodies Small *granulomas* composed of *macrophages*, *lymphocytes* and multinucleate cells grouped around eosinophilic *hyaline* material derived from collagen.

Characteristic of the *myocarditis* of rheumatic fever.

ascites Accumulation of fluid in the peritoneal cavity producing swelling; causes include infections, portal hypertension and various tumours.

ascites tumour Tumour which grows in, for example, the peritoneal cavity as a suspension of cells. Obviously such cells have lost *anchorage dependence*, and they can easily be isolated and passaged. *Hybridomas* are sometimes grown as ascites tumours, and the ascites fluid can then be used as the crude "antiserum".

ascorbic acid (vitamin C) A requisite in the diet of humans and guinea pigs. May act as a reducing agent in enzymic reactions, particularly those catalyzed by hydroxylases. See Table V1.

ascus Cell involved in sexual reproduction in ascomycete fungi. Forms an elongated spore case containing four or eight haploid spores (ascospores) arranged in a row.

Askanazy cells Abnormal thyroid epithelial cells found in autoimmune *thyroiditis*. The cubical cells line small *acini* and have *eosinophilic* granular cytoplasm and often bizarre nuclear morphology. Also known as Hurthle cells, oxyphil cells or oncocytes.

asialoglycoprotein The carbohydrate moiety of many vertebrate glycoproteins bears terminal residues of *neuraminic acid*. If such residues are removed, eg. by treatment with a *neuraminidase*, the resulting proteins are known as asialoglycoproteins. In the case of certain plasma proteins, the asialo-derivatives are specifically bound by a receptor on the surface of liver parenchymal cells (the clearance receptor).

asparaginase (EC 3.5.1.1) Enzyme which hydrolyses L-asparagine to L-aspartate and ammonia; used as an anti-tumour agent especially against lymphosarcoma and lymphatic leukaemia.

asparagine (β-asparagine; Asn; N; MW 132) The β-amide of aspartic acid; the L-form is one of the 20 amino acids directly coded in proteins. Coded independently of aspartic acid. See Table A2.

aspartate (aspartic acid; Asp; D; MW 133) L-aspartate is one of the 20 amino acids directly coded in proteins; the free amino acid is a neurotransmitter. See Table A2.

Aspergillus A genus of common ascomycete fungi found in soil. Industrially important in production of organic acids (*A. niger*), and a popular fungus for genetic study (*A. nidulans*).

aspirin (acetyl salicylic acid) An analgesic, antipyretic and anti-inflammatory drug. It is a potent *cyclo-oxygenase* inhibitor and blocks the formation of *prostaglandins* from *arachidonic acid*.

association constant (K_a; K_{ass}) Reciprocal of *dissociation constant*. A measure of the extent of a reversible association between two molecular species at equilibrium.

aster Star-shaped cluster of microtubules radiating from the polar organising centre at the start of mitosis.

astroblast An embryonic *astrocyte*.

astrocyte A *glial cell* found in vertebrate brain, named for its characteristic star-like shape. Astrocytes lend both mechanical and metabolic support for neurons, regulating the environment in which they function.

astrocytoma A neuro-ectodermal tumour (*glioma*) arising from *astrocytes*. Probably the commonest glioma, it has a tendency to show *anaplasia*.

astroglia See *astrocytes*.

astrogliosis Hypertrophy of the *astroglia*, usually in response to injury.

ataxia telangiectasia Louis Barr syndrome; a hereditary *autosomal* recessive disease in humans characterised by a high frequency of spontaneous chromosomal aberrations, neurological deterioration and susceptibility to various cancers. Ataxia: imbalance of muscle control; telangectasia: dilated capillary vessels. In part an immune deficiency disease and in part one of DNA repair; it is believed to be due to hypersensitivity to background ionising radiation.

atheroma Degeneration of the walls of the arteries because of the deposition of fatty plaques in the *intima* of the vessel wall, and scarring and obstruction of the lumen. Each plaque may represent the proliferating progeny of a single smooth-muscle cell; of monoclonal origin.

atherosclerosis General term for conditions in which arteries are thickened and hardened. See *atheroma*.

atopy General term referring to immediate *hypersensitivity*.

ATP (adenosine 5′ triphosphate) The synthesis of ATP in cells from ADP is driven by energy-yielding processes. Enzymic transfer of the terminal phosphate or pyrophosphate to a wide variety of substrates provides a means of transferring chemical free energy from *metabolic* to *anabolic* processes.

ATPase An enzyme capable of releasing the terminal (γ) phosphate from ATP, yielding *ADP* and inorganic phosphate. The description could mislead, because in most cases the enzymic activity is not a straightforward hydrolysis, but is part of a coupled system for achieving an energy-requiring process, such as ion-pumping or the generation of motility.

atrial natriuretic factor Obsolete name for *atrial natriuretic peptide*.

atrial natriuretic peptide (ANP) A polypeptide hormone found mainly in the atrium of many species of vertebrates, released in response to atrial stretching, and thus to elevated blood pressure. ANP acts to reduce blood pressure through stimulating the rapid excretion of sodium and water in the kidneys (reducing blood volume), by relaxing vascular smooth muscle (causing vasodilation) and through actions on the brain and adrenal glands.

atrophy Wasting away of tissue.

attachment constriction See *centromere*.

attachment plaques Specialised structures at the ends of a chromosome by which it is attached to the nuclear envelope at the *leptotene* stage of mitosis.

attenuation Viruses which have been *passaged* extensively may become attenuated (non-virulent), or their virulence may be markedly reduced; they can then be used as a vaccine.

AUG The *codon* in *messenger RNA* which specifies initiation of a polypeptide chain or, within a chain, incorporation of a *methionine* residue.

aurovertin Inhibitor of the *respiratory chain* that binds to ATPase.

Australia antigen See *hepatitis B*.

autoantibody Antibody which reacts with an antigen that is a normal component of the body. May lead to destruction of cells and tissues bearing that antigen. Autoimmunity is a causative factor in diseases such as *systemic lupus erythematosus*, Hashimoto's thyroiditis, *myasthenia gravis* and, more uncertainly, *rheumatoid arthritis*.

autocatalytic A compound that catalyses its own chemical transformation. More commonly a reaction that is catalyzed by one of its products or an enzyme-catalyzed reaction in which one of the products functions as an enzyme activator.

autocrine Secretion of a substance, such as a *growth factor*, which affects the secretory cell itself. One route to independence of *growth control* is by autocrine growth factor production.

autogamy Self-fertilisation, common in plants and also in some ciliate protozoa, where gametic nuclei from a single micronucleus subsequently fuse to form the zygote nucleus.

autoimmunity A state of immunity to self antigens. Response may be humoral or cellular. If sufficiently severe or prolonged, may give rise to autoimmune disease.

autolysis Spontaneous *lysis* (rupture) of cells or organelles produced by the release of internal hydrolytic enzymes. Normally associated with the release of lysosomal enzymes.

autonomic nervous system Part of the nervous system which is not under conscious control, comprising two *antagonistic* components, the *sympathetic* and *parasympathetic* divisions. Together, they control the heart, viscera, smooth muscle, glands and skin.

autophagy Removal of cytoplasmic components, particularly membrane bounded organelles, by digesting them within *secondary lysosomes* (autophagic vacuoles). Particularly common in embryonic development and senescence.

autoradiography Technique in which a specimen containing radioactive atoms is overlaid with a photographic emulsion, which is subsequently developed, revealing the localisation of radioactivity as a pattern of silver grains. Resolution is determined by the path length of the radiation, and so the low-energy beta-emitting isotope, tritium, is usually used.

autosomes Chromosomes other than the sex chromosomes.

autotrophs Organisms which synthesise all their organic molecules from inorganic carbon compounds (carbon dioxide, bicarbonate or carbonate). May be photo-autotrophs or chemo-autotrophs, depending upon the source of the energy; the latter are also known as lithotrophic organisms.

auxins A group of *plant growth substances*, the most common example being *indole acetic acid* (IAA), responsible for raising the pH around cells, making the cell wall less rigid and allowing elongation.

auxotroph Mutant which differs from the wild-type in requiring a nutritional supplement for growth; a deficiency mutant.

auxotyping Method for strain-typing *Neisseria* by checking their requirements for specific nutrients in defined media.

avian leukaemia virus Group of C-type RNA tumour viruses (Oncovirinae) which cause various leukaemias and other tumours in birds. The acute leukaemia viruses, which are replication-defective and require helper viruses, include avian erythroblastosis (AEV), myeloblastosis (AMV) and myelocytomatosis viruses. AEV carries two transforming genes, *v-erbA* and *v-erbB*; the cellular homologue of the latter is the structural gene for the *epidermal growth factor* receptor. AMV carries *v-myb* and causes a myeloid leukaemia; avian myelocytomatosis virus carries *v-myc*. The avian lymphatic leukaemia viruses (ALV) are also *Retroviridae* but are replication-competent and induce neoplasia only after several months; they often occur in conjunction with replication-defective leukaemia viruses.

avian sarcoma viruses *Retroviridae* rapidly causing tumours in chickens; best known is *Rous sarcoma virus*.

avidin *Biotin*-binding protein from egg-white. Binding is irreversible — a diet of raw egg-white leads to biotin deficiency.

avidity Strength of binding, usually of a small molecule with multiple binding sites by a larger; particularly the binding of a complex antigen by an antibody. (*Affinity* refers to simple receptor–ligand systems).

axenic Of an organism or cell; raised without contaminating, perhaps symbiotic, micro-organisms.

axokinin Axonemal protein (56kD) which, when phosphorylated by a cAMP-dependent protein kinase, reactivates the *axoneme*.

axolemma Plasma membrane of an axon.

axon The process, or processes, of a neuron which conducts impulses (*action potentials*), usually over long distances. See *dendrite*.

axon hillock Tapering region between a *neuron's* cell body and its axon. This region is responsible for summating the graded inputs from the *dendrites*, and producing *action potentials* if the threshold is exceeded.

axoneme The central *microtubule* complex of eukaryotic *cilia* and *flagella* with the characteristic "9 + 2" arrangement of tubules when seen in cross-section.

axoplasm The cytoplasm of a *neuron*.

axopod (*pl.* axopodia) Thin processes (a few μm in diameter but up to 500μm long), supported by complex arrays of *microtubules*, which radiate from the bodies of *Heliozoan* protozoa.

axostyles Ribbon-like bundles of *microtubules* found in certain parasitic flagellate protozoa; some axostyles may generate bending waves by *dynein*-mediated sliding of microtubules.

5-azacytidine (β-ribofuranosyl 5-azacytosine) The ribonucleoside of *5-azacytosine*.

5-azacytosine An analogue of the *pyrimidine* base cytosine, in which carbon-5 is replaced by a nitrogen. In DNA, unlike cytosine, it cannot be methylated.

azide Usually the sodium salt NaN_3, an inhibitor of electron transport which blocks electron flow from cytochrome oxidase to oxygen.

azo dyes Dyes which contain the –N=N– linkage. They are easily prepared from diazo compounds.

azurin Blue copper-containing protein from *Pseudomonas aeruginosa*; also the name of a dye used in histochemistry.

azurophil granules *Primary lysosomal* granules found in *neutrophil* granulocytes; contain a wide range of hydrolytic enzymes. Sometimes referred to as primary granules to distinguish them from the specific or secondary granules.

B

B cells (of pancreas) Cells within discrete endocrine islands (*Islets of Langerhans*) embedded in the major exocrine tissue of vertebrate pancreas. The B or β cells (originally distinguished by differential staining from A, C and D), are responsible for synthesis and secretion into the blood of the hormone insulin.

B chromosome Small acentric chromosome; part of the normal genome of some races and species of plants.

b-c₁ complex A part of the mitochondrial *electron transport chain*, which accepts electrons from *ubiquinone*, and passes them on to *cytochrome* c. The b-c₁ complex consists of 2 cytochromes.

B-DNA The structural form of *DNA* originally described by Crick and Watson. Normally found in hydrated DNA and is strictly an average, approximate structure for a family of B forms. In B-DNA, the double helix is a right-handed helix with about 10 residues per turn and has a major and a minor groove. The planes of the base pairs are perpendicular to the helix axis. Other structural forms are *A-DNA* and *Z-DNA*.

B-lymphocyte Lymphocyte that synthesises immunoglobulin. The B-lymphocytes are a maturational series from pre-B-cells to antibody-secreting cells (plasma cells). At various maturational steps B-lymphocytes may be detected by the presence of cytoplasmic immunoglobulin, of surface immunoglobulin and later by secretion of immunoglobulin. By virtue of the antibody that they make, B-lymphocytes mediate the humoral immune response. In birds, B-cells originate from the *Bursa of Fabricius*; in mammals they originate from the bone marrow.

Babes-Ernst granules Metachromatic intracellular deposits of polyphosphate found in *Corynebacterium diphtheriae* when the bacteria are grown on sub-optimal media. Stain reddish with methylene blue or toluidine blue.

bacillus Any cylindrical (rod-shaped) bacterium. Bacilli are usually 0.5–1.0μm long, 0.3–1μm wide. The genus *Bacillus* comprises a group of aerobic, spore-forming bacillus-shaped bacteria.

Bacillus Calmette-Guerin (BCG) An attenuated *mycobacterium* derived from **Mycobacterium tuberculosis**. The bacterium is used in vaccination against tuberculosis. Extracts of the bacterium have remarkable powers in stimulation of *lymphocytes* and *leucocytes*.

bacitracins Branched cyclic peptides produced by strains of *Bacillus licheniformis*. Interfere with murein (*peptidoglycan*) synthesis in Gram positive bacteria.

bacteraemia (USA bacteremia) The presence of living bacteria in the circulating blood; usually implies the presence of small numbers of bacteria which are transiently present without causing clinical effects, in contrast to septicaemia in which there are clinical symptoms such as fever.

bacteria One of the two major classes of prokaryotic organism (the other being the *Cyanobacteria*). Bacterial cells are small (linear dimensions of around 1μm), noncompartmentalised, with circular DNA, and ribosomes of 70S. Protein synthesis differs from that of eukaryotes; thus many antibacterial antibiotics that interfere with protein synthesis do not affect the eukaryotic host. Recently bacteria have been subdivided into *Eubacteria* and *Archaebacteria*, although some would consider the Archaebacteria to be a third kingdom, distinct from both Eubacteria and Eukaryotes. The Eubacteria can be further subdivided on the basis of their staining using *Gram stain*. Since the difference between Gram positive and Gram negative depends upon a fundamental difference in cell wall structure it is therefore more soundly based than classification on gross morphology alone (into cocci, bacilli, etc.).

bacterial chemotaxis The response of bacteria to gradients of attractants or repellents. In a gradient of attractant the probability of deviating from a smooth forward path is reduced if the bacterium is moving up-gradient. Since the opposite is true if moving down-gradient, the effect is to bias displacement towards the source of attractant. Strictly, should perhaps be considered a *klinokinesis* with adaptation.

bacterial flagella Thin filaments composed of *flagellin* subunits which are rotated by the basal motor assembly and act as propellors. If rotating anticlockwise (as viewed from the flagellar tip) the bacterium moves in a straight path, if clockwise the bacterium "tumbles".

bacteriochlorophyll Varieties of *chlorophyll* (bacteriochlorophylls a, b, c, d, e and g) found in *photosynthetic bacteria* and differing from plant chlorophyll in the substituents around the tetrapyrrole nucleus of the molecule, and in the absorption spectra.

bacteriocide A substance that kills bacteria.

bacteriocins Proteins (exotoxins), often *plasmid*-coded, produced by bacteria and which kill other bacteria (not eukaryotic cells). *Colicins* are produced by about 40% of *E. coli* strains: colicin E2 is a DNAase, colicin E3 an RNAase.

bacteriophaeophytin-b (USA bacteriopheophytin-b) One of the components of the bacterial photosynthetic *reaction centre*. (See also *ubiquinone*).

bacteriophages (phages) Viruses that infect bacteria. The bacteriophages that attack *Escherichia coli* are termed coliphages, examples are λ phage, the T phages (T1, etc.) and male-specific phages (M13, etc.). Basically, phages consist of a protein coat or *capsid* enclosing the genetic material, DNA or RNA, which is injected into the bacterium upon infection. In virulent phages all synthesis of host DNA, RNA and proteins ceases and the phage genome directs the synthesis of phage nucleic acids and proteins using the host's transcriptional and translational apparatus. These phage components then self-assemble to form new phage particles. The synthesis of a phage lysozyme leads to rupture of the bacterial cell wall releasing, typically, 100–2000 phage progeny. Temperate phages, such as λ, may also show this lytic cycle when they infect a cell, but more frequently they induce *lysogeny*. The study of bacteriophages has been important for our understanding of gene structure and regulation. Phage λ has been extensively used as a *vector* in recombinant DNA studies.

bacteriorhodopsin A light-driven proton-pumping protein, similar to *rhodopsin*, found in "purple patches" in the cytoplasmic membrane of the bacterium *Halobacterium halobium*. It is composed of 248 amino acids, comprising 7 transmembrane helices, and contains the light-absorbing *chromophore*, *retinal*. Light absorption maxima: 568nm (light-adapted); 558nm (dark-adapted). Each absorbed photon causes the movement of 2 protons from cytoplasmic to extracellular sides of the membrane. The resulting proton gradient is used (amongst other things) to drive *chemiosmotic* synthesis of ATP.

bacteriostatic Adjective applied to substances which inhibit the growth of bacteria without necessarily killing them.

bacteroid Small, often irregularly rod-shaped bacterium, eg. those found in root nodules of nitrogen-fixing plants.

bag cell neurons Cluster of electrically-coupled neurons in the abdominal ganglion of **Aplysia** that are homogeneous, easily dissected out and release peptides which stimulate egg laying.

Balbiani ring The largest *puffs* seen on the *polytene chromosomes* of diptera are called Balbiani rings after the nineteenth century microscopist who first described these chromosomes.

band cells Immature *neutrophils* released from the bone-marrow reserve in response to acute demand.

band III Protein (90kD) of the human erythrocyte membrane, identified as the major anion transport/exchange protein; analogous proteins exist in erythrocytes from other species. A dimeric transmembrane glycoprotein, with binding sites for many cytoplasmic proteins, including *ankyrin*, on its cytoplasmic domain.

banding patterns Chromosomes stained with certain dyes, commonly *quinacrine* or *Giemsa*, show a pattern of transverse bands of light and heavy staining which is characteristic for the individual chromosome. The basis of the differential staining, which is the same in most tissues, is not understood: each band generally represents 5–10% of the length, about 10^7 base pairs, although this is not true for *polytene chromosomes* which show more than 4000 bands per *Drosophila* genome.

Barr body Small dark-staining inactivated chromosome seen in female (XX) cells. According to the *Lyon hypothesis*, random inactivation occurs.

basal body Structure found at the base of eukaryotic *cilia* and *flagella* consisting of a continuation of the nine outer sets of axonemal *microtubules* but with the addition of a C-tubule (C-subfibre) to form a triplet (like the *centriole*). May be self-replicating and serves as a nucleating centre for axonemal assembly. Often anchored in the cytoplasm by the *rootlet system*. Synonymous with *kinetosome*.

basal cells General term for relatively undifferentiated cells in an epithelial sheet which give rise to more specialised cells (act as *stem cells*). In the *stratified squamous epithelium* of mammalian skin the basal cells of the epidermis (stratum basale) give rise by an unequal division to another basal cell and to cells which progress through the spinous, granular and horny layers, becoming progressively more *keratinised*, the outermost being shed as *squames*. In olfactory mucosa the basal cells give rise to olfactory and sustentacular cells. In the epithelium of epididymis their function is unclear, but they probably serve as stem cells.

basal lamina See *basement membrane*.

base-pairing The specific hydrogen-bonding between *purines* and *pyrimidines* in double-stranded nucleic acids. In DNA the pairs are *adenine* and *thymine*, and *guanine* and *cytosine*, while in RNA they are adenine and *uracil*, and guanine and cytosine. It is the basis for the formation of a DNA double-helix from two complementary single strands.

basement membrane Extracellular matrix characteristically found under epithelial cells. There are two distinct layers: the basal lamina, immediately adjacent to the cells, is a product of the epithelial cells themselves and contains collagen type IV, and the reticular lamina is produced by fibroblasts of the underlying *connective tissue* and contains fibrillar collagen.

baseplate A hypothetical cell adhesion molecule possibly involved in sponge cell adhesion, existence unproven.

basidium The organ which carries the sexually-derived exospores of the basidiomycetous fungi. In smuts, the basidium is linear and produces four crops of basidiospores; in rusts, the linear basidium produces a single crop of basidiospores; in the remainder, with a few exceptions such as the cultivated mushroom, a club-shaped basidium gives a single crop of four spores.

basilar membrane A thin layer of tissue covered with mesothelial cells that separates the cochlea from the scala tympani in the ear.

basket cells Cerebellar neurons with many small dendritic branches which enclose the cell bodies of adjacent *Purkinje cells* in a basket-like array.

basolateral plasma membrane The plasma membrane of epithelial cells which is adjacent to the *basement membrane* or the adjoining cells of the sheet. Differs both in protein and phospholipid composition from the *apical plasma membrane* from which it is isolated by the *zonula occludens*.

basophil Mammalian bone marrow-derived *granulocyte* with large heterochromatic basophilic granules that contain *histamine* bound to a protein and heparin-like mucopolysaccharide matrix. They are not phagocytic. Very similar to *mast cells*, but it is not clear whether basophils migrate into tissues and give rise to mast cells.

basophilia 1. Having an affinity for basic dyes. 2. Condition in which there is an excess of *basophils* in the blood.

batrachotoxin A *neurotoxin* from the Columbian poison frog, *Phyllobates*, which causes *sodium channels* to open. Batrachotoxin R is more effective than related batrachotoxin A.

Bayer's junctions Sites of adhesion between the outer and cytoplasmic membranes of Gram negative bacteria; each cell has 200–400.

Becquerel (Bq) The Systeme Internationale (SI; MKS) unit of radioactivity, named after the discoverer of radioactivity, and equal to 1 disintegration per second. Use is fairly recent, superceding the Curie (Ci). 1Ci = 37GBq.

beige mouse A mouse strain typified by beige hair and *lymphadenopathy*, reticulum cell neoplasms and giant lysosomal granules in leucocytes. May be the murine equivalent of *Chediak-Higashi syndrome* of humans.

belt desmosome Another name for the *zonula adhaerens* or adhaerens junction.

Bence-Jones proteins Dimers of *immunoglobulin* light chains, frequently produced by *myelomas* Bence-Jones proteins are sufficiently small to be excreted by the kidney.

benign tumour A clone of cells which does not invade locally or *metastasise*, having lost *growth control* but not *positional control*. Usually surrounded by a fibrous capsule of compressed tissue.

benzopyrene (benzapyrene) Polycyclic aromatic hydrocarbon. Potent mutagen and carcinogen.

Berk-Sharp technique A technique of genetic mapping in which *messenger RNA* is hybridised with *single-stranded DNA* and the non-hybridised DNA then digested with S1-*endonuclease*; the residual DNA which hybridised with the messenger is then characterised by electrophoresis. Also known as S1-mapping.

Bernard-Soulier syndrome Genetic deficiency in platelet membrane glycoprotein Ib; platelets aggregate normally (see *Glanzmann's thrombasthenia*) but do not stick to collagen of sub-endothelial basement membrane.

betacyanin Heterocyclic purple-red water-soluble pigment found in members of the plant Order Centrospermae. Beetroot possesses the pigment which is often incorrectly referred to as an *anthocyanin*.

beta emitter A radionuclide whose decay is accompanied by the emission of beta particles, most commonly negatively-charged electrons. Many isotopes used in biology, such as ^{3}H, ^{14}C, ^{35}S and ^{32}P are pure beta emitters.

β-cells (pancreas) See *B cells (pancreas)*.

β$_2$-microglobulin Immunoglobulin-like polypeptide (12kD, homologous with the constant region of the immunoglobulin heavy chain) which is found on the surfaces of most cells, associated non-covalently with Class I *histocompatibility antigens*.

beta-pleated sheet (β-pleated sheet) Beta secondary structure in proteins consists of two almost fully-extended polypeptide chains lying side by side, linked by interchain hydrogen bonds between peptide C=O and N–H groups. The chains may run in the same or opposite directions (yielding parallel or antiparallel beta structures, respectively). When multiple chains are involved, an extended sheet, the beta-pleated sheet, is formed.

BFU-E See *burst forming unit (erythrocytic)*.

BHK cells (Baby Hamster Kidney cells) A quasi-diploid established line of Syrian hamster cells, descended from a clone (Clone 13) isolated by Stoker and McPherson from an unusually rapidly-growing primary culture of new-born hamster kidney tissue. Usually described as fibroblastic, although smooth muscle-like in that they express the muscle intermediate filament protein *desmin*. Widely used in studies of viral transformation and of cell physiology.

bindin Molecule of around 30kD normally sequestered in the *acrosome* of a sea-urchin spermatozoon, and which through its specific *lectin*-like binding to the *vitelline layer* of the egg confers species-specificity in fertilisation.

bioassay An assay for the activity or potency of a substance which involves testing its activity on living material.

bioblasts When Altmann first observed mitochondria he considered them to be intracellular parasites and christened them bioblasts.

biogenic amines Amines found in both animals and plants that are frequently involved in signalling. There are several groups: ethanolamine derivatives include *choline*, *acetylcholine* and muscarine; catecholamines include *adrenaline*, *noradrenaline* and *dopamine*; polyamines include *spermine*; indolylalkylamines include tryptamine and *serotonin* (5-HT); betaines include carnitine; polymethylene diamines include cadaverine and *putrescine*.

bioluminescence Light produced by a living organism. The best known system is firefly luciferase (an ATPase), which is used routinely as a sensitive ATP assay system. Many other organisms, particularly deep-sea organisms, produce light and even leucocytes emit a small amount of light when their oxidative metabolism is stimulated. Does not really differ from *chemiluminescence*, except that the light-emitting molecule occurs naturally and is not a synthetic compound like *luminol* or *lucigenin*.

biosynthesis Synthesis by a living system (as opposed to chemical synthesis).

biotin (vitamin H; co-carboxylase) A prosthetic group (MW 244.3) for carboxylase enzymes. Important in fatty acid biosynthesis and catabolism and has found widespread use as a covalent label for (macro)molecules which may then be detected by high-affinity binding of labelled *avidin*.

bipolar cells A class of retinal *interneurons*, named after their morphology, which receive input from the photoreceptors and send it to the *ganglion cells*. Bipolar cells are *non-spiking*; their response to light is evenly graded, and shows *lateral inhibition*.

bipolar filaments Filaments which have opposite polarity at the two ends; classic example is the *thick filament* of striated muscle.

birefringence Optical property of a material in which the refractive index is different for light polarised in one plane compared to the orthogonal plane. May arise from molecular organisation of the material (form birefringence), alignment of molecules due to tension (stress birefringence) or alignment of rod-like particles in flow (flow birefringence). With crossed Nicoll prisms, a birefringent material appears bright against a dark background. See also *polarisation microscopy*.

bithorax complex A group of *homeotic mutations* of **Drosophila** that map to the bithorax region on chromosome III. The mutations all cause the third thoracic segment to develop like the second thoracic segment to varying extents, the most severe being ultrabithorax which is lethal if the fly is homozygous for it. Flies which are homozygous for the three less severe mutant alleles, bithorax, anterior bithorax and posterior bithorax, have an extra pair of wings on the third thoracic segment where there is normally one pair of wing rudiments that act as a balancing organ, the halteres. The genes of the bithorax complex are thought to determine the differentiation of the posterior thoracic segments and the abdominal segments.

Bittner agent Earlier name, now superceded, for the mouse *mammary tumour virus*.

bivalent Used of two homologous chromosomes when they are in *synapsis* during *meiosis*.

black membrane An artificial (phospho)lipid membrane formed by "painting" a solution of phospholipid in organic solvent over a hole in a hydrophobic support immersed in water. Drainage of the solvent from the film produces diffraction colours until the thickness falls below the wavelength of light — it then appears to be black. The structure is an extended bimolecular leaflet.

black widow spider venom Potent neurotoxin which induces catastrophic release of acetylcholine from presynaptic terminals of cholinergic *chemical synapses*.

blast cells Cells of a proliferative compartment in a cell lineage.

blastema A group of cells in an organism that will develop into a new individual by asexual reproduction, or into an organised structure during regeneration.

blastocoel (USA blastocele) The cavity formed within the mass of cells of the *blastula* of many animals during the later stages of cleavage.

blastocyst In mammalian development, cleavage produces a thin-walled hollow sphere, whose wall is the *trophoblast*, with the embryo proper being represented by a mass of cells at one side. The blastocyst is formed before implantation and is equivalent to the *blastula*.

blastoderm In many eggs with a large amount of yolk, cell division (cleavage) is restricted to a superficial layer of the fertilised egg (meroblastic cleavage). This layer is termed the blastoderm. In birds it is a flat disc of cells at one pole of the egg, and in insects an outer layer of cells surrounding the yolk mass.

blastomere One of the cells produced as the result of cell division, cleavage, in the fertilised egg.

blastopore During *gastrulation* cells on the surface of the embryo move into the interior to form the *mesoderm* and *endoderm*. The opening formed by this invagination of cells is the blastopore; it opens from the archenteron, the primitive gut, to the exterior. In some animals this opening becomes the anus, whilst in others it closes up and the anus opens at the same spot or nearby. In some animals, eg. chick, invagination occurs without a true blastopore and the site at which the cells move in (primitive streak) may be termed a virtual blastopore.

blastula Stage of embryonic development of animals near the end of cleavage but before *gastrulation*. In animals where cleavage (cell division) involves the whole egg, the blastula usually consists of a hollow ball of cells.

bleb Protrusion from the surface of a cell, usually approximately hemispherical; may be filled with fluid or supported by a mesh-work of microfilaments.

blepharoplast Term originally used for the large compound centrioles formed during spermatogenesis of certain ferns and cycads. Later (but not now) adopted by many protistologists as an alternative name for the *basal body*, or even (in Germany) the kinetoplast (mitochondrial DNA) of **Trypanosoma** spp.

blocking antibody An antibody used in a reaction to prevent some other reaction taking place, for example one antibody competing with another for a cell surface receptor. See also *desensitisation*.

blood–brain barrier The blood vessels of the brain (and the retina) are much more impermeable to large molecules (like antibodies) than blood vessels elsewhere in the body. This has important implications for the ability of the organism to mount an immune response in these tissues, although the basis for the difference in endothelial permeability is not well understood.

blood group antigens The set of cell surface antigens found chiefly, but not solely, on blood cells. A human of blood group A will generally possess antibodies against group B antigens so that when *erythrocytes* bearing blood group B are transfused into the first individual, immune reactions take place. Blood groups are divided into a number of sub-sets, eg. the ABO group, the Rhesus group, the H group, the M and N group, etc. In most cases the specificity of these antigens is determined by the polysaccharide sequences of membrane glycoproteins or glycolipids. Unlike the ABO system, the other blood group antigens do not have natural antibodies for the allelic forms.

blood vessels All the vessels lined with *endothelium* through which blood circulates.

Bloom's syndrome Rare human autosomal recessive defect associated with increased incidence of cancer. Thought to be due to a deficiency in DNA ligase I which functions in DNA replication in association with polymerase α.

blots From blotting. Any technique whereby a sample of material is transferred from a damp solid phase to an absorbent material, thereby preserving the spatial arrangement. The samples can be stabilised on the blot and analysed for the presence of specific components using radiolabelled or fluorescent affinity reagents. See *Southern blot*, *Northern blot*, *Western blot*.

blue-green algae Group of prokaryotes which should now be referred to as *Cyanobacteria*.

blue naevus A non-malignant accumulation of highly-pigmented *melanocytes* deep in the *dermis*.

bombesin Tetradecapeptide *neurohormone* with *paracrine/autocrine* effects, first isolated from skin of fire-bellied toad (*Bombina bombina*); mammalian equivalent is *gastrin-releasing peptide* (GRP). Bombesin cross-reacts with GRP receptors. Both are *mitogenic* for *Swiss 3T3* fibroblasts at nanomolar levels. Neuropeptides of this type are found in many tissues and at high levels in pulmonary (small cell carcinoma) and thyroid tumours.

bone marrow Tissue found in the centre of most bones; site of *haematopoiesis*. The most radiation-sensitive tissue of the body.

Bordetella pertussis A small, aerobic, Gram negative bacillus, causative organism of whooping cough. Produces a variety of toxins including a dermonecrotising toxin, an adenyl cyclase, an *endotoxin* and *pertussis toxin*, as well as surface components such as *fimbrial* haemagglutinin.

bottle cells The first cells to migrate inwards at the *blastopore* during amphibian *gastrulation*. The "neck" of the bottle is at the outer surface of the embryo.

Boyden chamber Simple chamber used to test for *chemotaxis*, especially of leucocytes. Consists of two compartments separated by a Millipore filter (3–8μm pore size); chemotactic factor is placed in one compartment and the gradient develops across the thickness of the filter (ca 150μm). Cell movement into the filter is measured after an incubation period less than the time taken for the gradient to decay. See also *checkerboard assay*.

bradykinin Vasoactive nonapeptide (RPPGFSPFR) formed by action of proteases on kininogens. Very similar to *kallidin* (which has the same sequence but with an additional N-terminal lysine). Bradykinin is a very potent vasodilator and increases permeability of post-capillary venules; it acts on endothelial cells to activate phospholipase A_2. It is also spasmogenic for some smooth muscle and will cause pain.

bright-field microscopy Optical microscopy, in which absorption to a great extent and diffraction to a minor extent give rise to the image, as opposed to *phase contrast* or *interference* methods of microscopy.

bromelain Thiol protease (EC 3.4.22.4) from pineapple.

brown fat cells Brown fat is specialised for heat production and the *adipocytes* have many mitochondria in which an inner-membrane protein can act as an uncoupler of *oxidative phosphorylation* allowing rapid thermogenesis.

brush border The densely packed *microvilli* on the apical surface of, for example, intestinal epithelial cells.

Bruton's disease Sex-linked recessive agammaglobulinaemia caused by a deficiency in B-lymphocyte function.

BUdR (bromo-deoxyuridine; deoxynucleoside of 5-bromo-uracil) Analogue of thymidine that induces point mutation because of its tendency to tautomerisation: in the enol form it pairs with guanine instead of adenine. It is used as a mutagen, and also as a marker for DNA synthesis (the incorporation of BUdR can be recognised because the staining pattern differs; an even more sensitive new method uses a monoclonal antibody staining procedure). When incorporated into DNA it causes the DNA to have increased density, which can be detected in a *caesium chloride* gradient.

buffer A system which acts to minimise the change in concentration of a specific chemical species in solution against addition or depletion of this species. pH buffers consist of weak acids or weak bases in aqueous solution; the working range of a given buffer is given by $pK_a \pm 1$. Metal ion buffers contain a metal ion chelator (eg. *EDTA*), partially saturated by the metal ion and acting as a buffer for the metal ion.

buffy coat Thin yellow-white layer of leucocytes on top of the mass of red cells when whole blood is centrifuged.

bufotenine (3-(2-(dimethylamino)ethyl)-1H-indol-5-ol; mappine) An indole alkaloid with hallucinogenic effects, isolated from *Piptadenia* spp. (Mimosidae); first isolated from skin glands of toad (*Bufo* sp.).

bullous pemphigoid Form of pemphigoid (which also affects mucous membranes), in which blisters (bulli) form on the skin. Patients have circulating antibody (usually IgG) to *basement membrane* of *stratified epithelium*, although the antibody titre does not correlate with the severity of the disease.

Bunyaviridae Family of enveloped viruses with single-stranded RNA genome which infect vertebrates and arthropods (and are thus grouped within the *arboviruses*); cause, for example, viral haemorrhagic fevers. Virion is spherical or oval, 90–100nm diameter.

Burkitt's lymphoma Malignant tumour of *lymphoblasts* derived from B-lymphocytes (B-cell lymphoma). Most commonly affects children in tropical Africa: both *Epstein Barr* virus and immunosuppression due to malarial infection are involved. Not only do cells contain multiple copies of the Epstein Barr virus genome, but there are characteristic reciprocal translocations between chromosomes 8 and 14 or between 8 and 2 or 22. The effect of the translocation is probably to bring the cellular oncogene, *c-myc*, next to an active immunoglobulin *promoter*, leading to over-expression of *myc*.

burr cells Triangular helmet-shaped cells found in blood, usually indicative of disorders of small blood vessels.

Bursa of Fabricius *Lymphoid tissue* found at the junction of the cloaca and the gut of birds giving rise to the so-called *B-lymphocyte* series.

burst forming unit (erythrocytic) (BFU-E) A bone marrow *stem cell* lineage detected in culture by its mitotic response to *erythropoietin*, and subsequent erythrocytic differentiation in about 12 mitotic cycles into erythrocytes.

C

C1 – C9 Proteins of the mammalian *complement* system.

C1 First component of *complement*; actually three sub-components, C1q, C1r and C1s, which form a complex in the presence of calcium ions. C1q, the recognition subunit, has an unusual structure of collagen-like triple helices forming a stalk for the immunoglobulin-binding globular heads. Upon binding to immune complexes the C1 complex becomes an active protease which cleaves and activates C4 and C2.

C2 Second component of *complement*. A β_2-globulin.

C3 Third component of *complement*, present in plasma at around 0.5–1mg/ml. Both classical and alternate pathways converge at C3 which is cleaved to yield C3a, an anaphylatoxin, and C3b which acts as an opsonin and is bound by *CR1*: C3b in turn can be proteolytically cleaved to iC3b (ligand for *CR3*) and C3dg by C3b-inactivator. C3b complexed with Factor B (to form C3bBb) will cleave C3 to give more C3b, although the C3bBb complex is unstable unless bound to *properdin* and a carbohydrate-rich surface. The C3b/C4b2a complex and C3bBb are both C5 convertases (cleave *C5*). Cobra venom factor is homologous with C3b but the complex of cobra venom factor, properdin and Factor Bb is insensitive to C3b inactivator.

C4 Fourth component of *complement* to be discovered, although the third to be activated in the classical pathway. Becomes activated by cleavage (by C1) to C4b which complexes with C2a to act as a C3 convertase, generating C3a and C3b. The C4b2a3b complex acts on C5 to continue the cascade.

C5 Fifth component of *complement*, which is cleaved by C5-convertase to form C5a, a 74-residue anaphylatoxin and potent chemotactic factor for leucocytes, and C5b. C5a rapidly loses a terminal arginine to form C5a desarg, which retains chemotactic but not anaphylatoxin activity. C5b combines with C6, C7, C8 and C9 to form a membranolytic complex.

C6 Sixth component of *complement*. See *C5* and *C9*.

C7 Seventh component of *complement*. See *C5* and *C9*.

C8 Eighth component of *complement*: three peptide chains α, β, and γ. See *C5* and *C9*.

C9 Ninth component of *complement*. Complexed with C5b,6,7,8 it forms a potent membranolytic complex (sometimes referred to as the membrane attack complex). Membranes which have bound the complex have toroidal "pores"; a single pore may be enough to cause lysis.

C banding (centromeric banding) Method of defining chromosome structure by staining with Giemsa and looking at the *banding pattern* in the heterochromatin of the centromeric regions. Giemsa banding (*G banding*) gives higher resolution. See also *quinacrine*.

C3 plants Plants which fix CO_2 in photosynthesis by the *Calvin–Benson cycle*. The enzyme responsible for CO_2 fixation is *ribulose bisphosphate carboxylase/oxidase*, whose products are compounds containing three carbon atoms. C3 plants are typical of temperate climates. *Photorespiration* in these plants is high.

C4 plants Plants found principally in hot climates whose initial fixation of CO_2 in photosynthesis is by the *HSK pathway*. The enzyme responsible is *PEP carboxylase*, whose products contain four carbon atoms. Subsequently the CO_2 is released and re-fixed by the *Calvin–Benson cycle*. The presence of the HSK pathway permits efficient photosynthesis at high light intensities and low CO_2 concentrations. Most species with this type of respiration have little or no *photorespiration*.

C polysaccharide (C substance) Polysaccharide released by pneumococci which contains galactosamine-6-phosphate and phosphoryl choline. *C-reactive protein* is so called because it will precipitate this polysaccharide through an interaction with the phosphoryl choline.

C2-kinin A kinin-like fragment generated from complement *C2*; causes vasodilation and increased vascular permeability. Distinct from *bradykinin*.

C-reactive protein (CRP) A protein found

in serum in various disease conditions. C-reactive protein is synthesised by *hepatocytes* and its production may be triggered by *prostaglandin* E1. It consists of five polypeptide subunits forming a molecule of total molecular weight 105kD. It binds to polysaccharides present in a wide range of bacterial, fungal and other cell walls or cell surfaces and to *lecithin* and to phosphoryl- or choline-containing molecules. It is related in structure to *serum amyloid*. See also *acute phase proteins*.

C-region The parts of the heavy or light chains of *immunoglobulin* molecules which are of constant sequence, in contrast to variable or V regions. The constancy of sequence is relative because there are certain constant region genes and alleles thereof (see *allotypes*), but within one animal homozygous at the light and heavy chain constant region genes, all immunoglobulin molecules of any one class have constant sequences in their C-regions. Note that the constant region sequences for the various types of immunoglobulin, eg. IgG, IgA, etc. differ from each other.

C-subfibre The third partial microtubule associated with the A- and B-tubules of the outer axonemal doublets in the *basal body* (and in the centriole) to form a triplet structure.

C value paradox Comparison of the amount of DNA present in the haploid genome of different organisms (the C value) reveals two problems: the value can differ widely between two closely-related species, and there seems to be far more DNA in higher organisms than could possibly be required to code for the modest increase in complexity.

cachectin Protein produced by macrophages that is responsible for the wasting (cachexia) associated with some tumours. Now known to be identical to *tumour necrosis factor* (TNF). Has two or three 17kD subunits, all derived from a single highly-conserved protein.

cadherins Molecules involved in calcium-dependent cell adhesion. N-cadherin in mouse has a tissue distribution distinct from E-cadherin which is associated with epithelial cells (and is equivalent to uvomorulin and L-CAM). See *cell adhesion molecules*.

Caenorhabditis elegans Nematode much used in lineage studies since the number of nuclei is determined, and the nervous system is relatively simple. The organism can be maintained *axenically* and there are mutants in behaviour, in muscle proteins, and in other features. Sperm are amoeboid and move by an unknown mechanism which does not seem to depend upon actin or tubulin.

caesium chloride (USA cesium chloride) Salt which yields aqueous solutions of high density. When equilibrium has been established between sedimentation and diffusion during ultracentrifugation, a linear density gradient is established in which macromolecules such as DNA band at a position corresponding to their own buoyant density.

calbindin-D-28k Vitamin D-induced calcium-binding protein (28kD) found in primate striate cortex and other neuronal tissues.

calcineurin Calmodulin-stimulated protein-phosphatase (EC 3.1.3.16), the major calmodulin-binding protein in brain. It is also thought to be involved in the control of motility in sperm.

calcitonin A polypeptide hormone produced by C-cells in the thyroid that causes a reduction of calcium ions in the blood.

calcitonin gene related peptide (CGRP) Neuropeptide of 37 amino acids with structural homology to salmon calcitonin. Co-localises with *substance P* in neurons. Intracerebral administration of CGRP leads to a rise in noradrenergic sympathetic outflow, a rise in blood pressure, and a fall in gastric secretion.

calcium ATPase Usually used of the calcium-pumping ATPase present in high concentration as an integral membrane protein of the sarcoplasmic reticulum of muscle. This pump lowers the the cytoplasmic calcium level and causes contraction to stop. Normal function of the pump seems to require a local phospholipid environment from which cholesterol is excluded.

calcium binding proteins There seem to be two major groups of calcium-binding proteins, those which are similar to *calmodulin*, such as troponin C and parvalbumin, and those which bind calcium and phospholipid (eg. *lipocortin*) and which have been grouped under the generic name of annexins. Many other proteins will bind calcium, although the binding site usually has considerable homology with the calcium binding domains of calmodulin.

calcium channel Membrane channel which is specific for calcium. Probably the best characterised is the *voltage gated ion-channel* of the sarcoplasmic reticulum which is ryanodine-sensitive.

calcium dependent regulator protein (CDRP) Early name for *calmodulin*.

calcium pump A *transport protein* responsible for moving calcium out of the cytoplasm. See *calcium ATPase*.

calcivirus Genus of *Picornaviridae*.

caldesmon Protein (150kD, normally dimeric) from the smooth muscle of bird gizzard: a calcium-sensitive F-actin crosslinker.

calelectrin Calcium-binding protein (34kD) from ray (*Torpedo marmorata*).

callose A plant cell-wall polysaccharide (a β(1–3)-*glucan*) found in phloem *sieve plates*, wounded tissue, pollen tubes, cotton fibres and certain other specialised cells.

callus 1. *Bot.* Undifferentiated plant tissue produced at wound edge — callus tissue can be grown *in vitro* and induced to differentiate by varying the ratio of the hormones *auxin* and *cytokinin* in the medium. 2. *Path.* Mass of new bony trabeculae and cartilaginous tissue formed by *osteoblasts* early in the healing of a bone fracture.

calmodulin Ubiquitous and highly conserved calcium binding protein (17kD) with four binding sites for calcium. Ancestor of *troponin* C, *leiotonin* C, and *parvalbumin*.

calpactins Calcium-binding proteins from cytoplasm. Calpactin I is identical to *lipocortin* II, calpactin II to lipocortin I; one of the major targets for phosphorylation by pp60^{src} (see **src** *gene*).

calpain Calcium-activated cytoplasmic proteases. Calpain I is activated by micromolar calcium, calpain II by millimolar calcium. Calpain has two subunits; the larger (80kD) has 4 domains, one homologous with *papain*, one with *calmodulin*; the smaller (30kD) has one domain homologous with calmodulin. First isolated from erythrocytes, but now described from other cells.

calpastatin Cytoplasmic inhibitor of calcium activated protease, *calpain*.

calretinin Neuronal protein (29kD) of the *calmodulin* family isolated from chick retina.

Has 58% sequence homology with calbindin, the intestinal cell isoform.

calsequestrin Protein (44kD) found in the cisternae of sarcoplasmic reticulum: sequesters calcium.

caltractin Calcium-binding protein (20kD) from **Chlamydomonas reinhardtii**, that is associated with the striated *rootlet system* of the basal body. It shares sequence homology with *calmodulin* and with the *cdc31* gene product that is associated with spindle pole body duplication in the yeast *Saccharomyces cerevisiae*.

caltrin Inhibitor of calcium ion transport found in bovine seminal plasma (47 amino acids; MW 5411, on gels ca 10kD) and that resembles seminal antibacterial protein (confusingly called plasmin, though not related to the protease).

calvarium One of the bones which makes up the vault of the skull (in humans these are the frontal, 2 parietals, occipital, and 2 temporals). Calvaria are often used in organ culture to investigate bone catabolism or synthesis.

Calvin–Benson cycle (Calvin cycle) Metabolic pathway responsible for photosynthetic CO_2 fixation in plants and bacteria. The enzyme which fixes CO_2 is *ribulose bisphosphate carboxylase/oxidase*. The cycle is the only photosynthetic pathway in *C3 plants* and the secondary pathway in *C4 plants*. The enzymes of the pathway are present in the stroma of the chloroplast.

CAM See *crassulacean acid metabolism* or *cell adhesion molecule*.

cambium 1. *Bot. Meristematic* plant tissue, commonly present as a single layer of cells, which forms new cells on both sides. Located either in vascular tissue (vascular cambium), forming xylem on one side and phloem on the other, or in cork (cork cambium or phellogen). 2. *Path.* Inner region of the periosteum from which *osteoblasts* differentiate.

canaliculi In bone, channels which run through the calcified matrix between lacunae containing *osteocytes*. In liver, small channels between *hepatocytes* through which bile flows to the bile duct and thence to the intestinal lumen.

cancer Originally descriptive of breast carcinoma, now a general term for diseases caused by any type of malignant tumour.

Candida albicans A dimorphic fungus that is an opportunistic pathogen of humans.

candidiasis Infection by **Candida albicans**, common on mucous membranes ("Thrush"), but in immunosuppressed patients can opportunistically infect many tissues.

cap binding protein Protein (24kD) with affinity for cap structure at 5′-end of mRNA and probably assists, together with other *initiation factors*, in binding the mRNA to the 40S ribosomal subunit. Translation of mRNA *in vitro* is faster if it has a cap binding protein.

capacitation A process occurring in mammalian sperm after exposure to secretions in the female genital tract. Surface changes take place, probably involved with the *acrosome*, which are necessary before the sperm can fertilise an egg.

capillary The small blood vessels which link arterioles with venules. Lumen may be formed within a single endothelial cell, and have a diameter less than that of an erythrocyte, which must deform to pass through. Blood flow through capillaries can be regulated by precapillary sphincters, and each capillary probably only carries blood for part of the time.

capnine Sulphonolipid isolated from the envelope of the *Cytophaga-Flexibacter* group of Gram negative bacteria. The acetylated form of capnine seems to be necessary for gliding motility.

capping 1. Movement of cross-linked cell-surface material to the posterior region of a moving cell, or to the perinuclear region. 2. The intracellular accumulation of intermediate filament protein in the pericentriolar region following microtubule disruption by colchicine. 3. The blocking of further addition of subunits by binding of a cap protein to the free end of a linear polymer such as actin. 4. mRNA molecules in eukaryotes are modified (capped) at their 5′ ends by methylation; this region is later recognised by *cap binding protein*.

capsid A protein coat which covers the nucleoprotein core or nucleic acid of a virion. Commonly shows icosahedral symmetry and may itself be enclosed in an envelope (as in the *Togaviridae*). The capsid is built up of subunits (some integer multiple of 60, the number required to give strict icosahedral symmetry) which self-assemble in a pattern typical of a particular virus. The subunits are often packed, in smaller capsids, into 5- or 6-membered rings (pentamers or hexamers) which constitute a morphological unit (capsomere). The packing of subunits is not perfectly symmetrical in most cases, and some units may have strained interactions and are said to have quasi-equivalence of bonding to adjacent units.

capsomeres See *capsid*.

capsule 1. *Bact.* Thick gel-like material attached to the wall of Gram positive or Gram negative bacteria, giving colonies a "smooth" appearance. May contribute to pathogenicity by inhibiting phagocytosis. Mostly composed of very hydrophilic acidic polysaccharide, but considerable diversity exists. 2. *Path.* Cellular response in invertebrate animals to a foreign body too large to be phagocytosed. A multicellular aggregate of *haemocytes* or coelomocytes isolates the foreign object. In some insects the capsule is apparently acellular and composed of *melanin*.

carbachol (carbamyl choline) Parasympathomimetic drug formed by substituting the acetyl of acetylcholine with a carbamyl group; acts on both *muscarinic* and *nicotinic* receptors and is not hydrolysed by acetylcholine esterase.

carbohydrates Very abundant compounds with the general formula $C_n(H_2O)_n$. The smallest are monosaccharides like glucose; polysaccharides (eg. starch, cellulose, glycogen) can be large and indeterminate in length.

γ-carboxyglutamate An amino acid found in some proteins, particularly those which bind calcium. Formed by post-translational carboxylation of glutamate.

carboxypeptidases Enzymes (particularly of pancreas) which remove the C-terminal amino acid from a protein or peptide. Carboxypeptidase B (EC 3.4.17.2) is specific for terminal lysine or arginine, whereas carboxypeptidase A will remove any amino acid.

carcinoembryonic antigen Antigen found in blood of patients suffering from cancer of colon and some other diseases, which is otherwise normally found in foetal gut tissue.

carcinogen An agent capable of initiating development of malignant tumours. May be a chemical, a form of electromagnetic radiation, or an inert solid body.

carcinoid Tumour arising from specialised cells with paracrine functions (*APUD cells*), also known as argentaffinoma. The primary tumour is commonly in the appendix, where it is clinically benign; hepatic secondaries may release large amounts of vasoactive amines such as serotonin to the systemic circulation.

carcinoma Malignant neoplasia of an epithelial cell: by far the commonest tumour. Those arising from glandular tissue are often called *adenocarcinomas*. Carcinoma cells tend to be irregular with increased basophilic staining of the cytoplasm, have an increased nuclear/cytoplasmic ratio and polymorphic nuclei. As with most generalisations, however, this is not true for all.

cardiac cell Strictly speaking any cell of or derived from the cardium of the heart, but often used loosely of heart cells.

cardiac jelly Gelatinous extracellular material which in the embryo lies between endocardium and myocardium.

cardiac muscle Specialised striated muscle of vertebrate heart. Individual cells branch and are connected by intercalated discs (similar to *adhaerens junctions*) and *gap junctions*. Cardiac muscle is involuntary and the beat is intrinsic (myogenic), being regulated and coordinated by *Purkinje cells* and the atrioventricular pacemaker region. The frequency and force of beat is increased by noradrenaline released by sympathetic neurons and diminished by acetylcholine released by the parasympathetic innervation.

cardiolipin A diphosphatidyl glycerol that functions as the antigen used to detect antibodies in the Wasserman test for syphilis.

carditis Inflammation of the heart, including pericarditis, myocarditis and endocarditis, according to whether the enveloping outer membrane, the muscle or the inner lining is affected.

carotenoids Accessory lipophilic photosynthetic pigments in plants and bacteria, including *carotenoids* and *xanthophylls*; red, orange or yellow, with broad absorption peaks at 450–480nm. Act as secondary light-harvesting pigments, passing energy to *chlorophyll*, and as protective agents, preventing photo-oxidation of chlorophyll. Found in chloroplasts and also in plastids in some non-photosynthetic tissues, eg. carrot root.

carotid body cell Cell derived from the neural crest, involved in sensing pH and oxygen tension of the blood.

carrageenan (carrageenin) Sulphated cell-wall polysaccharide found in certain red algae. Contains repeating sulphated disaccharides of galactose and (sometimes) anhydrogalactose. It is used commercially as an emulsifier and thickener in foods, and is also used to induce an inflammatory lesion when injected into experimental animals (probably activates *complement*).

cartilage Connective tissue dominated by extracellular matrix containing collagen type II and large amounts of proteoglycan, particularly chondroitin sulphate. Cartilage is more flexible and compressible than bone and often serves as an early skeletal framework, becoming mineralised as the animal ages. Cartilage is produced by *chondrocytes* which come to lie in small lacunae surrounded by the matrix they have secreted.

casein Group of proteins isolated from milk. α_s- and β-caseins are amphipathic polypeptides of around 200 amino acids with substantial hydrophobic C-terminal domains that associate to give micellar polymers in divalent cation-rich medium. κ-casein is a glycoprotein rather different from α- and β-casein.

caseous necrosis (caseation) The development of a necrotic centre (with a cheesy appearance) in a tuberculous lesion.

Casparian band (Casparian strip) A band of ligno-suberin found on the radial walls of the primary endodermal cell in plant roots and stems, which acts as a seal to prevent back-leakage of secreted material (analogous to the *zonula occludens* between epithelial cells). Found particularly where root parenchymal cells secrete solutes into xylem vessels.

castanospermine Plant alkaloid from seeds of the Australian chestnut tree. Inhibitor of α-glucosidase I; the effect is to leave N-linked oligosaccharides in their "high-mannose", unmodified state.

catabolin Obsolete term for a form of *interleukin-1* which stimulates the breakdown of connective tissue extracellular matrix.

catabolism Degradative processes, the opposite of anabolism.

catalase Tetrameric haem enzyme (EC 1.11.1.6; 245kD) which breaks down hydrogen peroxide.

cataract Opacity of the lens of the eye.

catecholamine A type of *biogenic amine* derived from tyramine, characterised as alkylamino derivatives of *o*-dihydroxybenzene. Catecholamines include *adrenaline*, *noradrenaline* and *dopamine*, with roles as *hormones* and *neurotransmitters*.

cathepsins Intracellular proteolytic enzymes of animal tissues, such as cathepsin B (EC 3.4.22.1), a lysosomal thiol proteinase; C, dipeptidyl peptidase, (EC 3.4.14.1); D (EC 3.4.23.5), which has pepsin-like specificity; G (EC 3.4.23.5), similar to chymotrypsin; H, which possesses aminopeptidase activity; N, which attacks N-terminal peptides of collagen, and so on.

cationic proteins Proteins of azurophil granules of neutrophils, rich in arginine. A chymotrypsin-like protease found in azurophil granules is also very cationic, as is cathepsin G and neutrophil elastase. Eosinophil cationic protein (21kD) is particularly important because it damages *schistosomula in vitro*.

cationised ferritin *Ferritin*, treated with dimethyl propanediamine, and used to show, in the electron microscope, the distribution

Table C1. CD antigens. Designations of CD (cluster determinant) antigens are being extended rapidly, and this table is not comprehensive. The inclusion of a "w" indicates that the designation derives from the workshop on CD antigens and has not been formally ratified.

Antigen	MW (kD)	Distribution	Comment
CD1a, b, c	49, 45, 43	T	-
CD2	50	T	Binds LFA-3
CD3	20, 26	T	Associated with T cell receptor
CD4	60	T subset	T helper and inducer
CD5	67	T	
CD6	120	T	
CD7	40	T	Fc_μ receptor
CD8	32	T subset	Cytotoxic T cells (MHC restricted)
CD9	24	M, P, preB	
CD10	100	preB, cALL	
CD11a	180 (95)	L	α subunit of LFA-1. (formerly CDw18)
CD11b	160 (95)	M, G	CR3 (Mac-1)
CD11c	150 (95)	M, (G)	gp150, 95
CD12		M, G, P	
CD13	150	G, M	
CD14		M, (G), FDRC	
CD15		G, (M)	
CD16	50–60	G	Fc_γ RIII
CD17		G, M, P	
CD18	95	L	β subunit of LFA-1 integrins
CD19	95	B	
CD20	35	B, FDRC	
CD21	140	B, FDRC	C3d receptor
—			
CD25	55	activated T	IL-2 receptor
CD31	130–140	M, G, P, (T)	gpIIa (?)
CD32	40	M, G, P, B	FcRII
CD35	220	G, M, FDRC	CR1
CDw40	50	B, IRC, carcinoma	
CDw41	150, 95	P	gpIIb/IIIa (integrin)
CDw42		P	gpIb
CD45		L	Leucocyte common antigen

B = B cells; cALL = common-type acute lymphoblastic leukaemia; FDRC = follicular dendritic reticulum cells; G = granulocytes; IRC = interdigitating reticulum cells; L = leucocytes; M = monocytes; P = platelets; T = T cells.

of negative charge on the surface of a cell. The amount of cationic ferritin binding is very approximately related to the surface charge.

CCCP (m-chloro-carbonylcyanide-phenyl-hydrazine) An *uncoupler* which dissipates proton gradients across membranes.

CD nomenclature The CD (cluster of differentiation) system is an internationally agreed system for naming differentiation antigens present on cell surfaces. Originally applied to leucocytes, it is being extended to other cell types as antigens shared with the immune system are identified. See Table C1.

***cdc* genes** Cell division cycle genes, of which several have now been defined, especially in yeasts.

cDNA (complementary DNA; copy DNA) DNA produced by a *reverse transcriptase* acting on *messenger RNA*. Has the advantage, for purposes of genetic engineering, that the *introns* are missing.

CDP See *cytidine 5'-diphosphate*.

CDRP See *calmodulin*.

cecropin One of a group of inducible antibacterial proteins purified and characterised from *Hyalophora cecropia* (silkmoth) pupae, and now found in several other species of endopterygote insects. Small basic proteins which cause lysis of both Gram positive and Gram negative bacteria.

celiac disease See *coeliac disease*.

cell adhesion See *adhesins, cadherins, cell adhesion molecules (CAMs), contact sites A, DLVO theory, integrins, sorting out, uvomorulin* and various specialised junctions (*adhaerens junctions, desmosomes, focal adhesions, gap junction* and *zonula occludens*).

cell adhesion molecule (CAM) Although this could mean any molecule involved in cellular adhesive phenomena, it has acquired a more restricted sense, namely a molecule on the surface of animal tissue cells, antibodies (or Fab fragments) against which specifically inhibit some form of intercellular adhesion. Examples are LCAM (Liver Cell Adhesion Molecule) and NCAM (Neural Cell Adhesion Molecule), both named from tissues in which first detected, although their occurrence is not in fact restricted to these.

cell behaviour General term for activities of whole cells such as movement and adhesion, by analogy with animal behaviour.

cell body Used in reference to neurons; the main part of the cell around the nucleus excluding long processes such as axons and dendrites.

cell centre *Microtubule organising centre* (MTOC) of the cell, the pericentriolar region.

cell culture General term referring to the maintenance of cell strains or lines in the laboratory.

cell cycle The sequence of events between mitotic divisions. The cycle is conventionally divided into G0, G1 (G standing for gap), S (synthesis phase during which the DNA is replicated), G2 and M (mitosis). Cells which will not divide again are considered to be in G0, and the transition from G0 to G1 is thought to commit the cell to completing the cycle and dividing.

cell death Cells die (non-accidentally) either when they have completed a fixed number of division cycles (around 60; the Hayflick limit) or at some earlier stage when programmed to do so, as in digit separation in vertebrate limb morphogenesis. Whether this is due to an accumulation of errors or a programmed limit is unclear; some transformed cells have undoubtedly escaped the limit.

cell division The separation of one cell into two daughter cells, involving both nuclear division (*mitosis*) and subsequent cytoplasmic division (*cytokinesis*).

cell electrophoresis Method for estimating the surface charge of a cell by looking at its rate of movement in an electrical field; almost all eukaryotic cells have a net negative surface charge. Measurement is complicated by the streaming potential at the wall of the chamber itself, and by the fact that the cell is surrounded by a layer of fluid (see *double layer*). The electrical potential measured (the zeta potential) is actually some distance away from the plasma membrane. One of the more useful modifications is to systematically vary the pH of the suspension fluid to determine the pK of the charged groups responsible (mostly carboxyl groups of sialic acid).

cell fractionation Strictly this should mean the separation of homogeneous sets from a

Table C2. Common cell lines

Table C2. Common cell lines. Although there are a great many cell lines available through the cell culture repositories and from trade suppliers, there are a few "classic" lines that will be met fairly frequently. Many of these well known lines are listed below, but the table is not comprehensive.

Name	Tissue of origin	Cell type	Comment
Human			
MRC5	Embryonic lung	Fib	Diploid, susceptible to virus infection
WI38	Embryonic lung	Fib	Diploid, finite division potential
HeLa	Cervical carcinoma	Epi	Established line
HEp2	Laryngeal carcinoma	Epi	
Raji	Burkitt lymphoma	Lym	Grows in suspension. EB virus undetectable
Daudi	Burkitt lymphoma	Lym	
HL60	—	Myl	Will differentiate to granulocytes or macrophages
J111	Monocytic leukaemia	Myl	
U937	Monocytic leukaemia	Myl	Will differentiate to macrophages
Monkey			
BSC-1	Kidney	Fib	Derived from African Green monkey, often used in
Vero	Kidney	Fib	virus studies
Hamster			
BHK21	Baby kidney	Fib	Syrian hamster. Usually C13 (clone 13)
CHO	Ovary	Epi	Chinese hamster ovary
Don	Lung	Fib	Chinese hamster
Potoroo			
PtK1	Female kidney	Epi	Small number of large chromosomes,
PtK2	Male kidney	Epi	Cells stay flat during mitosis
Dog			
MDCK	Kidney	Epi	Madin–Darby canine kidney
Mouse			
A9	From L929	Fib	HGPRT negative
L1210	Ascites fluid	Lym	Grows in suspension; DBA/2 mouse
MOPC31C	Plasmacytoma	Lym	Grows in suspension; secretes IgG
L929	Connective tissue	Fib	Clone of L cell
3T3	Whole embryo	Fib	Swiss or Balb/c types; very density dependent
SV40-3T3	From 3T3	Fib	Transformed by SV40 virus
P388D1	—	Lym	Grows in suspension
EAT	Ascites tumour	Fib	
S180	Sarcoma	Fib	Invasive; maintained *in vivo*
MCIM	Sarcoma	Fib	Methylcholanthrene-induced
B16	Melanoma	Mel	High and low metastatic variants (F1 and F10)
Rat			
GH1	Pituitary tumour	Fib	Secrete growth hormone
WRC-256	Carcinoma	—	Walker carcinoma; many variants
PC12	Adrenal	Neur	Phaeochromocytoma; can be induced to produce neurites

Epi = epithelial; Fib = fibroblastic; Lym = lymphocytic; Mel = melanin containing; Myl = myeloid; Neur = neural.

heterogeneous population of cells (by a method such as *flow cytometry*), but the term is more frequently used to mean subcellular fractionation ie. the separation of different parts of the cell by differential centrifugation, to give nuclear, mitochondrial, microsomal and soluble fractions.

cell fusion Fusion of two previously separate cells occurs naturally in fertilisation, in the formation of vertebrate skeletal muscle and certain giant cells, but can be induced artificially by the use of *Sendai virus* or fusogens such as polyethylene glycol. A cell formed by the fusion of dissimilar cells is often referred to as a heterokaryon.

cell growth Usually used to mean increase in the size of a population of cells though strictly should be reserved for an increase in cytoplasmic volume of an individual cell.

cell junctions Specialised junctions between cells. See *adhaerens junctions*, *desmosome*, *zonula occludens*, *gap junction*.

cell line A cell line is a permanently established cell culture which will proliferate indefinitely given appropriate fresh medium and space. Lines differ from cell strains in that they have escaped the Hayflick limit (see *cell death*) and become immortalised. Some species, particularly rodents, give rise to lines relatively easily, whereas other species do not. No cell lines have been produced from avian tissues, and the establishment of cell lines from human tissue is difficult. Many cell biologists would consider that a cell line is by definition already abnormal and that it is on the way towards becoming the culture equivalent of a neoplastic cell. See Table C2.

cell lineage The lineage of a cell relates to its derivation from the undifferentiated tissues of the embryo. Substances used to trace cell lineage by micro-injection include horseradish peroxidase and fluorescently-conjugated dextrans.

cell locomotion Active movement of a cell from one place to another.

cell mediated immunity Immune response which involves effector T-lymphocytes and not the production of humoral antibody. Responsible for *allograft* rejection, delayed *hypersensitivity* and in defence against viral infection and intracellular protozoan parasites.

cell migration Implies movement of a population of cells from one place to another — as in the movement of neural crest cells during morphogenesis.

cell movement A more general term than locomotion, which can include shape-change, cytoplasmic streaming etc.

cell plate Region in which the new cell wall forms after the division of a plant cell. In the plane of the equator of the spindle a disc-like structure, the *phragmoplast* forms, into which are inserted pole-derived microtubules. Golgi-derived vesicles containing pectin come together and fuse at the plate which eventually fuses with the plasma membrane thereby separating the daughter cells.

cell polarity 1. In epithelial cells the differentiation of apical and basal specialisations. In simple epithelia the apical and baso-lateral regions of plasma membrane differ in lipid and protein composition, and are isolated from one another by tight junctions. The apical membrane, for example, may be the only region where secretory vesicles fuse, or where there is a particular ionic pumping system. 2. A motile cell must have some internal polarity in order to move in one direction at a time: a region in which protrusion will occur (the front) must be defined. Locomotory polarity may be associated with the pericentriolar *microtubule organising centre*, and can be perturbed by drugs which interfere with microtubule dynamics.

cell proliferation Increase in cell number by division.

cell renewal Replacement of cells, for example those in the skin, by the proliferative activity of basal stem cells.

cell signalling Release by one cell of substances which transmit information to other cells.

cell sorting See *sorting out* or *flow cytometry*.

cell strain Cells adapted to culture, but with finite division potential. See *cell line*.

cell synchronisation A process of obtaining (either by selection, or imposition of a reversible blockade) a population of growing cells which are to a greater or lesser extent in phase with each other in the cycle of growth and division.

cell wall Extracellular material serving a structural role. In plants the primary wall is mainly cellulosic; the secondary wall, laid down after the cell ceases to enlarge, has additional materials such as lignin and suberin incorporated. In bacteria cell wall structure is complex: the walls of Gram positive and Gram negative bacteria are distinctly different. Removal of the wall leaves a *protoplast* or *spheroplast*.

cell–cell recognition Interaction between cells that is possibly dependent upon specific adhesion. Since the mechanism is not entirely clear in most cases, the term should be used with caution.

cell-free system Applied specifically to *in vitro* translation systems for mRNA. The most common systems are based upon a lysate of rabbit reticulocytes or on wheat germ.

cell-surface marker Any molecule characteristic of the plasma membrane of a cell or in some cases of a specific cell type. *5'-nucleotidase* and *sodium-potassium ATPase* are often used as plasma membrane markers.

cellobiose Reducing disaccharide composed of two D-glucose moieties β(1–4)-linked; the disaccharide subunit of cellulose, though not found as a free compound *in vivo*.

cellular immunity Immune response which involves enhanced activity by phagocytic cells and does not imply lymphocyte involvement. Since the term is easily confused with *cell mediated immunity*, its use in this sense should be avoided.

cellular slime mould See *Acrasidae*.

cellulases Enzymes which break down *cellulose*, and are involved in cell-wall breakdown in higher plants, especially during abscission. Produced in large amounts by certain fungi and bacteria. Degradation of cellulose *microfibrils* requires the concerted action of several cellulases.

cellulitis Inflammation of the subcutaneous connective tissues (dermis), mostly affecting face or limbs. *Streptococcus pyogenes* is commonly the causative agent. Also known as erysipelas.

cenocyte See *coenocyte*.

central lymphoid tissue Term defining the lymphoid organs necessary for ontogeny of immune cells, eg. the bone marrow, thymus (and Bursa of Fabricius in birds), in contrast to peripheral lymphoid organs, eg. lymph nodes, spleen etc. which are populated by lymphocytes that have undergone the antigen-independent (and in some cases, the antigen-dependent) stages of maturation.

centriolar region See *pericentriolar region* or *centrosome*.

centriole Organelle of animal cells which is made up of two orthogonally-arranged cylinders each with nine microtubule triplets composing the wall. Almost identical to *basal body* of cilium. The pericentriolar material, but not the centriole itself, is the major microtubule organising centre of the cell. Centrioles replicate (by induction of a daughter at right-angles to the parent) prior to mitosis and the daughter centrioles and their associated pericentriolar material come to lie at the poles of the spindle.

centrosome The microtubule organising centre which, in animal cells, surrounds the centriole, and which replicates to organise the two poles of the mitotic spindle.

centrosphere See *centrosome*.

cephalosporin Group of tetracyclic triterpene antibiotics isolated from culture filtrates of the fungus *Cephalosporium* sp.. Effective against penicillin-resistant staphylococci and other Gram positive bacteria.

ceramide An N-acyl *sphingosine*, the lipid moiety of *glycosphingolipids*.

cerebroside Glycolipid found in brain (11% of dry matter). Sphingosine core with fatty amide or hydroxy fatty amide and a single monosaccharide on the alcohol group (either glucose or galactose).

cereolysin Cytolytic (haemolytic) toxin released by *Bacillus cereus*. Inactivated by oxygen, reactivated by thiol reduction (hence thiol-activated cytolysin). Binds to cholesterol in the plasma membrane; rearrangement of the toxin–cholesterol complexes in the membrane leads to altered permeability.

ceruloplasmin A blue, copper-containing glycoprotein found in serum and long regarded as a transport protein for copper; also has oxidase activity.

cesium chloride See *caesium chloride*.

CFU-E Colony forming unit for *erythrocyte* cell lines.

chalone Cell-released tissue-specific inhibitor of cell proliferation thought to be responsible for regulating the size of a population of cells. Contentious.

channel forming ionophore An *ionophore* which makes an amphipathic pore with hydrophobic exterior and hydrophilic interior. Most known types are cation selective.

channel protein A protein which facilitates the diffusion of molecules/ions across lipid membranes by forming a hydrophilic pore. Most frequently multimeric with the pore formed by subunit interactions.

Chaos chaos Giant multinucleate freshwater amoeba (up to 5mm long) much used for studies on the mechanism of cell locomotion.

Chara See *Characean algae*.

Characean algae (Charophyceae) Class of filamentous green algae exemplified by the genus *Chara*, in which the mitotic spindle is not surrounded by a nuclear envelope. Probably the closest relatives, among the algae, to higher plants. The giant internodal cells (up to 5cm long) exhibit dramatic *cyclosis* and have been much used for studies on ion transport and cytoplasmic streaming.

chartins *Microtubule-associated proteins* of 64, 67 and 80kD, distinct from *tau* protein. Isolated from neuroblastoma cells. They are regulated by *nerve growth factor* and may influence microtubule distribution.

checkerboard assay Variant of the Boyden chamber assay for leucocyte chemotaxis introduced by Zigmond. By testing different concentrations of putative chemotactic factor in non-gradient conditions, it is possible to calculate the enhancement of movement expected due simply to *chemokinesis* and to compare this with the distances moved in positive and negative gradients. Good experimental design thus allows chemotaxis to be distinguished from chemokinesis, but it must be emphasised that the calculation is based upon an estimate of velocity and acceleration, and cannot be used when cell numbers are being counted.

Chediak-Higashi syndrome Autosomal recessive disorder characterised by the presence of giant lysosomal vesicles in phagocytes and in consequence poor bactericidal function. Some perturbation of microtubule dynamics seems to be involved. Reported from humans, albino Hereford cattle, mink, beige mice and killer whale.

chemical potential The work required (in J mol^{-1}) to bring a molecule from a standard state (usually infinitely separated in a vacuum) to a specified concentration. More usually employed as chemical potential difference, the work required to bring one mole of a substance from a solution at one concentration to another at a different concentration, $\Delta\mu = RT.\ln(c2/c1)$. This definition is useful in studies of *active transport*; note that, for charged molecules, the electrical potential difference must also be considered (see *electrochemical potential*).

chemical synapse A nerve–nerve or nerve–muscle junction where the signal is transmitted by release, from the membrane of the presynaptic cell, of a chemical transmitter which binds to a receptor on the membrane of the postsynaptic cell. An important feature is that signals can only pass in one direction.

chemiluminescence Light emitted as a reaction proceeds. Becoming used increasingly to assay ATP (using firefly *luciferase*) and the production of toxic oxygen species by activated phagocytes (using *luminol* or *lucigenin* as bystander substrates which release light when oxidised). See also *bioluminescence*.

chemiosmosis A mechanism (proposed by Mitchell) to explain energy transduction in the mitochondrion. As a general mechanism it is the coupling of one enzyme-catalysed reaction to another using the transmembrane flow of an intermediate species; eg. cytochrome oxidase pumps protons across the mitochondrial inner membrane and ATP synthesis is "driven" by re-entry of protons through the ATP-synthesising protein complex. The alternative model is production of a chemical intermediate species, but no compound capable of coupling these reactions has ever been identified.

chemiosmotic hypothesis See *chemiosmosis*.

chemoattractant A diffusible substance which elicits accumulation of cells.

chemoattraction Non-committal description of cellular response to a diffusible chemical, not necessarily by a tactic response. The term is preferable to "chemotaxis" when the mechanism is unknown.

chemodynesis Induction of cytoplasmic

streaming in plant cells by chemicals rather than by light (photodynesis).

chemokinesis A response by a motile cell to a soluble chemical which involves an increase or decrease in speed (positive or negative *orthokinesis*) or of frequency of movement, or a change in the frequency or magnitude of turning behaviour (*klino-kinesis*).

chemoreceptor A cell or group of cells specialised for responding to chemical substances in the environment.

chemorepellent Opposite of *chemoattractant*.

chemotactic See *chemotaxis*.

chemotaxis A response of motile cells or organisms, in which the direction of movement is affected by the gradient of a diffusible substance. Differs from chemokinesis in that the gradient alters probability of motion in one direction only, rather than rate or frequency of random motion.

CHF See *chick heart fibroblasts*.

chiasma (*pl.* chiasmata) Cytologically-visible junction points between non-sister chromatids at diplotene of meiosis, the consequence of a crossing-over event between maternally- and paternally-derived chromatids. A chiasma also serves a mechanical function and is essential for normal equatorial alignment at meiotic metaphase I and segregation at anaphase I in many species. Frequency of chiasmata is very variable between species.

chick heart fibroblasts The cells that emigrate from an explant of embryonic chick heart maintained in culture. Often used as archetypal normal cells.

chimera Organism composed of two genetically distinct types of cells, formed by the fusion of two early blastula-stage embryos, by the reconstitution of the bone marrow in an irradiated recipient, or by somatic segregation.

chitin A cross-linked polymer of N-acetyl-D-glucosamine which is the major structural component of arthropod exoskeletons. Widely distributed in plants and fungi. An inhibitor of chitin formation, diflubenzuron, has been evaluated for insecticidal potential (see also *juvenile hormone*).

Chlamydia Genus of minute prokaryotes which replicate in cytoplasmic vacuoles within susceptible eukaryotic cells. Genome about one-third that of *E. coli*. *C. trachomatis* causes trachoma in man; *C. psittaci* causes economically important diseases of poultry.

Chlamydomonas Genus now re-named *Dunalliela*. Unicellular green alga with two flagella; can easily be grown in the laboratory and has therefore been popular as an experimental organism: a range of mutants in flagellar function, for example, have been studied in detail.

chloramphenicol An antibiotic from *Streptomyces venezuelae* which inhibits protein synthesis in prokaryotes, mitochondria and chloroplasts by acting on the 50S subunit. It is relatively toxic but finds wide application in medicine.

chlorenchyma Any form of *parenchyma* tissue containing chloroplasts; found especially in leaf *mesophyll* and outer cortex of green stems.

chlorophylls The photosynthetic pigments of higher plants, but closely related to bacteriochlorophylls. Magnesium complexes of tetrapyrolles.

Chlorophyta (Green algae) Division of algae containing photosynthetic pigments similar to those in higher plants and having a green colour. Includes unicellular forms, filaments and leaf-like thalluses (eg. *Ulva*). Some members form *coenobia*, and the *Characean algae* have branched filaments.

chloroplast Photosynthetic organelle of higher plants. Lens-shaped and rather variable in size but approximately 5μm long. Surrounded by a double membrane and contains circular DNA (though not enough to code for all proteins in the chloroplast). Like the mitochondrion, it is semi-autonomous. It resembles a cyanobacterium from which, on the endosymbiont hypothesis, it might be derived. The photosynthetic pigment, chlorophyll, is associated with the membrane of vesicles (thylakoids) which are stacked to form grana.

chloroquine Antimalarial drug which has the interesting property of increasing the intralysosomal pH when added to intact cells in culture.

chlorosis Yellowing or bleaching of plant tissues due to the loss of chlorophyll or failure of chlorophyll synthesis. Symptomatic of many plant diseases, also of deficiencies of light or certain nutrients.

chlorpromazine Neuroleptic aliphatic phenothiazine, thought to act primarily as dopamine antagonist, but also antagonist at α-adrenergic, H1 histamine, muscarinic and serotonin receptors. Used clinically as an anti-emetic and anti-psychotic. Has been shown to alter fibroblast behaviour.

CHO cells Established line of fibroblasts from Chinese Hamster Ovary.

cholate In practice, the sodium salt of cholic acid, which has strong detergent properties and can replace membrane lipids to generate soluble complexes of membrane proteins.

cholecalcin See *calbindin-D-28k*.

cholecystitis Inflammatory condition of the wall of the gallbladder caused by *Salmonella typhi*.

cholecystokinin Hormone released by intestinal cells which induces enzyme release from pancreatic acinar cells.

cholera toxin A multimeric protein toxin from *Cholera vibrio*. The toxic A subunit activates adenyl cyclase irreversibly by ADP-ribosylation of a Gs protein. The B subunit facilitates passage of the A subunit across the cell membrane.

cholesterol The major sterol of higher animals. An important component of cell membranes, especially of the plasma (outer) membrane, most notably the myelin sheath. Transported in the esterified form via plasma lipoproteins.

choline Esterified in the head group of phospholipids (phosphatidyl choline and sphingomyelin) and in the neurotransmitter acetylcholine; otherwise a biological source of methyl groups.

cholinergic neurons Neurons in which actylcholine is the neurotransmitter.

chondro- Prefix indicating cartilage related/associated.

chondroblast Embryonic cartilage-producing cell.

chondrocyte Differentiated cell responsible for secretion of extracellular matrix of *cartilage*.

chondroitin sulphates Major components of the extracellular matrix and connective tissue of animals. They are repeating polymers of glucuronic acid and sulphated N-acetyl glucosamine residues that are highly hydrophilic and anionic. Found in association with proteins.

chondronectin Protein (180kD) isolated from chick serum which specifically favours attachment of *chondrocytes* to Type II *collagen* if present with the appropriate cartilage *proteoglycan*.

chorioallantoic membrane Extraembryonic membrane formed in birds and reptiles by the apposition of the allantois to the inner face of the *chorion*. The chorioallantoic membrane is highly vascularised, and is used experimentally as a site upon which to place pieces of tissue in order to test their invasive capacity.

choriocarcinoma Malignant tumour of trophoblast.

chorion 1. Protective membrane around the eggs of insects and fishes. 2. Extraembryonic membrane surrounding the embryo of amniote vertebrates. The outer epithelial layer of the chorion is derived from the trophoblast.

chorionic gonadotrophin A glycoprotein hormone synthesised in the placenta which controls the size of the gonads and the synthesis of sex hormones.

choroid Middle layer of the vertebrate eye, between retina and sclera. Ideal source of fibroblasts with which to confront pigmented retinal epithelium from embryonic chick retina. Well vascularised and also pigmented to throw light back onto the retina (the tapetum is an iridescent layer in the choroid of some eyes). Not to be confused with the choroid plexus, a highly vascularised region of the roof of the ventricles of the vertebrate brain which secretes cerebrospinal fluid.

Christmas disease Congenital deficiency of blood-clotting factor IX (first described in the Christmas issue of British Medical J., 1952). Inherited in similar sex-linked way as classical haemophilia.

chromaffin tissue Tissue in medulla of adrenal gland containing two populations of cells, one producing adrenaline, the other noradrenaline. The catecholamine is associated with carrier proteins (chromogranins) in membrane vesicles (chromaffin granules).

chromatic aberration When using white

light through a lens system, it is inevitable that different wave-lengths (colours) are brought to a focus at slightly different points. As a consequence, there are chromatic aberrations in the image; good microscope objectives are therefore corrected for this at two wave-lengths (achromats) or at three wave-lengths (apochromats), as well as for *spherical aberration*.

chromatid Single chromosome containing only one DNA duplex. Although in *G2* and early in mitosis each chromosome actually has two chromatids, they do not become visible until late metaphase.

chromatin Stainable material of interphase nucleus consisting of nucleic acid and associated histone protein packed into *nucleosomes*. Euchromatin is loosely packed and accessible to RNA polymerases, whereas heterochromatin is highly condensed and probably transcriptionally inactive.

chromatin body Barr body; condensed X chromosome in female mammalian cell.

chromatography Technique for separating molecules based on differential absorption and elution. Term for separation methods involving flow of a fluid carrier over a non-mobile absorbing phase.

chromatophores 1. Pigment-containing cells of the dermis, particularly in teleosts and amphibians; by controlling the intracellular distribution of pigment granules the animal can blend with the background. See *melanocytes*, *melanophores*. 2. Term occasionally used for chloroplasts in the chromophyte algae.

chromocentre Generally a condensed heterochromatic region of a chromosome which stains particularly strongly, but in the polytene chromosomes of *Drosophila* the chromocentre is of under-replicated heterochromatin and stains lightly.

chromogranins See *chromaffin tissue*.

chromomere Granular region of condensed *chromatin*. Used of chromosomes at leptotene and zygotene stages of meiosis, of the condensed regions at the base of loops on lampbrush chromosomes, and of condensed bands in polytene chromosomes of Diptera.

chromophores See *chromatophore*.

chromoplast Chromatophore filled with red/orange or yellow carotenoid pigment.

Responsible for colour of carrot and of many petals.

chromosome The DNA of eukaryotes is subdivided into chromosomes, presumably for convenience of handling, each of which has an unbroken length of DNA associated with various proteins. The chromosomes become more tightly packed at mitosis and meiosis. Each chromosome has a characteristic length and *banding* pattern.

chronic Persistent, long-lasting (as opposed to *acute*).

chronic inflammation Inflammatory response to a persistent antigenic stimulus. Dominant cells are macrophages and lymphocytes; very few granulocytes are present.

chronic granulomatous disease (CGD) Disease, usually fatal in childhood, in which the production of hydrogen peroxide by phagocytes does not occur because of a lesion in an NAD- or NADP-dependent oxidase. Catalase-negative bacteria are not killed and there is no luminol-enhanced *chemiluminescence* when the cells are tested. The absence of the oxygen-dependent killing mechanism is not itself fatal but seriously compromises the primary defence system. At least three separate lesions can cause the syndrome, the commonest being a defect in plasma membrane cytochrome.

Chrysophyta (Chrysophyceae: golden algae) Division or Class of algae, coloured golden-brown due to high levels of the *xanthophyll*, fucoxanthin. Mostly single-celled or colonial. Also called Chrysomonadida by protozoologists.

chymostatin Low-molecular weight oligopeptide–fatty acid compound of microbial origin which inhibits *chymotrypsin* and *papain*.

chymotrypsin Serine proteases from pancreas. Preferentially hydrolyse Phe, Tyr, or Trp peptide and ester bonds.

cicatrisation Contraction of fibrous tissue, formed at a wound site by fibroblasts, reducing the size of the wound but causing tissue distortion and disfigurement. Once thought to be due to contraction of collagen but now known to be due to cellular activity.

CIG (cold insoluble globulin) Obsolete synonym for *fibronectin*.

ciliary body Tissue which includes the group of muscles that act on the eye lens to produce accommodation and the arterial circle of the iris. The inner ciliary epithelium is continuous with the *pigmented retinal epithelium*, the outer ciliary epithelium secretes the aqueous humour.

ciliary ganglion *Neural crest*-derived ganglion acting as relay between parasympathetic neurons of the oculomotor nucleus in the midbrain and the muscles regulating the diameter of the pupil of the eye.

cilium (*pl.* cilia) Motile appendages of eukaryotic cells which contain an *axoneme*, a bundle of microtubules arranged in a characteristic fashion with nine outer doublets and a central pair (9 + 2 arrangement). Active sliding of doublets relative to one another generates curvature, and the asymmetric stroke of the cilium drives fluid in one direction (or the cell in the other direction).

circular dichroism (CD) Differential absorption of right-hand and left-hand circularly polarised light resulting from molecular asymmetry involving a chromophore group. CD is used to study the conformation of proteins and nucleic acids in solution.

cirrhosis Irreversible condition affecting the whole liver involving fibrosis, scarring, loss of parenchymal cells, inflammation, disruption of the normal tissue architecture, and eventually hepatic failure.

cirri Large motor organelles of hypotrich ciliate protozoa: formed from fused *cilia*.

***cis* -activation** Activation of a gene by an activator located on the same chromosome ie. not by a diffusible product.

cisternae Membrane-bounded saccules of the smooth and rough endoplasmic reticulum and Golgi apparatus. Operationally might be considered as an extra-cytoplasmic compartment since substances in the cisternal space will eventually be released to the exterior.

cistron A genetic element defined by means of the *cis–trans* **complementation** test for functional allelism; broadly equivalent to the sequence of DNA which codes for one polypeptide chain, including adjacent control regions.

citric acid cycle See *tricarboxylic acid cycle*.

clathrin Protein (180kD plus 34 and 36kD light chains) which forms the basketwork of triskelions around a *coated vesicle*.

cleavage The early divisions of the fertilised egg to form blastomeres. The cleavage pattern is radial in some phyla, spiral in others.

Cleland's reagent Dithiothreitol.

climacteric A particular stage of fruit ripening, characterised by a surge of respiratory activity, and usually coinciding with full ripeness and flavour in the fruit. Its appearance is hastened by ethylene at low concentrations.

Clonal Selection The process whereby one or more clones, ie. cells expressing a particular gene sequence, are selected by naturally-occurring processes from a mixed population. Usually the clonal selection is for general expansion by mitosis, particularly with reference to *B-lymphocytes* where selection with subsequent expansion of clones occurs as a result of antigenic stimulation only of those lymphocytes bearing the appropriate receptors.

clone A propagating population of organisms, either single cell or multicellular, derived, in the case of single cells or cells in culture, from a single progenitor cell. In the case of multicellular organisms, the clone is derived asexually from a single ancestor; as a result, such organisms should be genetically identical, though mutation events may abrogate this.

cloning The process whereby clones are established. In simple systems single cells may be isolated without precise knowledge of their genotype. In other systems (see *gene cloning*) the technique requires partial or complete selection of chosen genotypes. In plants, the term refers to artificial or natural vegetative reproduction.

Clostridium Genus of Gram positive anaerobic spore-forming bacilli commonly found in soil. Many species produce exotoxins of great potency, the best known being *C. botulinum* and *C. tetani*.

cnidoblast Developing form of *cnidocyte*.

cnidocyst See *nematocyst*.

cnidocyte Ectodermal cell of Cnidaria (coelenterates) specialised for defence or capturing prey. Each cell has a *nematocyst* which can be replaced once discharged.

coacervate Colloidal aggregate containing a mixture of organic compounds. One theory of the evolution of life is that the formation of coacervates in the primaeval soup was a step towards the development of cells.

coated pit First stage in the formation of a *coated vesicle*.

coated vesicle Vesicle formed as an invagination of the plasma membrane (a coated pit), and which is surrounded by a basket of *clathrin*. Associated with receptor-mediated pinocytosis and receptor recycling.

cobalamin Vitamin B_{12} (extrinsic factor). See Table V1.

cobra venom In general highly toxic venoms with a predominantly neurotoxic action (cf the necrotic action of viper venoms). Bind strongly and irreversibly to the acetylcholine receptor. See also *venoms*.

cobra venom factor See *C3*.

cocarcinogens Substances which, though not carcinogenic in their own right, potentiate the activity of a carcinogen. Strictly speaking they differ from tumour promotors in requiring to be present concurrently with the carcinogen.

cocci Bacteria with a spherical shape.

co-culture In order to get some cells to grow at low (clonal) density it is sometimes helpful to grow them together with a feeder layer of macrophages or irradiated cells. The mixing of different cell types in culture is otherwise normally avoided, although it is possible that this could prove an informative approach to modelling interactions *in vivo*.

cochlear hair cell The sound-sensing cell of the inner ear. The cells have modified ciliary structures (hairs), which enable them to produce an electrical (neural) response to mechanical motion caused by the effect of sound waves on the cochlea. Frequency is detected by the position of the cell in the cochlea and amplitude by the magnitude of the disturbance.

codon The coding unit of DNA which specifies the function of the corresponding messenger RNA. A triplet of bases recognised by anticodons on *transfer RNA* and hence specifying an amino acid to be incorporated into a protein sequence. Codons are degenerate, ie. each amino acid has more than one codon. The stop-codon determines the end of a polypeptide. See Table C3.

coeliac disease (USA celiac disease) Gluten enteropathy: atrophy of villi in small intestine leads to impaired absorption of nutrients. Caused by sensitivity to gluten (protein of wheat and rye). Sufferers have serum antibodies to gluten and show delayed hypersensitivity to gluten; the risk

Table C3. The codon assignments of the genetic code.

First Position (5' end)		Second Position				Third Position (3' end)
		U	C	A	G	
U		Phe, F	Ser, S	Tyr, Y	Cys, C	U
		Phe, F	Ser, S	Tyr, Y	Cys, C	C
		Leu, L	Ser, S	Stop (ochre)	Stop : (Trp)[a]	A
		Leu, L	Ser, S	Stop (amber)	Trp, W	G
C		Leu, L	Pro, P	His, H	Arg, R	U
		Leu, L	Pro, P	His, H	Arg, R	C
		Leu, L	Pro, P	Gln, Q	Arg, R	A
		Leu, L	Pro, P	Gln, Q	Arg, R	G
A		Ile, I	Thr, T	Asn, N	Ser, S	U
		Ile, I	Thr, T	Asn, N	Ser, S	C
		Ile, I : (Met)[a]	Thr, T	Lys, K	Arg, R : (stop)[a]	A
		Met, M (start)	Thr, T	Lys, K	Arg, R : (stop)[a]	G
G		Val, V	Ala, A	Asp, D	Gly, G	U
		Val, V	Ala, A	Asp, D	Gly, G	C
		Val, V	Ala, A	Glu, E	Gly, G	A
		Val, V (Met)[b]	Ala, A	Glu, E	Gly, G	G

[a]Unusual codons used in human mitochondria.
[b]Normally codes for valine but can code for methionine to initiate translation from an mRNA chain.

factor is ten times greater in HLA-B8 positive individuals.

coenobium Colony of cells formed by certain green algae, in which little or no specialisation of the cells occurs. The cells are often embedded in a mucilaginous matrix. Examples: **Volvox**, *Pandorina*.

coenocyte (USA cenocyte) Organism which is not subdivided into cells but has many nuclei within a mass of cytoplasm (a syncytium), as for example some fungi and algae, and the acellular slime mould *Physarum*.

coenzyme Either: low-molecular weight intermediate which transfers groups between reactions (eg. NAD), or: catalytically active low-molecular weight component of an enzyme (eg. haem). Coenzyme and apoenzyme together constitute the holoenzyme.

coenzyme A Coenzyme derived from pantothenic acid which carries acyl groups as thioesters. Acetyl-CoA reacts with oxaloacetate to produce citrate, or can be reduced to acetaldehyde in some bacterial fermentations.

cofactor Inorganic complement of an enzyme reaction, usually a metal ion. See *coenzyme*.

colcemid Methylated derivative of *colchicine*.

colchicine Alkaloid (MW 400) isolated from the Autumn crocus (*Colchicum autumnale*) which blocks microtubule assembly by binding to the *tubulin* heterodimer (but not to tubulin). As a result of interfering with microtubule reassembly will block mitosis at *metaphase*.

cold antibody (cold agglutinin) Antibodies which *agglutinate* particles with greater activity below 32°C. They are IgM antibodies specifically reactive with blood groups I and i in humans, and agglutinate red blood cells on cooling, causing Raynaud's phenomenon *in vivo*.

cold insoluble globulin (CIG) Name, now obsolete, originally given to fibronectin prepared from *cryoprecipitate*.

coleoptile Closed hollow cylinder or sheath of leaf-like tissue surrounding and protecting the plumule (shoot axis and young leaves) in grass seedlings.

coleorhiza Closed hollow cylinder or sheath of tissue surrounding and protecting the radicle (young root) in grass seedlings.

colicins Antibacterial proteins produced by *E. coli* and capable of killing other enteric bacteria. Colicins E2 and E3 are *AB toxins* with DNAase and RNAase activity respectively. Coded on plasmids which can be transferred at conjugation.

collagen Major structural protein (285kD) of extracellular matrix. An unusual protein both in amino acid composition (very rich in glycine (30%), proline, *hydroxyproline*, lysine, and *hydroxylysine*; no tyrosine or tryptophan), structure (a triple helical arrangement of 95kD polypeptides giving a *tropocollagen* molecule, dimensions 300nm × 0.5nm) and resistance to proteases. Most types are fibril-forming with characteristic quarter-stagger overlap between molecules producing an excellent tension-resisting fibrillar structure. Type IV, characteristic of *basement membrane*, does not form fibrils. Many different types of collagen are now recognised. Some are glycosylated (glucose-galactose dimer on the hydroxylysine), and nearly all types can be crosslinked through lysine side-chains. See *dermatosparaxis*, *scurvy*.

collagenase Proteolytic enzyme capable of breaking native collagen. Once the initial cleavage is made, less specific proteases will complete the degradation. Collagenases from mammalian cells are metallo-enzymes and are collagen-type specific. May be released in latent (proenzyme) form into tissues and require activation by other proteases before they will degrade fibrillar matrix. Bacterial collagenases are used in tissue disruption for cell harvesting.

collenchyma Plant tissue in which the *primary cell walls* are thickened, especially at the cell corners. Acts as a supporting tissue in growing shoots, leaves and petioles. Often arranged in cortical "ribs", as seen prominently in celery and rhubarb petioles. *Lignin* and *secondary walls* are absent; the cells are living and able to grow.

colonisation factors The pili on enteropathogenic forms of *E. coli* facilitate adhesion of the bacteria to receptors (probably GM1 gangliosides) on gut epithelial cells and are often referred to as colonisation or adherence factors. Colonisation factor antigens may be plasmid coded, are essential for pathogenicity and are strain-specific, for example K88 (diarrhoea in piglets), CFA/I and CFA/II on strains causing similar disease in man.

colony forming unit (CFU-S) Stem cells that will reconstitute the immune system of, for example, irradiated mice. Bone marrow cells from a non-irradiated donor will, when injected, form colonies in the spleen (hence -S), each colony representing the progeny of a pluripotent stem cell. Operationally, therefore, the number of colony forming units is a measure of the number of stem cells.

colV A plasmid of *E. coli* coding for *colicin* V that confers resistance to complement-mediated killing, for a siderophore to scavenge iron, and for F-like pili which permit conjugation.

comb plates Large flat organelles formed by the fusion of many cilia. Vertical rows of comb plates form the motile appendages of Ctenophores (comb-jellies).

combined immunodeficiency (severe c.i. syndrome) A congenital immunological deficiency of both humoral and cell-mediated immunity. There is thymic agenesis, lymphocyte depletion and hypogammaglobulinaemia and life expectancy is low unless marrow transplantation is successful.

communicating junction Another name for a *gap junction*.

companion cell Relatively small plant cell, with little or no vacuole, found adjacent to a phloem *sieve tube* and originating with the latter from a common mother cell; a single sieve tube may have more than one companion cell. Thought to be involved in translocation of sugars in and out of the sieve tube.

compartment In the insect wing, for example, there are two compartments, anterior and posterior, each containing several clones, but clones do not cross the boundary. It seems from studies with *homeotic mutants* that cells in different compartments are expressing different sets of genes. The evidence for such developmental compartments in vertebrates is sparse at present.

competitive inhibitor Inhibitor which occupies the active site of an enzyme or the binding site of a receptor and prevents the normal substrate or ligand from binding. At sufficiently high concentration of the normal ligand inhibition is lost: the K_m is altered by the competitive inhibitor, but the V_{max} remains the same.

complement A heat-labile cascade system of enzymes in plasma associated with response to injury. Activation of the complement cascade occurs through two convergent pathways. In the classical pathway the formation of antibody/antigen complexes leads to binding of *C1*, the release of active esterase which activates *C4* and *C2* which in turn bind to the surface. The C42 complex splits *C3* to produce C3b, an opsonin, and C3a (anaphylatoxin). C423b acts on *C5* to release C5a (anaphylatoxin and chemotactic factor) leaving C5b which combines with C6789 to form a cytolytic *membrane attack complex*. In the alternate pathway C3 cleavage occurs without the involvement of C142, and can be activated by IgA, endotoxin, or polysaccharide-rich surfaces (eg. yeast cell wall, zymosan). Factor B combines with C3b to form a C3 convertase which is stabilised by Factor P, generating a positive feedback loop. The alternate pathway is presumably the ancestral one upon which the sophistication of antibody recognition has been superimposed in the classical pathway. The enzymatic cascade amplifies the response, leads to the activation and recruitment of leucocytes, increases phagocytosis and induces killing directly. It is subject to various complex feedback controls which terminate the response. See also *C1 — C9*.

complement fixation Binding of *complement* as a result of its interaction with immune complexes (the classical pathway) or particular surfaces (alternative pathway).

complementation The ability of a mutant chromosome to restore normal function to a cell which has a mutation in the homologous chromosome when a hybrid or heterokaryon is formed: the explanation being that the mutations are in different functional genetic units, so that a complete set of non-mutant information is present.

complementary base pairs The crucial property of DNA is that the two strands are complementary; guanine and cytosine pair through three hydrogen bonds, adenine and thymine only form two hydrogen bonds. In RNA adenine and uracil are paired.

Con A See *concanavalin A*.

Con A binding sites See *Con A receptors*.

Con A receptors A common misuse of the

term receptor. Con A binds to the mannose residues of many different glycoproteins and glycolipids and the binding is therefore not to a specific site. It could be argued that the receptor is the Con A and cells have Con A ligands on their surfaces: certainly this would be less confusing.

concanavalin A (Con A) A lectin isolated from the jack bean. See Table L1 (Lectins).

condensing vacuole Vacuole formed from the *trans* face of the Golgi network during regulative secretion by the fusion of smaller vesicles. Within the condensing vacuole the contents are concentrated and may become semi-crystalline (*zymogen granules* or *secretory vesicles*).

conditioned medium Cell culture medium that has already been partially used by cells. Although depleted of some components, it is enriched with cell-derived material, probably including small amounts of *growth factors*; such cell-conditioned medium will support the growth of cells at much lower density and, mixed with some fresh medium, is therefore useful in cloning.

cone cell See *retinal cone*.

confluent culture A cell culture in which all the cells are in contact and the entire surface of the culture vessel is covered. The term is also often used with the implication that the cells have also reached their maximum density, though confluence does not necessarily mean that division will cease or that the population will not increase in size.

conformational change Alteration in the shape, usually the tertiary structure of a protein, as a result of alteration in the environment (pH, temperature, ionic strength) or the binding of a ligand (to a receptor) or binding of substrate (to an enzyme).

congenic Organisms which differ in *genotype* at (ideally) only one specified locus. Strictly speaking these are conisogenics. Thus one homozygous strain can be spoken of as being congenic to another.

conidium Asexual exospore of a fungus, borne at the tip of a specialised *hypha* (conidiophore).

conjugation Union between two gametes or between two cells leading to the transfer of genetic material. In eukaryotes the classic examples are in *Paramecium* and *Spirogyra*.

Conjugation between bacteria involves an F+ bacterium (with F-pili) attaching to an F−; transfer of the F-plasmid then occurs through the sex pilus. In Hfr mutants the F-plasmid is integrated into the chromosome and so chromosomal material is transferred as well. Conjugation occurs in many Gram negative bacteria (*Escherichia*, *Shigella*, *Salmonella*, *Pseudomonas*) and in Gram positive bacteria (*Bacillus*, *Staphylococcus* and *Streptomyces*).

connectin Cell surface protein (70kD) from mouse fibrosarcoma cells that binds *laminin* and *actin*.

connective tissue Rather general term for mesodermally derived tissue which may be more or less specialised. Cartilage and bone are specialised connective tissue, as is blood, but the term is probably better reserved for the less specialised tissue which is rich in extracellular matrix (collagen, proteoglycan etc.) and which surrounds other more highly ordered tissues and organs.

connective tissue diseases A group of diseases including rheumatoid arthritis, systemic lupus erythematosus, rheumatic fever, scleroderma and others, that are sometimes referred to as rheumatic diseases. They probably do not affect solely connective tissues but the diseases are linked in various ways and have interesting immunological features which have led some workers to suggest that the diseases may be autoimmune in origin.

connexon The functional unit of *gap junctions*; an assembly of six membrane-spanning proteins having a water-filled gap in the centre. Two connexons in juxtaposed membranes link to form a continuous pore through both membranes.

Conn's syndrome Uncontrolled secretion of *aldosterone* usually by an adrenal *adenoma*.

consensus sequence DNA sequence shared by several genes; need not be absolutely identical.

constitutive Constantly present, whether there is demand or not. Thus some enzymes are constitutive, whereas others are inducible.

constriction ring The equatorial ring of microfilaments which diminishes in diameter probably both by contraction and disassembly as *cytokinesis* proceeds.

contact following Behaviour shown by individual slime mould cells when they join a stream moving towards the aggregating centre. In *Dictyostelium*, *contact sites A* at front and rear of cell may be involved.

contact guidance Directed locomotory response of cells to an axial anisotropy of the environment, for example the tendency of fibroblasts to align along ridges or parallel to the alignment of collagen fibres in a stretched gel.

contact inhibition of growth/division See *density dependent inhibition*.

contact inhibition of locomotion/movement Reaction in which the direction of motion of a cell is altered following collision with another cell. In heterologous contacts both cells may respond (mutual inhibition), or only one (non-reciprocal). Type I contact inhibition involves paralysis of the locomotory machinery, Type II is a consequence of adhesive preference for the substratum rather than the dorsal surface of the other cell.

contact inhibition of phagocytosis Phenomenon described in sheets of kidney epithelial cells which, when confluent, lose their weak phagocytic activity, probably because of a failure of adhesion to the dorsal surface in the absence of ruffles.

contact sensitivity Response to contact with irritant; usually an immunological *hypersensitivity*.

contact sites A (csA) Developmentally-regulated adhesion sites which appear on the ends of aggregation-competent *Dictyostelium discoideum*, at the stage when the starved cells begin to come together to form the grex. Originally detected by the use of Fab fragments of polyclonal antibodies, raised against aggregation-competent cells and adsorbed against vegetative cells, to block adhesion in EDTA-containing medium. (Cell–cell adhesion mediated by contact sites A, unlike that mediated by contact sites B, is not divalent cation-sensitive). The fact that a mutant deficient in csA behaves perfectly normally in culture is puzzling.

contact sites B See *contact sites A*.

contact-induced spreading The response in which contact between two epithelial cells leads to a stabilised contact and the increased spreading of the cells, so that the area covered is greater than that covered by the two cells in isolation.

contractile ring See *constriction ring*.

contrapsin Trypsin inhibitor (*serpin*) from rat.

Coomassie blue Tradename for Kenacid Blue, a stain commonly used non-specifically for proteins on gels.

cooperativity Phenomenon displayed by enzymes or receptors which have multiple binding sites. Binding of one ligand alters the affinity of the other site(s). Both positive and negative cooperativity are known; positive cooperativity gives rise to a sigmoidal binding curve. Cooperativity is often invoked to account for non-linearity of binding data, although it is by no means the only possible cause.

cord blood Blood taken post-partum from the umbilical cord.

cord factor Glycolipid (trehalose-6,6'-dimycolate) found in the cell walls of Mycobacteria (causing them to grow in serpentine cords) and important in virulence, being toxic and inducing granulomatous reactions identical to those induced by the whole organism.

cornea Transparent tissue at the front of the eye. The cornea has a thin outer squamous epithelial covering and an endothelial layer next to the aqueous humour, but is largely composed of avascular collagen laid down in orthogonal arrays with a few fibroblasts. Transparency of the cornea depends on the regularity of spacing in the collagen fibrils.

Coronaviridae Family of single-stranded RNA viruses responsible for respiratory and perhaps gastro-intestinal diseases of humans. The outer envelope of the virus has club-shaped projections which radiate outwards and give a characteristic corona appearance to negatively-stained virions.

cortex 1. *Bot.* Outer part of stem or root, between the vascular system and the epidermis; composed of *parenchyma*. Secondary cortex (*phelloderm*) is produced internally by the cork cambium. 2. Region of cytoplasm adjacent to the plasma membrane. 3. *Histol.* Outer part of organ.

cortical granule Specialised secretory vesicles lying just below the plasma membrane of the egg, which fuse and release their contents immediately after fertilisation (activation) to prevent polyspermy.

cortical layer See *cortical meshwork*.

cortical meshwork Sub-plasmalemmal layer of tangled microfilaments anchored to the plasma membrane by their barbed ends. This meshwork contributes to the mechanical properties of the cell surface and probably restricts the access of cytoplasmic vesicles to the plasma membrane.

corticosteroids *Steroid hormones* produced in the adrenal cortex. Formed in response to *adrenocorticotrophin* (ACTH). Regulate both carbohydrate metabolism and salt/water balance. Glucocorticoids (eg. cortisol, cortisone) predominantly affect the former and mineralocorticoids (eg. aldosterone) the latter.

corticotrophin (corticotropin) See *adrenocorticotrophin*.

corticotrophin releasing factor See *adrenocorticotrophin*.

cortisol The major adrenal glucocorticoid; stimulates conversion of proteins to carbohydrates, raises blood sugar levels and promotes glycogen storage in the liver.

cortisone Derived from *cortisol* and having similar physiological actions.

Corynebacteria Genus of Gram positive non-motile rod-like bacteria, often with a club-shaped appearance. Most are facultative anaerobes with some similarities to mycobacteria and nocardiae. *C. diphtheriae* is the causative agent of diphtheria and produces a potent exotoxin (*diphtheria toxin*).

costa Rod-shaped intracellular organelle lying below the undulating membrane of the flagellate protozoan *Trichomonas*. In some species, generates active bending associated with local loss of *birefringence* at the bending zone, probably as a result of conformational change in the longitudinal lamellae. Major protein approximately 90kD.

co-translational import Process whereby a protein is vectorially discharged into a membrane-bounded compartment (usually the cisterna of rough endoplasmic reticulum) whilst translation is still proceeding.

co-transport In membrane transport, describes tight coupling of the transport of one species (generally Na^+) to another (eg. a sugar or amino acid). The transport of Na^+ from high to low concentration can provide the energy for transport of the second species up a concentration gradient. See *secondary active transport*.

cotyledon Modified leaf ("seed leaf"), found as part of the embryo in seeds, involved in either storage or absorption of food reserves. Dicotyledonous seeds contain two, monocotyledonous seeds only one. May appear above ground and show photosynthetic activity in the seedling.

coumarin A γ-lactone formed by the cyclisation of O-hydroxycinnamic acid. Pleasant-smelling compound found in many plants and released on wilting (probably a major component of the smell of fresh hay). Dicoumarol has anticoagulant activity, probably competing with vitamin K.

counterstain Rather non-specific stain used in conjunction with another histochemical reagent of greater specificity to provide contrast and reveal more of the general structure of the tissue; for example, Light Green is used as a counterstain in the Mallory procedure.

coupling The linking of two independent processes by a common intermediate, eg. the coupling of electron transport to oxidative phosphorylation or the ATP–ADP conversion to transport processes.

coupling factors Proteins responsible for coupling transmembrane potentials to ATP synthesis in *chloroplasts* and *mitochondria*. Include ATP-synthesising enzymes (F_1 in mitochondrion), which can also act as ATPases.

Coxsackie viruses Species of enteroviruses of the *Picornaviridae* first isolated in Coxsackie, N.Y.. Coxsackie A produces diffuse myositis, Coxsackie B produces focal areas of degeneration in brain and skeletal muscle in mice. Similar to polioviruses in chemical and physical properties. A complication of Coxsackie virus infection can be a persistent myalgia and lassitude, which is now recognised as a distinct syndrome (fatigue syndrome).

CR1 Complement receptor-1; binds particles coated with C3b. Present on neutrophils, mononuclear phagocytes, B-lymphocytes and Langerhans' cells (the latter having only a few), and involved in the opsonic phagocytosis of bacteria and uptake of immune complexes. Also present on follicular dendritic cells and glomerular podocytes.

CR2 Receptor for complement factor C3dg, iC3b; present only on B-lymphocytes, follicular dendritic cells and lines of B- and T-cells. CR2 is the site to which the *Epstein Barr virus* binds.

CR3 Receptor for the complement component C3bi (iC3b), present on neutrophils and mononuclear phagocytes, follicular dendritic cells, natural *killer cells* and *ADCC* effector lymphocytes. CR3 is of the LFA-1 class, an *integrin*.

CR4 Receptor for C3dg, the complement fragment that remains when C3b is cleaved to C3bi. Thought to be present on monocytes, macrophages and neutrophils, but there is some disagreement at present. Possibly the same as gp150/95.

CR5 Receptor for complement fragments C3dg and C3d, which is present on neutrophils and platelets. Resembles *CR4* but binding is said to be divalent cation-dependent. Not yet firmly established.

crassulacean acid metabolism (CAM) Physiological adaptation, particularly of certain succulent plants, in which CO_2 can be fixed (non-photosynthetically) at night into malic and other acids. During the day the CO_2 is regenerated and then fixed photosynthetically in the *Calvin–Benson cycle*. This adaptation permits the stomata to remain closed during the day, conserving water.

creatine kinase (creatine phosphokinase; EC 2.7.3.2) Dimeric enzyme (82kD) which catalyses the reversible formation of ATP and creatine from ADP and creatine phosphate in muscle.

creatine phosphate (phosphocreatine) Storage compound of vertebrate muscle. See *creatine kinase*.

crenation Distortion of the erythrocyte membrane giving a spiky, echinocyte, morphology. Results from ATP depletion or an excess of lipid species in the external lipid layer of the membrane.

Creutzfeld–Jacob disease Rare fatal presenile dementia of humans, similar to *kuru* and other slow viruses. Induces neurological disorder in goats 3–4 years after inoculation with CJD brain extract. Classified pathologically as a subacute spongiform encephalopathy.

crinophagy Digestion of the contents of secretory granules following their fusion with lysosomes.

***cro*-protein** Protein synthesised by bacteriophage λ in the lytic state. The *cro*-protein blocks the synthesis of the λ repressor (which is produced in the lysogenic stage, and inhibits *cro*-protein synthesis). Production of the *cro*-protein in turn controls a set of genes associated with rapid virus multiplication.

Crohn's disease Chronic inflammatory disease which may involve any part of the alimentary tract, but most commonly the distal ileum. The disease seems to have both genetic and environmental causes; not well understood.

crossing over Recombination as a result of DNA breakage and reunion between homologous chromatids in meiosis, giving rise to *chiasmata*.

croton oil Oil from the seeds of the tropical plant *Croton tiglium* (Euphorbiaceae), causes severe skin irritation and contains a potent *tumour promoter* (co-carcinogen), phorbol ester.

crown gall Gall, or tumour, found in many dicotyledonous plants, caused by the bacterium **Agrobacterium tumefaciens**.

cryoglobulin Abnormal plasma globulin (IgG or IgM) which precipitates when serum is cooled.

cryoprecipitate The precipitate which forms when plasma is frozen and then thawed; particularly rich in *fibronectin* and blood-clotting *Factor VIII*.

cryoprotectant Substance which is used to protect from the effects of freezing, largely by preventing large ice-crystals from forming. The two compounds commonly used for freezing cells are *DMSO* or *glycerol*.

crypt Deep pit which protrudes down into the connective tissue surrounding the small intestine. The epithelium at the base of the crypt is the site of stem cell proliferation and the differentiated cells move upwards and are shed 3–5 days later at the tips of the villi.

crystallins Major proteins of the vertebrate

lens. Range from high MW oligomeric species to low MW monomeric species. Immunological cross-reactivity suggests that the sequences of crystallin subunits are relatively highly conserved in evolution.

CSAT Monoclonal antibody defining an integral plasma membrane component of chick fibroblasts; probably the avian homologue of *integrin*. The CSAT antigen is a complex of three proteins each of 140kD and has binding sites for *talin* and *fibronectin*.

CTL See *cytotoxic T-cells*.

culture To grow *in vitro*.

curare Curare alkaloids are the active ingredients of arrow poisons used by S.American Indians. The term generally means those alkaloids which have muscle-relaxant properties because they block motor end plate transmission, acting as competitive antagonists for *acetylcholine*.

CURL The compartment for uncoupling of receptors and ligands; internalised receptor–ligand complexes are stripped of the ligand and recycled.

Cushing's syndrome A type of hypertensive disease in humans due probably to the oversecretion of *cortisol* due in turn to excessive secretion of *adrenocorticotrophin* (ACTH). Adrenal tumours are the usual primary cause.

cutin Waxy hydrophobic substance deposited on the surface of plants. Composed of complex long-chain fatty esters and other fatty acid derivatives. Impregnates the outer wall of epidermal cells and also forms a separate layer, the cuticle, on the outer surface of the epidermis.

Cyanobacteria (Cyanophyta) Modern term for the blue-green algae, prokaryotic cells which use chlorophyll on intracytoplasmic membranes for photosynthesis. The blue-green colour is due to the presence of phycobiliproteins. Found as single cells, colonies or simple filaments. In *Anabaena*, in which the cells are arranged as a filament, heterocysts capable of nitrogen-fixation occur at regular intervals. According to the *endosymbiont theory* Cyanobacteria are the progenitors of chloroplasts.

cyanogen bromide (CNBr) Agent which cleaves peptide bonds at methionine residues. The peptide fragments so generated can then, for example, be tested to locate particular activities.

Cyanophyta (Blue-green algae) See *Cyanobacteria*.

cyclic AMP (cAMP; 3'5'-cyclic ester of AMP) The first second-messenger to be characterised. Generated from ATP by the action of adenylate cyclase which is coupled to hormone receptors by G-proteins (*GTP-binding proteins*). cAMP activates cAMP-dependent protein kinase(s) and is broken down by a phosphodiesterase to form 5'AMP. Also functions as an extracellular morphogen for some slime moulds.

cyclic GMP (cGMP; 3'5'-cyclic ester of GMP) May have second-messenger functions and was thought to act as an intracellular counter-balance to cAMP.

cyclic phosphorylation Any process in which a phosphatide ester forms a cyclic diester by linkage to a neighbouring hydroxyl group.

cyclic photophosphorylation Process by which light energy absorbed by *photosystem I* (PS-I) in the chloroplast can be used to generate ATP without concomitant reduction of *NADP*$^+$ or other electron acceptors. Energised electrons are passed from PS-I to ferredoxin, and thence along a chain of electron carriers and back to the reaction centre of PS-I, generating ATP *en route*.

cycloheximide An antibiotic (MW 281) isolated from *Streptomyces griseus*. Blocks eukaryotic (but not prokaryotic) protein synthesis by preventing initiation and elongation on 80S ribosomes. Commonly used experimentally.

cyclo-oxygenase Enzyme complex present in most tissues which produces various prostaglandins and thromboxanes from arachidonic acid; inhibited by aspirin-like drugs, probably accounting for their anti-inflammatory effects.

cyclophilin Cytoplasmic protein to which *cyclosporin* binds.

cyclophosphamide An alkylating agent and important immunosuppressant. Acts by alkylating SH and NH_2 groups, especially the N7 of guanine.

cyclosis Cyclical streaming of the cytoplasm of plant cells, conspicuous in giant internodal cells of algae such as *Chara*, in pollen tubes and in stamen hairs of *Tradescantia*. Term also used to denote cyclical movement of food vacuoles from mouth to *cytoproct* in ciliate protozoa.

cyclosporin (ciclosporine) Group of cyclic oligopeptides isolated from *Tolypocladium inflatum*, with potent immunosuppressant activity on both humoral and cellular systems. The use of cyclosporin has made transplant surgery much easier, although the long-term consequences of suppressing immune function are not yet clear.

cystatins A group of natural cysteine protease inhibitors (approximately 13kD) widely distributed both intra- and extra-cellularly.

cysteine (Cys: C; MW 121). The only amino acid to contain a thiol (SH) group. In intracellular enzymes the unique reactivity of this group is frequently exploited at the catalytic site. In extracellular proteins found only as 1/2 cystine in disulphide bridges or fatty acylated. See Table A2.

cystic fibrosis Generalised abnormality of exocrine gland secretion which affects pancreas (blockage of the ducts leads to cyst formation and to a shortage of digestive enzymes), bowel, biliary tree, sweat glands and lungs. The production of abnormally viscous mucus in the lung predisposes to respiratory infection, a major problem in children with the disorder. A fairly common (1 in 2000 live births in caucasians) autosomal recessive disease.

cytidine Nucleoside consisting of D-ribose and the pyrimidine base cytosine.

cytidine 5′diphosphate (CDP) CDP (derived from cytidine 5′triphosphate) is important in phosphatide biosynthesis; activated choline is CDP-choline.

cytocalbins *Calmodulin*-binding proteins associated with the cytoskeleton.

cytochalasins A group of fungal metabolites which inhibit the addition of G-actin to a nucleation site and therefore perturb labile microfilament arrays. Cytochalasin B inhibits at around 1μg/ml but at about 5μg/ml begins to inhibit glucose transport. Cytochalasin D affects only the microfilament system and is therefore preferable.

cytochemistry Branch of histochemistry associated with the localisation of cellular components by specific staining methods, as for example the localisation of acid phosphatases by the Gomori method. Immunocytochemistry involves the use of labelled antibodies as part of the staining procedure.

cytochrome Enzymes of the electron transport chain which are pigmented by virtue of their **haem** prosthetic groups. Very highly conserved in evolution.

cytochrome oxidase Terminal enzyme of the *electron transport chain* which accepts electrons from (ie. oxidises) cytochrome c and transfers electrons to molecular oxygen.

cytochrome P450 Mixed-function mitochondrial and microsomal oxidase of the cytochrome b group involved, among other things, in steroid hydroxylation reactions in the adrenal cortex.

cytokeratins Generic name for the intermediate filament proteins of epithelial cells.

cytokinesis Process in which the cytoplasm of a cell is divided after nuclear division (mitosis) is complete.

cytokinins Class of *plant growth substances* active in promoting cell division. Also involved in cell growth and differentiation and in other physiological processes. Examples: *kinetin*, *zeatin*.

cytology The study of cells. Implies the use of light or electron microscopic methods for the study of morphology.

cytolysis Cell lysis.

cytolysosome Membrane-bounded region of cytoplasm which is subsequently digested.

cytomegalovirus Probably the most widespread of the Herpetoviridae group, a large double-stranded DNA virus. Infected cells enlarge and have a characteristic inclusion body (composed of virus particles) in the nucleus. Opportunistic in immunocompromised hosts (*AIDS*, transplant recipients) and may cause disease *in utero* (leading to abortion or stillbirth or to various congenital defects). Latent infections are easily established both *in vitro* in cell culture and *in vivo*.

cytoplasm Substance contained within the plasma membrane excluding, in eukaryotes, the nucleus.

cytoplasmic bridge (*plasmodesma*) Thin strand of cytoplasm linking cells as in higher plants, *Volvox*, between *nurse cells* and developing eggs, and between developing sperm cells. Unlike gap junctions, allows the transfer of large macromolecules.

cytoplasmic inheritance Inheritance of parental characters through non-chromosomal means; thus mitochondrial DNA is cytoplasmically inherited since the information is not segregated at mitosis. In a broader sense the organisation of a cell may be inherited through the continuity of structures from one generation to the next. It has often been speculated that the information for some structures may not be encoded in the genomic DNA, particularly in protozoa which have complex patterns of surface organelles. See *maternal inheritance*.

cytoplasmic streaming Bulk flow of the cytoplasm of cells. Most conspicuous in large cells such as amoebae and the internodal cells of *Chara* where the rate of movement may be as high as 100μm/sec.

cytoplast Fragment of cell with nucleus removed (in *karyoplast*); usually achieved by cytochalasin B treatment followed by mild centrifugation on a step gradient.

cytoproct Cell anus: region at posterior of a ciliate where exhausted food vacuoles are expelled.

cytosine Pyrimidine base found in DNA and RNA; pairs with guanine. The glycosylated base is *cytidine*.

cytoskeleton Rather an imprecise term (with slightly unfortunate and probably incorrect implications for functional role) generally used to refer to the microfilament, microtubule and intermediate filament systems in cells. At one time the insoluble residue of detergent-extracted cells was referred to as a cytoskeletal preparation; in this case the specialised cell–cell junctions were also involved. Although intermediate filaments may have primarily a tension-resisting function, it is increasingly clear that both microtubules and microfilaments are involved with motor systems. The analogy with the vertebrate skeleton can be misleading.

cytosol That part of the cytoplasm that remains when organelles and internal membrane systems are removed.

cytotactin Site-restricted extracellular matrix protein involved in neuron-glia interactions, synthesised by glial cells (not neurons) of central and peripheral nervous system. Prominent in areas where cell movement is common.

cytotoxic T-cells (CTL) Subset of T-lymphocytes responsible for lysing target cells and for killing virus-infected cells (in the context of Class I *histocompatibility antigens*). In most cases, CD8+ (see Table C1).

D

D cells (δ cells) Cells of the pancreas; about 5% of the cells present in primate pancreas with small argentaffin-positive granules. Their function is unclear, but they may release somatostatin.

D-gene segment (diversity gene segment) Part of the gene for the immunoglobin *heavy chain*, it codes for part of the *hypervariable region* of the V_H domain and is located between the V_H and J_H segments. There are probably about 20 different D segments.

Dane particle 42nm particle, the complete infective virion of *hepatitis B*.

dansyl chloride (1-dimethyl-amino-naphthalene-5-sulphonyl chloride) A strongly fluorescent compound which will react with the terminal amino group of a protein. After acid hydrolysis of all the other peptide bonds, the terminal amino acid is identifiable as the dansylated residue.

dapsone Drug related to the sulphonamides (diaminodiphenyl sulphone) which is used to treat leprosy (of which the causative agent is *Mycobacterium leprae*). May act by inhibiting folate synthesis.

dark current (of retina) Current caused by constant influx of sodium ions into the *rod outer segment* of retinal photoreceptors, and which is blocked by light (leading to hyperpolarisation). The plasma membrane sodium channel is controlled through a cascade of amplification reactions initiated by photon capture by *rhodopsin* in the disc membrane.

dark-field microscopy A system of microscopy in which particles are illuminated at a very low angle from the side so that the background appears dark and the objects are seen by diffracted and reflected patches of light against a dark background.

dark reaction The reactions in photosynthesis which occur after NADPH and ATP production, and which take place in the stroma of the chloroplast; constitute the *Calvin–Benson cycle*. By means of the reaction, CO_2 is incorporated into carbohydrate.

deacetylase An enzyme that removes an acetyl group: one of the most active deacetylation reactions is the constant deacetylation (and reacetylation) of lysyl residues in histones (the half-life of an acetyl group may be as low as 10min). Acetylation (which removes a positive charge on the lysine ε-amino group) is thought to be increased in active genes, therefore deacetylation would be important in switching off genes.

deamination (of nucleic acids) The loss of the amino groups of cytosine (yielding uracil), methyl cytosine (yielding thymine), or of adenine (yielding hypoxanthine). It can be argued that the presence of thymine in DNA in place of the uracil of RNA stabilises genetic information against this lesion, since repair enzymes would restore a GU base pair to GC.

decay accelerating factor Plasma protein which regulates *complement* cascade by blocking the formation of the C3bBb complex (the C3 convertase of the alternate pathway). Widely distributed in tissues but deficient in paroxysmal nocturnal haemoglobinuria.

dedifferentiation Loss of differentiated characteristics. In plants, most living cells, including the highly differentiated haploid *microspores* (immature pollen cells) of angiosperms, can lose their differentiated features and give rise to a whole plant; in animals this is less certain, and there is still controversy as to whether the undifferentiated cells of the blastema that forms at the end of an amputated amphibian limb (for example) are derived by dedifferentiation, or by proliferation of uncommitted cells. Neither is it clear whether dedifferentiation in animal cells might just be the temporary loss of phenotypic characters, with retention of the *determination* to a particular cell type.

deep cells Cells (blastomeres) in the teleost blastula that lie between the outer cell layer and the yolk syncytial layer, and are the cells from which the embryo proper is constructed during gastrulation; much studied in the fish, *Fundulus*.

defective virus A virus genetically deficient in replication, but which may nevertheless be replicated when it co-infects a host cell in the presence of a wild-type "helper" virus. Most acute transforming *Retroviridae* are defective, since their acquisition of oncogenes seems to be accompanied by deletion of essential viral genetic information.

defensins Antimicrobial cationic peptides (29–34 residues) found in the azurophil granules of neutrophil leucocytes. At least ten kinds, constituting more than 5% of the total protein of the neutrophil, are known. Fairly well conserved in sequence.

defined medium Cell culture medium in which all components are known. In practice this means that the serum (which is normally added to culture medium for animal cells) is replaced by insulin, transferrin and possibly specific growth factors such as *platelet-derived growth factor*.

definitive erythroblast Embryonic erythroblast found in the liver; smaller than *primitive erythroblasts*, they lose their nucleus at the end of the maturation cycle and produce erythrocytes with adult haemoglobin.

degeneracy The coding of a single amino acid by more than one base triplet (*codon*). Of the 64 possible codons, three are used for stop signals, leaving 61 for only 20 amino acids. Since all codons can be assigned to amino acids, it is clear that some amino acids must be coded by several different codons, in some cases as many as six.

degranulation Release of secretory granule contents by fusion with the plasma membrane.

dehydration Removal of water as in preparing a specimen for embedding or a histological section for clearing and mounting.

dehydrogenase Enzyme which oxidises a substrate by transferring hydrogen to an acceptor which is either $NAD^+/NADP^+$ or a flavin enzyme.

delayed-type hypersensitivity See *hypersensitivity*.

deletion mutant Mutant which arises as a result of the loss of one or more base pairs of *DNA*.

delta chains The heavy chains of mouse and human *IgD immunoglobulins*.

demyelinating diseases Diseases in which the myelin sheath of nerves is destroyed and which often have an autoimmune component. Examples are multiple sclerosis, acute disseminated encephalomyelitis (a complication of acute viral infection), experimental allergic encephalomyelitis, Guillain–Barre syndrome.

denaturation Reversible or irreversible loss of function in proteins and nucleic acids resulting from loss of higher order (secondary, tertiary or quaternary structure) produced by non-physiological conditions of pH, temperature, salt or organic solvents.

dendrite A long, branching outgrowth from a *neuron*, a postsynaptic region which carries electrical signals from synapses to the cell body. This classical definition, however, lost some weight with the discovery of axo–axonal and dendro–dendritic synapses. Sensory dendrites, eg. of stretch receptor neurons, are processes specialised to act as a receptor. See *axon*.

dendritic cells 1. Follicular dendritic cells, found in germinal centres of spleen and lymph nodes, and that retain antigen for long periods. 2. Accessory (antigen-presenting) cells, positive for Class II *histocompatibility antigens*, found in the red and white pulp of the spleen and lymph node cortex and associated with stimulating T-cell proliferation, particularly in primary immune responses. Langerhans cells of dermis may be a precursor of lymph node dendritic cells. 3. Dopa-positive cells derived from neural crest and found in the basal part of epidermis: melanocytes.

dendritic spines Wine-glass or mushroom-shaped protrusions from dendrites that represent the principal site of termination of afferent neurons on interneurons, especially in the cortical regions.

dendritic tree The characteristic (tree-like) pattern of outgrowths of neuronal *dendrites*.

denervation Removal of nerve supply to a tissue, usually by cutting or crushing the *axons*.

dengue Tropical disease caused by a flavivirus (one of the *arboviruses*), transmitted by mosquitoes. A more serious complication is dengue shock syndrome, a haemorrhagic fever probably caused by an immune complex *hypersensitivity* after re-exposure.

dense bodies Areas of electron density associated with the thin filaments in smooth muscle cells. Some are associated with the plasma membrane, others are cytoplasmic. Because the term has been used very loosely for a range of unknown electron-dense areas, it should be interpreted with care and avoided where possible.

density dependent inhibition of growth The phenomenon exhibited by most normal (*anchorage dependent*) animal cells in culture that stop dividing once a critical cell density is reached. The cessation of population growth may depend upon restricted access of the cells to growth factors once the cells are constrained to a morphology in which their free apical surface area drops below a certain level, or the response may be due simply to restricted spreading (although how the cell would perceive this is not yet known). Certainly, the critical density is considerably higher for most cells than the density at which a monolayer is formed; for this reason, most cell behaviourists prefer the term density dependent inhibition of growth as this avoids any confusion with *contact inhibition of locomotion*, a totally different phenomenon that is contact dependent.

density gradient A column of liquid in which the density varies continually with position, usually as a consequence of variation of concentration of a solute. Such gradients may be established by progressive mixing of solutions of different density (as for example, sucrose gradients) or by centrifuge-induced redistribution of solute (as for *caesium chloride* gradients). Density gradients are widely used for centrifugal and gravity-induced separations of cells, organelles and macromolecules. The separations may exploit density differences between particles, or primarily differences in size, in which latter case the function of the gradient is chiefly to stabilise the liquid column against mixing.

deoxycholate A bile salt. The sodium salt is used as a detergent to make membrane proteins water-soluble.

2-deoxyglucose Analogue of glucose in which the the hydroxyl on C-2 is replaced by a hydrogen atom. Since it is often taken up by cells but not further metabolised, it can be used to study glucose transport, and also to inhibit glucose utilisation.

deoxyhaemoglobin Haemoglobin without bound oxygen.

deoxyribonuclease (DNAase) An *endonuclease* with preference for DNA. Pancreatic DNAase I yields di- and oligo-nucleotide 5'phosphates, pancreatic DNAase II yields 3'phosphates. In chromatin, the sensitivity of DNA to digestion by DNAase I depends on its state of organisation, transcriptionally active genes being much more sensitive than inactive genes.

deoxyribose (2-deoxy-D-ribose) The sugar which, when linked by 3'-5'phosphodiester bonds forms the backbone of DNA.

dephosphorylation Removal of a phosphate group. Phosphorylation is such an important control mechanism that it is easy to forget that there also have to be systems to restore the original dephosphorylated state.

depolarisation A positive shift in a cell's *resting potential* (which is normally negative) towards zero, thus making it numerically smaller and less polarised, eg. -90mV to -50mV. The opposite of *hyperpolarisation*.

depsipeptides Polypeptides that contain ester bonds as well as peptides. Naturally occurring depsipeptides are usually cyclic; they are common metabolic products of micro-organisms and often have potent antibiotic activity (examples are *actinomycins*, enniatins, *valinomycin*).

depurination (of DNA) The N-glycosidic link between purine bases and deoxyribose in DNA has an appreciable rate of spontaneous cleavage *in vivo*, a lesion which must be enzymically repaired to ensure stability of the genetic information. Occurs readily *in vitro* in the presence of dilute acid.

dermal tissue Outer covering of plants, which includes the *epidermis* and periderm (non-living bark). See also *dermis*.

dermatan sulphate Glycosaminoglycan (15–40kD) typical of extracellular matrix of skin, blood vessels and heart. Repeating units of D-glucuronic acid-N-acetyl D-galactosamine or L-iduronic acid-N-acetyl D-galactosamine with 1–2 sulphates per unit. Broken down by L-iduronidase, but accumulates intralysosomally in *Hurler's disease* and *Hunter syndrome*.

dermatitis Inflammation of the *dermis*, often a result of *contact sensitivity*.

dermatosparaxis Recessive disorder of cattle in which a procollagen peptidase is absent. In consequence the amino- and carboxy-terminal peptides of procollagen are not removed, the collagen bundles are disordered, and the dermis is fragile. Similar to Ehlers–Danlos syndrome in humans.

dermis Mesodermally-derived connective tissue underlying the epithelium of the skin.

DeSanctis–Cacchione syndrome A variant of *xeroderma pigmentosum* in which a different DNA repair enzyme is involved. Hybrid fibroblasts formed by *Sendai virus* fusion of the two types show normal repair (complementation).

desensitisation In a general sense: see *adaptation*. Immunologically, the term is used to mean the administration of a graded series of increasing doses of an antigen to which there is an immediate hypersensitivity response. The technique is used in the treatment of allergy and works by inducing the production of blocking antibody (IgG) which inhibits IgE production or blocks IgE binding.

desmin A protein of *intermediate filaments*, somewhat similar to *vimentin*, but characteristic of muscle cells.

desmocalmin See Table D1.

desmocollin See Table D1.

desmoglein See Table D1.

desmoplakin See Table D1.

desmosine Component of elastin, formed from four side-chains of lysine and constituting a cross-linkage.

desmosome Specialised cell junction characteristic of epithelia into which intermediate filaments (tonofilaments of cytokeratin) are inserted. The gap between plasma membranes is of the order of 25–30nm and the intercellular space has a medial band of electron-dense material. Desmosomes are particularly conspicuous in tissues such as skin which have to withstand mechanical stress. Also known as macula adhaerens junctions or spot desmosomes. The various proteins associated with desmosomes are shown in Table D1.

desmotubule Cylindrical membrane-lined channel through a *plasmodesma*, linking the cisternae of endoplasmic reticulum in the two cells.

desoxy- See *deoxy-*.

destruxins Cyclic depsipeptide fungal toxins which suppress the immune response in invertebrates.

desynapsis Separation of the paired homologous chromosomes at the *diplotene* stage of meiotic prophase I.

detergents Amphipathic, surface active molecules with polar (water soluble) and non-polar (hydrophobic) domains. They bind strongly to hydrophobic molecules or molecular domains to confer water solubility. Examples include:- sodium dodecyl sulphate, fatty acid salts, the Triton family, octyl glucoside.

determination A cell may be committed to

Table D1. Proteins and glycoproteins of desmosomes (Courtesy of C. Skerrow).

	Apparent MW (kD) in gels	Synonyms	Location
Proteins			
dp1	230–250	band 1 desmoplakin 1	Cytoplasm
dp2	210–220	band 2 desmoplakin 2	Cytoplasm
dp3	83–90	band 5 desmoplakin 3 plakoglobin	Cytoplasm and non-desmosomal
Glycoproteins			
dg1	140–160	band 3 desmoglein 1	Plasma membrane
dg2	110–120	band 4a desmoglein 2a desmocollin 1	Plasma membrane
dg3	97–105	band 4b desmoglein 2b desmocollin 2	Plasma membrane

following a particular path of differentiation, yet exhibit no features which reveal this determination. Generally irreversible, but in the case of *imaginal discs* of *Drosophila* which are maintained by serial passage, *transdetermination* may occur.

detoxification reactions Reactions taking place usually in the liver or kidney in order to inactivate toxins, either by degradation or by conjugation to a hydrophilic moiety to promote excretion.

deutan (deuteranope) Form of colour blindness in which the green factor is absent. The gene maps on the X chromosome close to the *haemophilia* A locus.

deuterium oxide (D$_2$O) Heavy water, in which the hydrogen is replaced by deuterium, and which is used to stabilise assembled microtubules.

dexamethasone Steroid analogue (glucocorticoid), used as an anti-inflammatory drug. It has little effect on *cyclo-oxygenase*, but probably induces *lipocortin* (phospholipase A$_2$ inhibitor) synthesis.

dextrans High-molecular weight polysaccharides synthesised by some micro-organisms. Consist of D-glucose linked by $\alpha(1-6)$- bonds (and a few $\alpha(1-3)$- and $\alpha(1-4)$- bonds). Dextran 75 (average molecular weight 75kD) has a colloid osmotic pressure similar to blood plasma, so dextran 75 solutions are used clinically as plasma expanders. They will also cause charge-shielding, and at the right concentrations induce flocculation of red cells, a trick that is used in preparing leucocyte-rich plasma for white cell purification in the laboratory. Cross-linked dextran is the basis of *Sephadex*.

diabetes insipidus X-linked recessive disorder in which either the renal tubules do not respond to *antidiuretic hormone* (ADH) or there is inadequate ADH production, leading to excessive production of dilute urine.

diabetes mellitus Relative or absolute lack of *insulin*, or absence of effect of insulin, leading to uncontrolled carbohydrate metabolism. In juvenile onset diabetes (which may be an autoimmune response to pancreatic *B cells*) the insulin deficiency tends to be almost total, whereas in adult onset diabetes there seems to be no immunological component but an association with obesity.

diakinesis The final stage of the first *prophase* of *meiosis*; the term derives from the vigorous movement of *bivalents* as they attach to the microtubules of the spindle. The chromosomes condense to their greatest extent during this stage and normally the nucleolus disappears and the fragments of the nuclear envelope disperse.

diapedesis A somewhat archaic term for the emigration of leucocytes across the endothelium.

dibutyryl cyclic AMP An analogue of *cyclic AMP* which shares some of the pharmacological effects of this nucleotide, but is generally believed to enter cells more readily on account of its greater hydrophobicity.

2,6-dichlorobenzonitrile Inhibitor of cellulose biosynthesis in higher plants.

2,4-dichlorophenoxyacetic acid (2,4-D) A synthetic *auxin*, also used as a selective herbicide.

dictyosome Organelle found in plant cells and functionally equivalent to the *Golgi apparatus* of animal cells.

Dictyostelium A genus of the order *Acrasidae*, the cellular slime moulds.

dictyotene Prolonged *diplotene* of meiosis: the stage at which oocyte nuclei remain during yolk production or during maturation arrest characteristic of most mammals. In women, dictyotene commences *in utero* and lasts (for some oocytes) until the menopause.

differential adhesion The differential adhesion hypothesis was advanced by Steinberg to explain the mechanism by which heterotypic cells in mixed aggregates sort out into isotypic territories. Quantitative differences in homotypic and heterotypic adhesion are supposed to be sufficient to account for the phenomenon without the need to postulate cell-type specific adhesion systems: fairly generally accepted, although some tissue-specific *cell adhesion molecules* are now known to exist.

differential interference contrast Method of image formation in the light microscope based on the method proposed by Nomarski (though strictly speaking all forms of optical microscopy rely to a greater or lesser extent on differential interference). The light beam is split by a Wollaston prism in the condenser, to form slightly divergent beams polarised at right angles. One passes through the specimen (and is retarded if the refractive index is greater), and one through the background nearby: the two are recombined in a second Wollaston prism in the objective and interfere to form an image. The image is spuriously "three-dimensional" — the nucleus, for example, appears to stand out above the cell (or be hollowed out) because it has a higher refractive index than the cytoplasm. The Nomarski system has the advantage that there is no phase-halo, but the contrast is low and image formation with crowded cells is poor because the background does not differ from the specimen.

differentiation Process in development of a multicellular organism by which cells become specialised for particular functions. Requires that there is selective expression of portions of the genome; the fully differentiated state may be preceded by a stage in which the cell is already programmed for differentiation but is not yet expressing the characteristic phenotype (*determination*).

differentiation antigen Any large structural macromolecule which can be detected by immune reagents and which also is associated with the differentiation of a particular cell type or types. Many cells can be identified by their possession of a unique set of differentiation antigens. There should be no implication that the antigens cause differentiation. For leucocytes, an agreed *CD nomenclature* for the (cluster of differentiation) antigens exists, and is being extended to cells of other tissues. See Table C1.

diffraction When a wave-train passes an obstacle secondary waves are set up which interfere with the primary wave and give rise to bands of constructive and destructive interference. Around a point source of light, in consequence, is a series of concentric light and dark bands (coloured bands with white light).

diffusion coefficient (diffusion constant) For the translational diffusion of solutes, diffusion is described by Fick's First Law, which states that the amount of a substance crossing a given area is proportional to the spatial gradient of concentration and the diffusion constant (D), which is related to molecular size and shape. A useful derived relationship is that the mean square distance moved by molecules in time t is 6Dt.

diffusion limitation The boundary layer hypothesis; the basis of the hypothesis is that the proliferation of cells in culture is limited by the rate at which some essential component (almost certainly a growth factor) diffuses from the bulk medium into the layer immediately adjacent to the plasma membrane. By spreading out, a cell obtains a supra-threshold level of the factor and can divide; if unable to spread (because of crowding or poor adhesion) then the cell will remain in the *G0* stage of the *cell cycle*. See also *density dependent inhibition of growth*.

DiGeorge syndrome Congenital absence of the thymus and parathyroid due to defective development of the 3rd and 4th branchial arches, as a result of which the T-lymphocyte system is absent.

digestive vacuole Intracellular vacuole into which lysosomal enzymes are discharged and digestion of the contents occurs. More commonly referred to as a *secondary lysosome*.

digitalis General term for pharmacologically active compounds from the foxglove (*Digitalis*). The active substances are the cardiac glycosides, digoxin, digitoxin, strophanthin and *ouabain*. Causes increased force of contraction of the heart, disturbance of rhythm and reduced beat frequency. Also causes arteriolar constriction, venous dilation, nausea and visual disturbances.

digoxin See *digitalis*.

digitoxin See *digitalis*.

dihydrofolate reductase (EC 1.5.1.3) An enzyme involved in the biosynthesis of *folic acid* coenzymes, which transfers hydrogen from NADH to dihydrofolate, yielding tetrahydrofolic acid. Standard source is liver, and the enzyme is used to assay dihydrofolate.

2,4-dinitrophenol A small molecule used as an uncoupler of oxidative phosphorylation. Also used after reaction with various proteins to provide a strong and specific identified *haptenic* group.

dinoflagellates Members of the order Dinoflagellida (for botanists, Dinophyceae). They are aquatic and have 2 *flagella* lying in grooves in an often elaborately sculptured shell or *pellicle* which is formed from plates of cellulose deposited in membrane vesicles. The pellicle gives some dinoflagellates very bizarre shapes. The nuclei of dinoflagellates contain permanently condensed chromosomes, whose DNA is not associated with histones, and which are associated with the persistent nuclear envelope during a nuclear division process in which spindle microtubules remain extranuclear; dinoflagellates represent a condition between that of prokaryotes and eukaryotes, often termed mesokaryotes. The organisms are very abundant in marine plankton. *Gymnodinium* and *Gonyaulax*, which causes "red tide", produce toxins which if accumulated by filter-feeding molluscs can be fatal to humans who ingest them. Another common genus is *Peridinium*.

diphtheria toxin An *AB toxin* (62kD) coded by β corynephage of virulent *Corynebacterium diphtheriae* strains (which can produce a repressor of toxin production). The B subunit binds to receptors on the surface of the target cell and facilitates the entry of the enzymically active A subunit (21kD) which ADP-ribosylates *elongation factor 2*, thereby halting mRNA translation.

diplococcus Bacterial form in which two spherical cells (cocci) are joined to form a pair like a dumb-bell or figure-of-eight.

diploid cell Cell having its *chromosomes* in homologous pairs, and thus having 2 copies of each *autosomal* genetic *locus*. The diploid number (2n) equals twice the *haploid* number and is the characteristic number for most cells other than gametes.

diplotene The final stage of the first prophase of meiosis. All four *chromatids* of a tetrad are visible and homologous chromosomes start to move away from one another except at *chiasmata*.

Diptera Order of insects with one pair of wings, the second pair being modified into balancing organs, the halteres; the mouthparts are modified for sucking or piercing. These insects show complete metamorphosis in that they have larval, pupal and imaginal stages. The order includes the flies and mosquitoes; best known genera are **Drosophila** and *Anopheles*.

discoidin A lectin, isolated from the cellular slime mould *Dictyostelium discoideum*, which has a binding site for carbohydrate residues related to galactose. The lectin, which consists of two distinct species (discoidins I and II) is synthesised as the cells differentiate from vegetative to aggregation phase, and was originally thought to be involved in intercellular adhesion; discoidin I is now thought to be involved in adhesion to the substratum by a mechanism resembling that of *fibronectin* in animals.

disjunction mutant *Drosophila* mutant in which chromosomes are partitioned unequally between daughter cells at *meiosis*, as a result of nondisjunction.

disseminated intravascular coagulation Complication of septic shock in which endotoxin (from Gram negative bacteria) induces clotting of the blood within the vascular system, probably indirectly through its effect on neutrophils. It may also develop in other situations where neutrophils become systemically hyperactivated.

dissociation Any process by which a tissue is separated into single cells. Enzymic dissociation with trypsin or other proteases is often used.

dissociation constant See also K_d. In a chemical equilibrium of form $A + B \rightleftharpoons AB$, the equilibrium concentrations (strictly, activities) of the reactants are related such that $[A] \times [B]/[AB] =$ a constant, K_d, the dissociation constant, which in this simplest case has the dimensions of concentration. When A is H^+, this is the acid dissociation constant often designated K_a, and expressed as pK_a $(-\log_{10}K_a)$.

disulphide bond The –S–S– linkage. A linkage formed between the SH groups of two cysteine moieties either within or between peptide chains. Each cysteine then becomes a half-cystine residue. –S–S– linkages stabilise, but do not determine, secondary structure in proteins. They are easily disrupted by –SH groups in an exchange reaction and are not present in cytosolic proteins (cytosol has a high concentration of *glutathione* which has a free -SH residue).

division septum The cell wall which forms between daughter cells at the end of mitosis in plant cells or just before separation in bacteria; more commonly called the cell plate in plants.

DLVO theory Theory of colloid flocculation advanced independently by Derjaguin and Landau and by Vervey and Overbeek and subsequently applied to cell adhesion. Because electrostatic repulsion forces diminish with the separation of two surfaces following a different power law than do the electrodynamic London–Van der Waals forces of attraction, there exist distances (primary and secondary minima) at which the forces of attraction exceed those of repulsion; an adhesion will thus be formed. For cells there is quite good correlation between the calculated separations of primary and secondary minima and the cell separations in tight junctions (1–2nm) and more general cell–cell appositions (12–20nm) respectively. Although physicochemical theories such as this account for many of the observed phenomena in cell-cell adhesion, it is now clear that other factors (particularly *cell adhesion molecules*) also play an important part.

DMSO (dimethyl sulphoxide) Much used as a solvent for substances which do not dissolve easily in water and which are to be applied to cells (for example cytochalasin B, formyl peptides), also as a cryoprotectant when freezing cells for storage. It is used clinically for the treatment of arthritis, although its efficacy is disputed.

DNA (deoxyribonucleic acid) The principal genetic material of all cells and many viruses. A polymer of *nucleotides*. The monomer consists of phosphorylated 2-deoxyribose N-glycosidically linked to one of four bases *adenine, cytosine, guanine* or *thymine*. These are linked together by 3′,5′-phosphodiester bridges. In the Watson–Crick double-helix model two complementary strands are wound in a right-handed helix and held together by hydrogen bonds between *complementary base pairs*. The sequence of bases encodes genetic information. Three major conformations exist *A-DNA*, *B-DNA* (which corresponds to the original Watson–Crick model) and *Z-DNA*.

DNAase See *deoxyribonuclease*.

DNA glycosylase (DNA glycosidase) Class of enzymes involved in *DNA repair*. They recognise altered bases in DNA and catalyse their removal by cleaving the glycosidic bond between the base and the deoxyribose sugar. At least 20 such enzymes occur in cells.

DNA gyrase A type II *topoisomerase* of *Escherichia coli*, which is essential for DNA replication. This enzyme can induce or relax *supercoiling*.

DNA helicase An enzyme that uses the hydrolysis of ATP to unwind the DNA helix, usually at the replication fork, to allow the resulting single strands to be copied. Two molecules of ATP are required for each nucleotide pair of the duplex. Also called unwindase.

DNA hybridisation See *hybridisation*.

DNA library See *genomic library*.

DNA ligase Enzyme involved in DNA replication, repair and recombination. The DNA ligase of *E. coli* seals nicks in one strand of double-stranded DNA, a reaction required for linking precursor fragments during discontinuous synthesis on the lagging strand. Nicks are breaks in the phosphodiester linkage which leave a free 3′-OH and 5′-phosphate. The ligase from phage T4 has the additional property of joining two DNA molecules having completely base-paired ends. DNA ligases are crucial in joining DNA molecules and preparing radioactive probes (by *nick-translation*) in recombinant DNA technology.

DNA polymerases (EC 2.7.7.7) Enzymes involved in template-directed synthesis of DNA from deoxyribonucleoside triphosphates. I, II and III are known in *E. coli*; III appears to be most important in genome replication and I is important for its ability to edit out unpaired bases at the end of growing strands. Animal cells have alpha, beta, gamma and delta polymerases, with alpha and delta apparently responsible for replication of nuclear DNA, and gamma for replication of mitochondrial. All these function with a DNA strand as template. Retroviruses possess a unique DNA polymerase (*reverse transcriptase*) which uses an RNA template.

DNA rearrangement Wholesale movement of sequences from one position to another in DNA, such as occur somatically, for example in the generation of antibody diversity.

DNA repair Enzymic correction of errors in DNA structure and sequence which protect genetic information against environmental damage and replication errors.

DNA replication The process whereby a copy of a DNA molecule is made, and thus the genetic information it contains is duplicated. The parental double-stranded DNA molecule is replicated semi-conservatively, ie. each copy contains one of the original strands paired with a newly synthesised strand which is complementary in terms of AT and GC base pairing and of opposite chemical polarity. Though conceptually simple, mechanistically a complex process involving a number of enzymes.

DNA synthesis The linking together of nucleotides (as deoxyribonucleotide triphosphates) to form DNA. *In vivo*, most synthesis is *DNA replication*, but incorporation of precursors also occurs in repair. In the special case of retroviruses, DNA synthesis is directed by an RNA template (see *reverse transcriptase*).

DNA topoisomerase An enzyme capable of altering the linking number of DNA. Usually measured by following a change in the degree of supercoiling of double-stranded DNA molecules. Various topoisomerases can increase or relax supercoiling, convert single-stranded rings to intertwined double-stranded rings, tie and untie knots in single-stranded and duplex rings, catenate and decatenate duplex rings. Topoisomerase II of *E. coli* = gyrase.

DNA transfection (Term is hybrid between transformation and infection). A technique originally developed to allow viral infection of animal cells by uptake of purified viral DNA rather than by intact virus particles. Term is used to describe applications of same methodology to introduction of DNA into cells, such as activated oncogenes from tumours into tissue culture cells.

DNA tumour virus Virus with DNA genome which can cause tumours in animals (including humans). Examples are *Papovaviridae*, *Adenoviridae* and *Epstein Barr virus*.

DNA virus A virus in which the nucleic acid is double- or single-stranded DNA (rather than RNA). Major groups of double-stranded DNA viruses are *Papovaviridae*, *Adenoviridae*, *Herpetoviridae*, large bacteriophages and *poxviruses*: of single-stranded, *Parvoviridae* and coliphages φX174 and M13.

DNA-annealing The reformation of double-stranded DNA from denatured DNA. The rate of re-association depends upon the degree of repetition, and is slowest for unique sequences (this is the basis of the C_0t value).

DNA-renaturation See *DNA-annealing*.

dolichol Terpenoids with 13–24 isoprene units and a terminal phosphorylated hydroxyl group. Function as transmembrane carriers for glycosyl units in the biosynthesis of glycoproteins and glycolipids.

domain Used to describe a part of a molecule or structure which shares common physicochemical features, eg. hydrophobic, polar, globular, alpha-helical domains, or enzymic properties, eg. DNA-binding domains, ATPase domains.

donor splice junction The junction between an *exon* and an *intron* at the 5′ end of the intron. When the intron is removed during processing of *hnRNA* the donor junction is spliced to the acceptor junction at the 3′ end of the intron.

dopamine A *catecholamine neurotransmitter* and hormone (MW 153), formed by decarboxylation of dihydroxyphenylalanine (dopa). A precursor of *adrenaline* and *noradrenaline*.

double helix Conformation of a DNA molecule — like a ladder twisted into a helix.

double layer The zone adjacent to a charged particle in which the potential falls effectively to zero. An excess or deficiency of electrons on the surface (charge; not to be confused with the transmembrane potential) leads to an equivalent excess of ions of the opposite charge in the surrounding fluid. For most cells, which have negative charges, there will be an excess of cations immediately adjacent to the plasma membrane, and at physiological ionic strength the double layer is likely to be around 2–3nm thick.

doublet microtubules Microtubules of the axoneme. The outer nine sets are often referred to as doublet microtubules, although only one (the A-tubule) is complete and has 13 protofilaments. The B-tubule has only 10 or 11 protofilaments, and shares the remainder with the A-tubule. A- and B-tubules differ in their stability and in the other proteins attached periodically to them; it is the dynein arms attached to the A-tubule attaching to and detaching from the B-tubule of the adjacent doublet that generates sliding movement in the axoneme.

doubling time The time taken for a cell to complete the *cell cycle*.

Down's syndrome Also known as mongolism, most frequently a consequence of trisomy of chromosome 21. Sadly common (1 in 700 live births); incidence correlates with maternal age. The cause is usually non-disjunction at meiosis but occasionally a translocation of fused chromosomes 21 and 14, in which case it is familial.

downstream 1. Portions of DNA or RNA which are more remote from the initiation sites and which will therefore be translated or transcribed later. 2. Shorthand term for things which happen at a late stage in a sequence of reactions.

down-regulation Reduction in the responsiveness of a cell to a stimulus following first exposure, often by a reduction in the number of receptors expressed on the surface (as a consequence of reduced recycling). Tends to be over-used in an inaccurate or loose way, and the term is thus losing precision.

Drosophila A genus of small, cosmopolitan flies, *Diptera*. The best known species is *D. melanogaster*, often called the fruit fly or the vinegar fly. First investigated genetically by T.H.Morgan and his group, it has been extensively used in genetic studies. More recently it has been used for studies of embryonic development.

dual recognition hypothesis An outmoded hypothesis which is known to be incorrect now that the structure of the T-cell receptor is known. The proposal was that viral (and some chemical) antigens were recognised in association with *histocompatibility antigens* by separate receptors on the T-cell. The generation of cytotoxic T-cells was by association with Class I MHC antigens, of T-helper cells by association with Class II MHC antigens. See *altered self hypothesis*.

Duchenne muscular dystrophy A sex-linked hereditary disease confined to young males and to females with *Turner's syndrome*. It is characterised by degeneration and *necrosis* of skeletal muscle fibres, which are replaced by fat and fibrous tissue. Affected individuals become unable to walk when they are 7–14 years old and die at about 18 years usually as a result of inability to breathe properly. The incidence of this disorder is about 1 in 4000 male births and of these a third are estimated to be new mutational events. Cause seems to be a deficiency of *dystrophin*.

dynein Large multimeric protein (600–800kD) with ATPase activity; constitutes the side arms of the outer microtubule doublets in the ciliary axoneme and is responsible for the sliding. Probably (together with *kinesin*) involved in microtubule-associated movement elsewhere. Cytoplasmic dynein is *MAP1C*.

dynorphin Opiate peptide derived from the hypothalamic precursor pro-dynorphin (which also contains the neoendorphin sequences). Contains the penta-peptide leu-enkephalin sequence. Its binding affinity is greater for the κ-type than for the μ-type opioid receptor.

dysgenic System of breeding or selection that is genetically deleterious or disadvantageous.

dyskinetoplasty Absence of an organised kinetoplast from a flagellate protozoan cell.

dystrophin Protein (400kD) from skeletal muscle which is missing in Duchenne muscular dystrophy. Its exact role is not yet clear, though it seems to be associated with the cytoplasmic face of the *sarcolemma* and *T tubules* and may form part of the membrane cytoskeleton. There are sequence homologies with non-muscle α-*actinin* and with *spectrin*.

E

EAA See *excitatory amino acid*.

E classification (Enzyme Classification) Classification of enzymes based on the recommendations of the Committee on Enzyme Nomenclature of the International Union of Biochemistry. The first number indicates the broad type of enzyme (1 = oxidoreductase; 2 = transferase; 3 = hydrolase; 4 = lyase; 5 = isomerase; 6 = ligase (synthetase)). The second and third numbers indicate subsidiary groupings, and the last number, which is unique, is assigned arbitrarily in numerical order by the Committee.

E-face In *freeze fracture* the plasma membrane cleaves between the acyl tails of membrane phospholipids, leaving a monolayer on each half of the specimen. The E-face is the inner face of the outer lipid monolayer. From within the cell this is the view that you would have of the outer half of the plasma membrane if the inner layer could be removed. The complementary surface is the P-face (the inner surface of the inner leaflet of the bilayer). According to some authors, E stands for ectoplasmic, P for protoplasmic: not terms that are in common usage!

E-rosettes The clustering of sheep erythrocytes (E) around T-lymphocytes due to binding of erythrocyte *LFA-3* to CD2 on the T-cell (for CD numbers see Table C1). E-rosette formation is used as a marker for T-lymphocytes of humans and most mammals; in this case E are untreated, compared with other rosette tests such as EA where erythrocytes have antibody bound to their surfaces.

EA-rosettes A rosetting test for the presence of *Fc-receptors*, using erythrocytes coated with antibody.

EAC-rosettes Rosettes (see *E-rosettes*) formed from erythrocytes (E) coated with antibody (A) and complement (C). A test for C3b or C3bi receptors (*CR1* or *CR3*). The rosettes are more stable than E- or EA-rosettes.

Eadie–Hofstee plot Linear transformation of enzyme kinetic data in which the velocity of reaction (v) is plotted on the ordinate, v/S on the abscissa, S being the initial substrate concentration. The intercept on the ordinate is V_{max}, the slope is $-K_m$. Preferable to the *Lineweaver–Burke plot*.

early antigens Virus-coded cell surface antigens which appear soon after the infection of a cell by virus, but before viral nucleic acid synthesis has begun.

early region Archaic term for that region of a viral genome in which genes that are transcribed and expressed early during infection of a cell are clustered.

EBV See *Epstein Barr virus*.

ecdysone Family of *steroid hormones* found in insects, crustaceans and plants. In insects, α-ecdysone stimulates moulting. The steadily maturing character of the moults is affected by steadily decreasing levels of *juvenile hormone*. β-ecdysone (ecdysterone) has a slightly different structure and is also found widely. Phytoecdysones are synthesised by some plants.

echinocytes Erythrocytes which have shrunk (in hypertonic medium) so that the surface is spiky.

Echinosphaerium Previously *Actinosphaerium*. Genus of *Heliozoan* protozoa. The organisms are multinucleate and have a starburst of radiating axopodia, of which the microtubules have been much studied in relation to *tubulin* polymerisation and microtubule depolymerisation.

echoviruses A group of human *picorna viruses*. "Echo" is derived from enteric cytopathic human orphan, where orphan implies that the viruses are not associated with any disease, though some are now known to cause aseptic meningitis or other disorders.

ecm Common abbreviation for *extracellular matrix*.

E. coli See **Escherichia coli**.

Eco RI Probably the first and one of the most commonly used type II *restriction endonucleases* isolated from *E. coli*. It cuts the sequence GAATTC between G and A thus generating *sticky ends* with a 5' overhang.

ectoderm The outer of the three germ layers of the embryo (the other two being mesoderm and endoderm). Ectoderm gives rise to epidermis and neural tissue.

ectopic Misplaced, not in the normal location.

ectoplasm Granule-free cytoplasm of amoeba lying immediately below the plasma membrane.

ectoplasmic tube contraction Model for amoeboid movement in which it is proposed that protrusion of a pseudopod is brought about by contraction of the sub-plasmalemmal region everywhere else in the cell thus squeezing the central cytoplasm forwards. See *frontal zone contraction theory*.

ectromelia Congenital absence or gross shortening of long bones of limb or limbs. This was the common *teratogenic* defect induced by *thalidomide*.

edema See *oedema*.

EDTA Ethylenediaminetetraacetic acid (often used as the disodium salt). Chelator of divalent cations; $\log_{10} K_{app}$ for calcium at pH 7 is 7.27 (5.37 for magnesium). See *EGTA*.

EDTA-light chain Myosin light chains (18kD) from scallop muscle (two per pair of heavy chains), easily extracted by calcium chelation. Although the EDTA-light chains do not bind calcium they confer calcium sensitivity on the myosin heavy chains.

E_1E_2-ATPase A class of plasma membrane-localised ion-motive pumps which includes *sodium–potassium ATPase*. The phosphoenzyme has two conformational states, E_1 and E_2, and ion exchange is inhibited by oligomycin, orthovanadate and *ouabain*.

eglin C A proteinase inhibitor (70 amino acids) from leech.

EGTA (ethyleneglycol-bis(β-aminoethyl) N,N,N',N'-tetraacetic acid) Like *EDTA* a chelator of divalent cations but with a higher affinity for calcium (log K_{app} 6.68 at pH 7) than magnesium (log K_{app} 1.61 at pH 7). Will also bind other divalent cations. Note: the "apparent association constant", K_{app}, is used because protons compete for binding and the association constant varies according to pH. Thus, EGTA has $\log_{10} K_{app}$ for calcium of 2.7 at pH 5, 10.23 at pH 9.

Ehlers-Danlos syndrome See *dermatosparaxis*.

Ehringhaus compensator Device used in *interference* or *polarisation microscopy* to reduce the brightness of the object to zero in order to measure the phase retardation (optical path difference). The compensator consists of a birefringent crystal plate which can be tilted. An alternative to *Sénarmont compensation* and has the advantage that it can be applied to retardations of more than one wavelength.

Ehrlich ascites A line of mouse ascites tumour cells maintained by passage in animals.

EHS cells (Englebreth-Holm-Swarm sarcoma cells) A line of mouse cells which produce large amounts of basement membrane-type extracellular matrix (ecm), rich in *laminin*, collagen type IV, *nidogen* and heparan sulphate. Often used as a source of these ecm molecules.

eicosanoids Useful generic term for compounds derived from arachidonic acid. Includes *leukotrienes*, *prostacyclin*, *prostaglandins* and *thromboxanes*.

ektacytometry Method in which cells (usually erythrocytes) are exposed to increasing shear-stress and the laser diffraction pattern through the suspension is recorded; it shifts from circular to elliptical as shear increases. From these measurements a deformability index for the cells can be derived.

elaioplast Unpigmented type of *plastid* modified as an oil-storage organelle.

elastase Serine protease which will digest *elastin* and *collagen* type IV; inhibited by α_1-*protease inhibitor* of plasma.

elasticoviscous Alternate form of the commoner term *viscoelastic*.

elastin Glycoprotein (70kD) randomly coiled and cross-linked to form elastic fibres which are found in connective tissue. Like *collagen*, the amino acid composition is unusual with 30% of residues being glycine and with a high proline content. Cross-linking depends upon formation of *desmosine* from four lysine side groups. The mechanical properties of elastin are poorer in old animals.

elastonectin Elastin-binding protein (120kD) found in extracellular matrix, produced by skin fibroblasts.

electrical coupling General term for an intimate cytoplasmic contact, mediated by *gap junctions*, between touching cells, such that electrical current injected into either cell changes the membrane potential of both. In neurons, arrays of gap junctions form *electrical synapses*, which allow *action potentials* to pass directly between cells. However, electrical coupling is not confined to excitable cells: many embryonic and adult *epithelia* are coupled, possibly to allow metabolic cooperation.

electrical synapse A connection between two electrically excitable cells, such as neurons or muscle cells, *via* arrays of *gap junctions*. This allows the cells to be *electrically coupled*, and so an action potential in one cell moves directly into the other, without the 1ms delay inherent in *chemical synapses*. Electrical synapses do not allow modulation of their connection, and so only occur in neuronal circuits where speed of conduction is paramount (eg. the crayfish escape reflex). A few electrical synapses are rectifying, implying a more specialised property than a simple gap junction. See also *synapse*.

electrochemical potential Defined as the work done in bringing 1 mole of an ion from a standard state (infinitely separated) to a specified concentration and electrical potential. Measured in joules/mole. More commonly used to measure the electrochemical potential difference between two points (eg. either side of a cell membrane), thus sidestepping the rather abstract concept of a standard state. If the molecule is uncharged or the electrical potential difference between two points is zero, the electrochemical potential reduces to the *chemical potential* difference of the species. At equilibrium, the electrochemical potential difference (by definition) is zero; the situation can then be described by the *Nernst equation*.

electrodynamic forces London–Van der Waals forces: see *DLVO theory*.

electrofocusing Any technique whereby chemical species are concentrated using an applied electric field. See *isoelectric focusing*.

electron microprobe A technique of elemental analysis in the electron microscope based on spectral analysis of the scattered X-ray emission from the specimen induced by the electron beam. Using this technique it is possible to obtain quantitative data on, for example, the calcium concentration in different parts of a cell; it is

necessary to use ultra-thin frozen sections for detection of diffusible ions.

electron microscopy Any form of microscopy in which the interactions of electrons with the specimens are used to provide information about the final structure of that specimen. In transmission electron-microscopy (TEM) the diffraction and adsorption of electrons as the electron beam passes normally through the specimen is imaged to provide information on the specimen. In scanning electron-microscopy (SEM) an electron beam falls at a non-normal angle on the specimen and the image is derived from the scattered and reflected electrons. Secondary X-rays generated by the interaction of electrons with various elements in the specimen may be used for *electron microprobe* analysis.

electron transport chain A series of compounds which transfer electrons to an eventual donor with concomitant energy conversion. One of the best studied is in the mitochondrial inner membrane, which takes NADH (from the *tricarboxylic acid cycle*) or FADH and transfers electrons *via ubiquinone*, cytochromes and various other compounds, to oxygen. Other electron transport chains are involved in *photosynthesis*.

electrophoresis Separation of molecules based on their mobility in an electric field. High resolution techniques normally use a gel support for the fluid phase. Examples of gels used are starch, polyacrylamide, agarose or mixtures of polyacrylamide and agarose. Frictional resistance produced by the support causes size, rather than charge alone, to become the major determinant of separation. The electrolyte may be continuous (a single buffer), or discontinuous (where the sample is stacked by means of a buffer discontinuity), before it enters the running gel/running buffer. The gel may be a single concentration or gradient in which pore size decreases with migration distance. In *SDS-PAGE* of proteins or electrophoresis of polynucleotides, mobility depends primarily on size and is used to determine molecular weight. In pulse-field electrophoresis, two fields are applied alternately at right-angles to each other to minimise diffusion-mediated spread of large linear polymers. See also *cell electrophoresis*, *electrofocusing*, *pulse-field electrophoresis*.

electroplax A stack of specialised muscle

fibres found in electric eels, arranged in series. The fibres have lost the ability to contract; instead they generate extremely high voltages (ca 500V) in response to nervous stimulation. They contain asymmetrically distributed *sodium potassium ATPases*, *acetylcholine* receptors and *sodium channels* (gates) at extraordinarily high concentrations. The tissue was vital in the sequencing studies of the sodium gate.

electroporation Method for temporarily permeabilising cell membranes so as to facilitate the entry of large or hydrophilic molecules (as in *transfection*). A brief (ca 1msec) electric pulse is given with potential gradients of about 700V/cm.

electrostatic forces Like charges in close proximity produce forces of repulsion between them. Consequently, if two surfaces bear appreciable and approximately equal densities of charged groups on their surfaces, appreciable forces of repulsion may occur between them. The range of these forces is determined in the main by the ionic strength of the intervening medium, forces being of minimal range at high ionic strength. The forces are effective over approximately twice the *double layer* thickness.

elicitor Substance which induces the formation of *phytoalexins* in higher plants. May be exogenous (often produced by potentially pathogenic micro-organisms) or endogenous (possibly cell-wall degradation products).

ELISA (Enzyme-Linked Immuno-Sorbent Assay) A very sensitive technique for the detection of small amounts of protein or other antigenic substance. The basis of the method is the binding of the antigen by an antibody which is linked to the surface of a plate (usually a well of a microtitre plate). Formation of an immune complex is detected by use of peroxidase or phosphatase coupled to antibody, the enzyme being used to generate an amplifying colour reaction. Various ways of carrying out the assay are possible — if the aim is to detect antibody production from a myeloma clone for example, then the antigen may be bound to the plate, and the formation of the antibody/antigen complex may be detected using peroxidase coupled to an anti-Ig antibody.

ellipsosome Membrane-bounded compartment containing cytochrome-like pigment and found in the *retinal cones* of some fish.

elongation factor Peptidyl transferase components of ribosomes which catalyse formation of the acyl bond between the incoming amino acid residue and the peptide chain. *Diphtheria toxin* inhibits protein synthesis in eukaryotes by adding an ADP-ribosyl group to a modified histidine residue (diphthamide) in elongation factor II.

Embden–Meyerhof pathway (glycolysis) The main pathway for anaerobic degradation of carbohydrate. Starch or glycogen is hydrolysed to glucose-1-phosphate and then through a series of intermediates, yielding two ATP molecules per glucose, and producing either pyruvate (which feeds into the *tricarboxylic acid cycle*) or lactate.

embedding Tissue is embedded in wax or plastic in order to prepare sections for microscopical examination. The embedding medium provides mechanical support.

embolus A clot formed by platelets or leucocytes that blocks a blood vessel.

embryo The developmental stages of an animal or, in some cases a plant, during which the developing tissue is effectively isolated from the environment by, for example, egg membranes, foetal membranes and various structures in plants.

embryogenesis The processes leading to the development of an embryo from egg to completion of the embryonic stage.

embryonic induction The induction of differentiation in one tissue as a result of proximity to another tissue arising, for example, during gastrulation. One of the best known examples is the induction of the neural tube in the ectoderm by the underlying chordomesoderm. Although the information to form the tube is present in the competent determined ectoderm, it must be elicited by the inducing tissue. In some cases it is known that cell-cell contact between epithelium and mesenchyme is necessary.

emphysema Pulmonary emphysema is associated with chronic bronchitis and is probably caused by excessive leucocyte elastase activity in the alveolar walls (possibly as a result of the inactivation of α_1-protease inhibitor by active oxygen species released by leucocytes in inflammation). As a result, there is destruction and dilation of alveoli.

endarteritis Chronic inflammation of the arterial *intima*, often a late result of syphilis.

endocarditis Inflammation of the inner

membrane of the heart, that over the valves being particularly susceptible. May be caused by viral or bacterial infection, or indirectly as a response to rheumatic fever, scarlet fever or tonsilitis.

endocrine gland Gland which secretes directly into blood and not through a duct. Examples are pituitary, thyroid, parathyroid, adrenal glands, ovary and testis, placenta and B cells of pancreas.

endocytosis Uptake of material into a cell by the formation of a membrane-bound vesicle.

endocytotic vesicle See *endocytosis*.

endoderm A germ layer lying remote from the surface of the embryo which gives rise to internal tissues such as gut. Contrast *mesoderm* and *ectoderm*.

endodermis Single layer of cells surrounding the central stele (vascular tissue) in roots. The radial and transverse walls contain the hydrophobic *Casparian band*, which prevents water flow in or out of the stele through the *apoplast*. In monocotyledonous stems each vascular bundle is surrounded by endodermal tissue. In young organs, endodermal cells often act as *statocytes*.

endogenous Product or activity arising in the body or cell — as opposed to agents coming from outside.

endogenous pyrogen Fever-producing substance released by mononuclear phagocytes which acts on the hypothalamic thermoregulatory centre. Almost certainly *interleukin-1* is the major pyrogen.

endomitosis Chromosome replication without mitosis, leading to polyploidy. Many rounds of endomitosis give rise to the giant polytene chromosomes of Dipteran salivary glands and other larval tissue, though in this case the daughter chromosomes remain synapsed.

endonexin Calcium-dependent membrane-binding protein located on the endoplasmic reticulum of fibroblasts. Isolated protein will bind to liposomes if $1–10\mu M$ calcium is present but not if the liposomes contain sphingomyelin or cholesterol. An analogous calcium-dependent membrane-binding protein, synexin, co-distributes with endonexin and binds particularly to phosphatidyl serine. Another of the same class is p36, a component of brush-border membrane, a

target for the src *gene* tyrosine-specific *protein kinase*, and which binds phosphatidyl serine or phosphatidyl inositol.

endonuclease One of a large group of enzymes which cleave nucleic acids at positions within the chain. Some act on both RNA and DNA (eg. S1 nuclease, EC 3.1.30.1, which is specific for single-stranded molecules). *Ribonucleases* such as pancreatic, T1 etc. are specific for RNA, *deoxyribonucleases* for DNA. Bacterial *restriction endonucleases* are crucial in recombinant DNA technology for their ability to cleave double-stranded DNA at highly specific sites.

endopeptidase Member of subset of peptide hydrolases (the other being exopeptidases (*peptidases*)) which cleave protein at positions within the chain. Formally, the enzymes are peptidyl-peptide hydrolases, more usually known as *proteases* or proteolytic enzymes.

endoplasm Inner, granule-rich cytoplasm of amoeba.

endoplasmic reticulum (ER) Membrane system which ramifies through the cytoplasm. The membranes of the ER are separated by 50–200nm and the *cisternal* space thus enclosed constitutes a separate compartment. The Golgi region is composed of flattened sacs of membrane which together with ER and lysosomes constitute the GERL system. See also *smooth ER*, *rough ER*.

endoplasmin Most abundant protein in microsomal preparations from mammalian cells (100-fold more concentrated in endoplasmic reticulum than elsewhere). A glycoprotein (100kD) with calcium-binding properties. Same as *GRP* (glucose related protein).

endorphins A family of peptide hormones which bind to receptors that mediate the actions of opiates. Released in response to neurotransmitters and rapidly inactived by peptidases. Physiological responses to endorphins include analgesia and sedation.

endosmosis Movement of water into a cell as a result of greater internal osmotic pressure. The *water potential* within the vascular sap of a plant cell must be lower than that in the bathing medium or sap of a neighbouring cell.

endosome 1. Endocytotic vesicle derived

from the plasma membrane. More specifically an acidic non-lysosomal compartment in which receptor–ligand complexes dissociate. 2. A chromatinic body near the centre of a vesicular nucleus in some protozoa.

endosperm Tissue present in the seeds of angiosperms, external to and in some cases surrounding the *embryo*, which it provides with nourishment in the form of *starch* or other food reserves. Formed by the division of the *endosperm mother cell* after fertilisation; may be absorbed by the embryo prior to seed maturation, or may persist in the mature seed, when it may have a regulatory rather then nutritive role.

endosperm mother cell Cell of the higher plant embryo sac. Contains two "polar nuclei" (a nomenclature best avoided), and fuses with the sperm cell from the pollen grain. Gives rise to the *endosperm*. The embryo-sac is equivalent to the highly developed gametophyte.

endospore An asexual spore formed within a cell.

endosymbiotic bacteria Bacteria which establish a symbiotic relationship within a eukaryotic cell, eg. the nitrogen-fixing bacteria of legume root nodules. See also *endosymbiont hypothesis*.

endosymbiont hypothesis The hypothesis that semi-autonomous organelles such as mitochondria and chloroplasts were originally endosymbiotic bacteria or cyanobacteria. The arguments are convincing and although the hypothesis cannot be proven it is widely accepted.

endothelium Simple squamous epithelium lining blood vessels, lymphatics and other fluid-filled cavities (such as the anterior chamber of the eye). Mesodermally derived, unlike most epithelia.

endotoxin Heat-stable cell-bound toxin; term is now used specifically to refer to the lipopolysaccharide (LPS) of the outer membrane of Gram negative bacteria. There are three parts to the molecule, the *Lipid A* (six fatty acid chains linked to two glucosamine residues), the core oligosaccharide (branched chain of ten sugars), and a variable length polysaccharide side-chain (up to 40 sugar units in smooth forms) which can be removed without affecting the toxicity (rough LPS). Some endotoxin is probably released into the medium and endotoxin is responsible for many of the virulent effects of Gram negative bacteria.

endplate The area of sarcolemma immediately below the synaptic region of the motor neuron in a vertebrate neuromuscular junction.

enhancement effect Property of higher plant photosynthesis, discovered by Robert Emerson. The *quantum yield* of red light (less than 680nm) and far-red light (700nm), when shone simultaneously on a plant, is greater than the sum of the yields of the light of the two wavelengths separately. This effect provides evidence for the cooperative interaction of two *photosystems* in photosynthesis.

enkephalins Natural opiate pentapeptides isolated originally from pig brain. Leu-enkephalin (YGGFL) and met-enkephalin (YGGFM) bind particularly strongly to δ-type opiate receptors.

enteric Relating to the intestine.

Enterobacter Genus of enteropathic bacilli of the *Klebsiella* group. Not to be confused with the Family *Enterobacteriaceae* of which they are members.

Enterobacteriaceae A large family of Gram negative bacilli that inhabit the large intestine of mammals. Commonest is *Escherichia coli*; most are harmless commensals but others can cause intestinal disease (*Salmonella*, *Shigella*).

enterobactin Alternative name for *enterochelin*.

enterochelin Iron-binding compound (*siderophore*) of *E. coli* and *Salmonella* spp. A cyclic trimer of 2,3-dihydroxybenzoyl-serine.

enterocytes Rare term for cells of the intestinal epithelium.

enterotoxins Group of bacterial *exotoxins* produced by enterobacteria and which act on the intestinal mucosa. By perturbing ion and water transport systems they induce diarrhoea. *Cholera toxin* is the best known example.

enterovirus A genus of *Picornaviridae* that preferentially replicate in the mammalian intestinal tract, though they subsequently infect other tissues. It includes the *polioviruses*, *Coxsackie viruses* and *echoviruses*.

enzyme induction An increase in enzyme

synthesis in response to an environmental signal. The classic example is the induction of β-galactosidase in *E. coli*.

eosinophil Polymorphonuclear leucocyte (granulocyte) usually considered to be of the myeloid series but not of the same lineage as *neutrophils*, of which the granules stain red with eosin. Poorly phagocytic, particularly associated with helminth infections and with hypersensitivity.

eosinophil cationic protein Arginine-rich protein (21kD) in granules of eosinophils, that damages schistosomula *in vitro*. Not the same as the MBP (major basic protein) of the granules.

eosinophil chemotactic peptide (ECF; ECF of anaphylaxis; ECF-C) Weakly chemotactic tetrapeptides (of which two are identified: val-gly-ser-glu and ala-gly-ser-glu) released by mast cells and which are said to attract and activate eosinophils.

eosinophilia Condition in which there are unusually large numbers of eosinophils in the circulation, usually a consequence of parasitism by helminths or of allergy.

eosinophilopoietin Small (1500D) peptide, possibly released by T-lymphocytes, which regulates eosinophil development in the bone marrow. Probably *interleukin-5*.

ependymal cells Cells which line cavities in the central nervous system — considered to be a type of glial cell.

epicotyl The first internode of the axial axis of plant seedlings lying above the point of insertion of the cotyledon(s) and the first node. Can be relatively long in some *etiolated* seedlings.

epidermal cell 1. Cell of *epidermis* in animals. 2. Plant cell on the surface of a leaf or other young plant tissue, where bark is absent. The exposed surface is covered with a layer of *cutin*, waxes and esters.

epidermal growth factor (EGF) A mitogenic polypeptide initially isolated from male mouse submaxillary gland. The name refers to the early bioassay, but EGF is active on a variety of cell types, especially but not exclusively epithelial. Human equivalent named urogastrone owing to its hormone activity.

epidermis Outer epithelial layer of a plant or animal. May be a single layer (as in the case of plants) and may produce an extracellular

material (as for example the cuticle of arthropods). In some cases, a complex stratified squamous epithelium, as in many vertebrates.

epididymis Convoluted tubule connecting the vas efferens, which comes from the seminiferous tubules of the mammalian testis, to the vas deferens. Maturation and storage of sperm occur in the epididymis.

epigenesis The theory that development is a process of gradual increase in complexity as opposed to the preformationist view which supposed that mere increase in size was sufficient to produce adult from embryo.

epigenetics The study of mechanisms involved in the production of phenotypic complexity in morphogenesis. According to the epigenetic view of differentiation, the cell makes a series of choices (some of which may have no obvious phenotypic expression, and are spoken of as *determination* events) that lead to the eventual differentiated state. Thus, selective gene repression or derepression at an early stage in differentiation will have a wide-ranging consequence in restricting the possible fate of the cell.

epiglycanin Mucin-like molecule on the surface of certain spontaneous murine mammary carcinoma cells that may mask *histocompatibility antigens*. See also *epitectin*.

epimer Diastereomeric monosaccharides which have opposite configurations of a hydroxyl group at only one position, eg. D-glucose and D-mannose.

epimorphosis Term referring to pattern of regeneration in which proliferation precedes the development of a new part. Opposite of *morphallaxis*.

epinephrine Synonym for *adrenaline*.

episome Piece of hereditary material (plasmid) which can exist as free, autonomously replicating DNA or be attached to and integrated into the chromosome of the cell, in which case it replicates along with the chromosome. Examples of episomes are many *bacteriophages* such as λ and the male sex factor of *E. coli*.

epitectin Mucin-like glycoprotein found on surface of human tumour cells (also known as CA antigen) but not non-tumourigenic cell lines. It is present on the surface of some specialised cells (sweat glands, Type II pneumocytes from lung, bladder epithelium) and may therefore be a normal *differentiation antigen*. Also present in normal urine.

epithelioid cells In a general sense, a cell which has an appearance that is similar to that of epithelial cells: used specifically of the very flattened macrophages found in granulomas (eg. in tubercular lesions).

epithelium One of the simplest types of tissues. A sheet of cells, one or several layers thick, organised above a *basement membrane*, and often specialised for mechanical protection or *active transport*. Examples include skin, and the lining of lungs, gut and blood vessels.

epitope That part of an antigenic molecule to which the T-cell receptor responds; a site on a large molecule against which an antibody will be produced and to which it will bind.

epizootic Veterinary equivalent of an epidemic.

Epstein Barr virus Species of *Herpetoviridae*, that binds to *CR2* and that causes infectious mononucleosis and, in the presence of other factors, tumours such as Burkitt's lymphoma and nasopharyngeal carcinoma.

equatorial plate Region of the mitotic spindle where chromosomes are aligned at metaphase: as its name suggests, it lies midway between the poles of the spindle.

equilibrium constant (equilibrium dissociation constant; dissociation constant) The ratio of the reverse and forward rate constants for a reaction of the type A + B $\rightleftharpoons$ AB. At equilibrium the equilibrium constant (K) equals the product of the concentrations of reactants divided by the concentration of product, and has dimensions of concentration. K = (concentration A.concentration B) / (concentration AB). The affinity (association) constant is the reciprocal of the equilibrium constant.

equivalence The situation where two interacting species are present in concentrations just sufficient to produce occupation of all binding sites. Only used to describe high *avidity* interactions, especially the antibody/antigen interaction.

erb Two oncogenes, *erbA* and *erbB*, associated with erythroblastosis virus (an acute transforming retrovirus). The cellular homologue of *erbB* is the structural gene for the cell-surface receptor for epidermal growth factor, and of *erbA* is a steroid hormone receptor.

ergastic substances Metabolically inert and non-functional components in plant cells representing storage materials, such as starch grains, fat globules and protein bodies.

erysipelas A spreading infection of the dermis possibly associated with an allergic reaction to products of the causative organism, *Streptococcus pyogenes*.

erythroblast Rather non-committal name for a nucleated cell of the bone marrow which gives rise to erythrocytes. See also *normoblast*, *BFU-E*, *CFU-E*, *primitive erythroblast* and *definitive erythroblast*.

erythrocyte A vertebrate red blood cell. For erythrocyte membrane proteins, see Table E1.

erythrocyte ghost The membrane and cytoskeletal elements of the erythrocyte devoid of cytoplasmic contents, but preserving the original morphology.

erythrogenic toxin Toxin produced by strains of *Streptococcus pyogenes* responsible for scarlet fever. Three antigenic variants of the toxin are known. It is a small protein which is complexed with hyaluronic acid and can intensify the effects of other toxins such as *endotoxin* and *streptolysin O*.

erythroid cell Cell which will give rise to erythrocytes.

erythroleukaemic cell Abnormal precursor (virally transformed) of erythrocytes which can be grown in culture and induced to differentiate by treatment with, for example, DMSO. See *Friend murine erythroleukaemia cell*.

erythrophores *Chromatophores* which have red pigment.

erythropoiesis Process of production of erythrocytes in the marrow in adult mammals. A pluripotent stem cell (CFU) produces, by a series of divisions, committed stem cells (*BFU-E*) which give rise to *CFU-E*, cells which will divide only a few more times to produce mature erythrocytes. Each stem cell product can give rise to 2^{11} mature red cells.

erythropoietin Glycoprotein (46kD) produced in the kidney and which regulates the production of red blood cells in the marrow. Higher concentrations are required to stimulate **BFU-E** than **CFU-E** to produce erythrocytes.

Escherichia coli The archetypal bacterium for biochemists, used very extensively in experimental work. A rod-shaped Gram negative bacillus (0.5×3–$5\mu m$) abundant in the large intestine (colon) of mammals. Normally non-pathogenic, although some strains cause diarrhoeas in human infants, calves and piglets.

essential amino acids Those amino acids which cannot be synthesised by an organism and must therefore be present in the diet. The term is often applied anthropocentrically to those amino acids required by humans (Isoleu, Leu, Lys, Met, Phe, Thr, Trp and Val), though rats need two more (Arg and His).

essential fatty acids The three fatty acids required for growth in mammals, arachidonic, linolenic and linoleic acids. Only linoleic acid needs to be supplied in the diet; the other two can be made from it.

esterase An enzyme which catalyzes the hydrolysis of organic esters to release an alcohol or thiol and acid. The term could be applied to enzymes which hydrolyse carboxylate, phosphate and sulphate esters, but is more often restricted to the first class of substrate.

estradiol See *oestradiol*.

ethidium bromide A dye which intercalates into DNA. Intercalation into linear DNA is easier than into covalently closed circular DNA and the addition of ethidium bromide to DNA prior to ultracentrifugation on a *caesium chloride* gradient is much used to separate nuclear and mitochondrial or plasmid DNA. Because less intercalates into the circular DNA, the density remains higher.

ethylene Plant growth substance (phytohormone, plant hormone), involved in promoting growth, epinasty, fruit ripening, senescence and breaking of dormancy. Its action is closely linked with that of *auxin*.

etiolation Growth habit adopted by germinating seedlings or other plants kept in the dark. Involves rapid extension of shoot and/or hypocotyl and suppression of chlorophyll formation and leaf growth.

etioplast Form of *plastid* present in plants grown in the dark. Lacks chlorophyll, but contains chlorophyll precursors and can develop into a functional chloroplast in the light.

Eubacteria A major subdivision of the prokaryotes (includes all except *Archaebacteria*). Most Gram positive bacteria, cyanobacteria, mycoplasmas, enterobacteria, pseudomonads and chloroplasts are Eubacteria. The cytoplasmic membrane

Table E1. Erythrocyte membrane proteins. The mammalian erythrocyte ghost consists of a lipid bilayer linked to a cytoskeletal network. The proteins of the ghost vary across species, but there are some common patterns. Components are identified as far as possible by comparison with the proteins of the human erythrocyte ghost, after electrophoretic separation on SDS polyacrylamide gel, and numbered according to the Steck classification. (*J. Cell Biol.* 1974. **62**, 1–29)

Band number after Steck.[a]	MW (kD)	Other name or function
1	240	Spectrin α
2	220	Spectrin β
2.1	200	Ankyrin. Links band 3 to spectrin
3	93	Anion transporter
4.1	82	Links spectrin to glycophorin
4.2	76	
4.5	46	Glucose transporter
4.9	48	
5	43	Actin; forms short oligomers, involved in gelation of spectrin and band 4.1
6	35	Glyceraldehyde 3-phosphate dehydrogenase
7	28	

[a]These bands are visible when the gel is stained with a typical "protein" dye, eg., Coomassie brilliant blue. Other bands are only detected when stained for carbohydrate with the Periodic Acid/Schiff reagent (PAS). Four bands are characterised:- PAS1, PAS2, PAS3 and PAS4. Of these PAS1 and PAS2 are the glycoprotein glycophorin (55kD) in different oligomeric states. PAS3 and PAS4 are minor components.

contains ester-linked lipids, there is **peptido-glycan** in the cell wall (if present) and no **introns** have been discovered.

euchromatin The chromosomal regions which are diffuse during interphase and condensed at the time of nuclear division. They show what is considered to be the normal pattern of staining (eu = true) as opposed to **heterochromatin**.

Euglena gracilis Phytoflagellate protozoon of the algal order Euglenophyta (zoological order Euglenida). An elongate cell with two **flagella**, one emerging from a pocket at the anterior end. The organism exhibits positive **phototaxis**, determined by a photoreceptive spot on the basal part of the flagellum shaft being shielded by a carotenoid-containing stigma ("eyespot") in the wall of the pocket.

Eukaryote Organism whose cells have (1) chromosomes with nucleosomal structure and separated from the cytoplasm by a two-membrane nuclear envelope, and (2) compartmentalisation of function in distinct cytoplasmic organelles.

euploidy The state in which the chromosome number is an integer multiple of the starting number, ie. one or more whole chromosome sets (genomes) is present.

excision repair Mechanism for the repair of environmental damage to one strand of DNA (loss of **purines** due to thermal fluctuations, formation of pyrimidine dimers by ultraviolet irradiation). The site of damage is recognised, excised by an **endonuclease**, the correct sequence is copied from the complementary strand by a **DNA polymerase** and the ends of this correct sequence are joined to the rest of the strand by a **ligase**. The term is sometimes restricted to bacterial systems where the polymerase also acts as endonuclease.

excitable cell A cell in which the membrane response to **depolarisations** is non-linear, causing amplification and propagation of the depolarisation (an "action potential"). Apart from neurons and muscle cells, electrical excitability can be observed in fertilised eggs, some plants and glandular tissue. Excitable cells contain **voltage gated ion-channels**.

excitation contraction coupling Name given to the chain of processes coupling excitation of a muscle by the arrival of a nervous impulse at the **motor end plate** to the contraction of the filaments of the **sarcomere**. The crucial link is the release of calcium from the sarcoplasmic reticulum, and the analogy is often drawn between this and stimulus–secretion coupling, which also involves calcium release into the cytoplasm.

excitatory amino acid (EAA) The naturally occurring amino acids L-glutamate and L-aspartate and their synthetic analogues, notably **kainate**, **quisqualate**, and **NMDA**. They have the properties of excitatory neurotransmitters in the central nervous system (CNS), may be involved in long term potentiation, and can act as **excitotoxins**. At least three classes of EAA receptor have been identified; the agonists of the N-type receptor are L-aspartate, NMDA, and ibotenate; the agonists of the Q-type receptor are L-glutamate and quisqualate; agonists of the K-type are L-glutamate and kainate. All three receptor types are found widely in the CNS, and particularly the telencephalon; N- and Q-type receptors tend to occur together, and may interact; their distribution is complementary to the K-type receptors. The ion fluxes through the Q and K receptors are relatively brief, whereas the flux through the N-type is longer, and carries a significant amount of calcium. Additionally the N-type receptor is blocked by magnesium near the resting potential, and thus shows **voltage gated ion-channel** properties, leading to a regenerative response; this is why N-type receptors have been linked to long term potentiation. Invertebrate glutamate receptors may not have the same properties as those described above.

excitatory synapse A synapse (either **chemical** or **electrical**) in which an action potential in the presynaptic cell increases the probability of an action potential occurring in the postsynaptic cell. See **inhibitory synapse**.

excitotoxin Class of substances that damage neurons through paroxysmal overactivity. The best known excitotoxins are the **excitatory amino acids**, that can produce lesions in the central nervous system similar to those of Huntingdon's chorea or Alzheimer's disease.

exfoliatin Epidermolytic toxin produced by some strains of *Staphylococcus aureus*; causes detachment of the outer layer of skin by disrupting **desmosomes** of the stratum granulosum.

exine External part of pollen wall that is often elaborately sculptured in a fashion characteristic of the plant species. Contains *sporopollenin*. The term is also used for the outer part of a spore wall.

exocrine Exocrine glands release their secreted products into ducts which open onto epithelial surfaces.

exocytosis Release of material from the cell by fusion of a membrane-bounded vesicle with the plasma membrane.

exocytotic vesicle Vesicle, for example a secretory vesicle or *zymogen granule*, which can fuse with the plasma membrane to release its contents.

exons The sequences of the primary RNA transcript (or the DNA which encodes them) that exit the nucleus as part of a *messenger RNA* molecule. In the primary transcript or the *split gene* neighbouring exons are separated by *introns*.

exotoxins Toxins released from Gram positive and Gram negative bacteria — as opposed to *endotoxins* which form part of the cell wall. Examples are *cholera*, *pertussis* and *diphtheria toxins*. Usually specific and highly toxic. See Table E2.

experimental allergic encephalomyelitis An autoimmune disease which can be induced in various experimental animals by the injection of homogenised brain or spinal cord in *Freund's adjuvant*. The antigen appears to be a basic protein present in myelin, and the response is characterised by focal areas of lymphocyte and macrophage infiltration into the brain, associated with demyelination and destruction of the blood–brain barrier. Sometimes used as a model for demyelinating diseases, although whether this is entirely justifiable is not clear.

Table E2. Exotoxins

Name	Source	Target/mode of action
α-toxin	*Clostridium perfringens*	Phospholipase C
Anthrax toxin	*Bacillus anthracis*	Three components, one a soluble adenyl cyclase
Botulinum toxins	*Clostridium botulinum*	Inhibits acetylcholine release
Cholera toxin	*Vibrio cholerae*	ADP-ribosylation of G_s
Diphtheria toxin	*Corynebacterium diphtheriae*	ADP-ribosylation of EF-2
δ-toxin	*Clostridium perfringens*	Binds to cholesterol
Enterotoxins	*Staphyloccus aureus*	Neurotoxic
	Pseudomonas aeruginosa	Causes diarrhoea
Erythrogenic toxin	*Streptococcus pyogenes*	Skin hypersensitivity
Exfoliatin	*Staphyloccus aureus*	Disrupts desmosomes
Haemolysin	*Staphyloccus aureus*	
α-Haemolysin		Unknown mechanism
β-Haemolysin		Acts as sphingomyelinase C
γ-Haemolysin		Haemolytic
δ-Haemolysin		Surfactant, haemolytic
Haemolysin	*Pseudomonas aeruginosa*	Toxic for macrophages
Heat-labile toxin	*Bordetella pertussis*	Dermonecrotic
Heat-labile toxin	*Escherichia coli*	Similar to cholera toxin
Kanagawa haemolysin	*Vibrio haemolytica*	Haemolytic, cardiotoxic
Leucocidin	*Staphyloccus aureus and Pseudomonas aeruginosa*	Lyses neutrophils and macrophages
Pertussis toxin	*Bordetella pertussis*	ADP-ribosylates G_i
Pneumolysin	*Streptococcus pneumoniae*	Binds to cholesterol
Stable toxin	*Escherichia coli*	Activates guanylate cyclase
Streptolysin D	*Streptococcus pyogenes*	Binds cholesterol
Streptolysin S	*Streptococcus pyogenes*	Membranolytic
Subtilysin	*Bacillus subtilis*	Haemolytic surfactant
Tetanolysin	*Clostridium tetani*	Binds cholesterol
Tetanus toxin	*Clostridium tetani*	Inhibits glycine release at synapse
Toxin A	*Pseudomonas aeruginosa*	ADP-ribosylates EF-2

G_s, G_i: see GTP binding proteins
EF-2: elongation factor 2

expression vector A *vector* which results in the expression of inserted DNA sequences when propagated in a suitable host cell, ie. the protein coded for by the DNA is synthesised by the host's system.

extensin Glycoprotein of the plant cell wall, characterised by its high hydroxyproline content. Carbohydrate side-chains are composed of simple galactose residues and oligosaccharides containing 1 to 4 arabinose residues. Part of a larger class of *hydroxyproline-rich glycoproteins*. Function uncertain.

extracellular matrix (ecm; ECM) Any material produced by cells and secreted into the surrounding medium, but usually applied to the non-cellular portion of animal tissues. The ecm of connective tissue is particularly extensive and the properties of the ecm determine the properties of the tissue. In broad terms there are three major components: fibrous elements (particularly *collagen*, *elastin* or *reticulin*), link proteins (eg. *fibronectin*, *laminin*) and space-filling molecules (usually *glycosaminoglycans*). The matrix may be mineralised to resist compression (as in bone) or dominated by tension-resisting fibres (as in tendon). The basal lamina (*basement membrane*) of epithelial cells is another commonly encountered ecm. Although ecm is produced by cells, it has recently become clear that the ecm can influence the behaviour of cells quite markedly, an important factor to consider when growing cells *in vitro*: removing cells from their normal environment can have far-reaching effects.

extrachromosomal element Any heritable element not associated with the chromosome(s). It is usually a *plasmid* or the DNA of organelles such as mitochondria and chloroplasts.

exudate cells Leucocytes which enter tissues (exude from the blood vessels) during an inflammatory response. See also *peritoneal exudate*.

F

F_1-ATPase A subunit of the *F_1F_0-ATPase*.

F_1F_0-ATPase A large enzyme complex found in mitochondria, chloroplasts and bacteria. In mitochondria, the enzyme uses the proton gradient which is generated across the organelle membrane by the *electron transport chain* to generate *ATP*. The F_1F_0-ATPase has MW of approximately 500kD and is visible by electron microscopy as 10nm-diameter spheres. It is widely believed to be an ATPase being driven in reverse (an "ATP synthase"). See also *chemiosmosis*.

F-duction See *sex-duction*.

F-factor *Plasmid* which confers the ability to conjugate (ie. fertility) on bacterial cells, and carries the *tra* (transfer) genes; first described in *E. coli*.

Fab Fragment of immunoglobulin prepared by *papain* treatment. Fab fragments (45kD) consist of one light chain linked through a disulphide bond to a portion of the heavy chain, and contain one antigen-binding site. They can be considered as univalent antibodies.

F(ab)$_2$ The fragment (90kD) of an immunoglobulin produced by *pepsin* treatment. These fragments have two antigen-combining sites and contain two light chains and two variable region heavy chains plus one constant region domain in each heavy chain. The fragment is divalent but lacks the complement-fixing (Fc) domain.

Fabry disease *Storage disease* due to deficiency of ceramide trihexosidase: also known as angiokeratoma.

facilitated diffusion (passive transport) A process by which substances are conveyed across cell membranes faster than would be possible by diffusion alone. This is generally achieved by proteins which provide a hydrophilic environment for polar molecules throughout their passage through the plasma membrane, acting as either shuttles or pores. See *symport*, *antiport*, *uniport*.

facilitation Greater effectiveness of synaptic transmission by successive presynaptic impulses, usually due to increased transmitter release.

facilitator neuron A neuron whose firing enhances the effect of a second neuron on a third. This allows the effects of neuronal activity to be modulated. See *facilitation*.

FACS See *flow cytometry*.

Factors I — XIII Blood clotting factors, especially from humans. These factors form a cascade in which the activation of the first factor leads to enzymic attack on the next factor and so on, finally resulting in blood clotting. See Table F1.

facultative heterochromatin That *heterochromatin* which is condensed in some cells and not in others, or condensed at some times and not at others, presumably representing stable differences in the activity of genes in different cells. The best known example results from the random inactivation of one of the pair of X chromosomes in the cells of female mammals (*Lyonisation*).

Table F1. Blood clotting factors

Factor	Name	MW	Function
I	Fibrinogen	340kD	Cleaved to form fibrin
II	Prothrombin	70kD	Converted to thrombin by Factor X
III	Thromboplastin	—	Lipoprotein which acts with VII to activate X
IV	Calcium ions	—	Needed at various stages
V	Proaccelerin	—	Product, accelerin, promotes thrombin production
VII	Proconvertin	—	Activated by trauma to tissue
VIII	Antihaemophilic factor	>10^6D	Acts with IXa to activate X
IX	Christmas factor	55kD	See VIII
X	Stuart factor	55kD	When activated converts II to thrombin
XI	Thromboplastin antecedent	124kD	Converts IX to active form
XII	Hagemann factor	76kD	Activated by surface contact
XIII	Fibrin-stabilising factor	350kD	Transglutaminase which cross-links fibrin

FAD Flavine adenine dinucleotide, a prosthetic group of many flavine enzymes. See *flavine nucleotides*.

Falconisation Trade name for the treatment of polystyrene to make it appropriate for use in cell culture. The main commercial process is probably corona discharge in air or other gas mixtures at low pressure. Treatment of polystyrene with sulphuric acid will produce the same effect.

familial hypercholesterolemia Excess of cholesterol in plasma as a result of defects in the recycling process which leads to reduced uptake of LDL (*low density lipoprotein*) into *coated vesicles*. Several different defects in the LDL-receptor system have been identified.

Fanconi syndrome Transport disease (recessive defect) in which the renal reabsorption of several substances (phosphate, glucose, amino acids) is impaired.

Fanconi's anaemia Defect in thymine-dimer excision from DNA predisposing to development of leukaemia.

Farmer's lung Type III *hypersensitivity* response to *Micropolyspora faeni*, a thermophilic bacterium found in mouldy hay.

Farr-type assay Method of radioimmunoassay in which free antigen remains soluble and antibody–antigen complexes are precipitated.

fascicle Literally, a bundle. In particular, this is used to describe the tendency of *neurites* to grow together (fasciculate).

fascicular cambium Form of *cambium* present in the vascular bundles of higher plants.

fascin Actin filament-bundling protein (58kD) from sea-urchin (*echinoderm*) eggs.

fat cell See *adipocyte*.

fat droplets Micro-aggregates of (mainly) triglycerides visible within cells.

fate map Diagram of an early embryo (usually a *blastula*) showing which tissues the cells in each region will give rise to (ie. their developmental fate). Fate maps are normally constructed by labelling small groups of cells in the blastula with *vital stains* and seeing which tissues are stained when the embryo develops.

fats A term largely applied to storage lipids in animal tissues. The primary components are triglyceride esters of long-chain fatty acids.

fatty acids Chemically R–COOH where R is an aliphatic moiety. The common fatty acids of biological origin are linear chains with an even number of carbon atoms. Free fatty acids are present in living tissues at low concentrations. The esterified forms are important both as energy storage molecules and structural molecules. See *triglycerides*, *phospholipid*.

fatty streak Superficial fatty patch in the artery wall caused by the accumulation of cholesterol and cholesterol oleate in distended smooth muscle cells (foam cells).

favism Haemolytic anaemia induced in individuals who are glucose-6-phosphate dehydrogenase-deficient by eating fava beans (from *Vicia fava*).

Fc That portion of an immunoglobulin molecule ("Fragment crystallisable") which is produced by *papain* hydrolysiis. Consists of the C-terminal domains of the two heavy chains linked by disulphide bonds. The Fc of IgG binds to a cell when the antigen binding sites (*Fab*) of the antibody are occupied or the antibody is aggregated; the Fc portion is also important in complement activation. Monomeric IgE binds to the mast cell Fc receptor by its Fc portion. Fc moieties from different antibody classes and subclasses have different properties.

Fc-receptor (FcR) Receptor for the Fc portion of immunoglobulins, belonging to the Ig superfamily of proteins. Several distinct types of FcR are recognised, each binding particular classes or sub-classes of Ig:- FcRγ I: 72kD protein on monocyte surface that binds monomeric IgG with high affinity; FcRγ II: (CDw32) 40kD, a low-affinity receptor present on neutrophils, monocytes, eosinophils, platelets and B-cells; FcRγ III: (CD16) 50–70kD, commonest FcR in blood though rather low affinity, present on neutrophils, large granular lymphocytes and macrophages (but not monocytes); has a glyco-phosphatidyl inositol tail (see *glypiation*). In humans, all FcR bind IgG1 and IgG3 much better than IgG2 and IgG4; in mice, FcR are Ig subclass-specific and not necessarily equivalent to the human FcR. The FcR for IgE (FcRϵ) on mast cells binds monomeric IgE; clustering of the IgE leads to exocytosis of granule contents.

feedback regulation Control mechanism

which uses the consequences of a process to regulate the rate at which the process occurs: if, for example, the products of a reaction inhibit the reaction from proceeding (or slow down the rate of the reaction), then there is negative feedback, something that is very common in metabolic pathways. Positive feedback is liable to lead to exponential increase and may be explosively dangerous in some cases. Other examples are the action of voltage-dependent *sodium channels* in generating action potentials and the activation of blood clotting *Factors V and VIII* by *thrombin*. Without damping, feedback can lead to resonance ("hunting") and oscillation in the system.

fermentation Breakdown of organic substances, especially by micro-organisms such as bacteria and yeasts, yielding incompletely oxidised products. Some forms can take place in the absence of oxygen, in which case *ATP* is generated in reaction pathways in which organic compounds act as both donors and acceptors of electrons. Historically, the production of ethyl alcohol or acetic acid from *glucose*. Also applied to anaerobic *glycolysis* as in *lactate* formation in muscle.

ferredoxins Low-molecular weight iron-sulphur proteins which transfer electrons from one enzyme system to another without themselves having enzyme activity.

ferrichromes Siderochromes: ligands for iron-binding secreted by micro-organisms to sequester and transport iron.

ferritin An iron-storage protein of mammals, found in liver, spleen, and bone marrow. Morphologically a shell of apoferritin (protein) with a core of ferrous hydroxide/phosphate. It is much used as an electron-dense marker in electron microscopy.

fertilisation An essential process in sexual reproduction, involving the union of two specialised *haploid* cells, the gametes, to give a diploid cell, the zygote, which then develops to form a new organism.

ferulic acid Phenolic compound present in the plant cell wall that may be involved in cross-linking polysaccharide.

fetal See *foetal*.

fibre cell Greatly elongated cell in plant axes or on surface (flax and cotton respectively). Usually dead at maturity, this cell type is specialised for the provision of mechanical strength, and has walls thickened with cellulose or lignin.

fibrillar centres Location of the nucleolar ribosomal chromatin at telophase: as the nucleolus becomes active the ribosomal chromatin and associated ribonucleoprotein transcripts compose the more peripherally-located dense fibrillar component.

fibrillar region Dense-staining region of the nucleolus composed of 5nm fibres, RNA transcripts.

fibrillin Widely-distributed connective tissue protein (350kD) associated with microfibrils (10nm diameter).

fibrin Monomeric fibrin (323kD) is produced from *fibrinogen* by proteolytic removal of the highly-charged (aspartate- and glutamate-rich) *fibrinopeptides* by thrombin, in the presence of calcium ions. The monomer readily polymerises to form long insoluble fibres (23nm periodicity; half-staggered) which are stabilised by covalent crosslinking (by *Factor XIII*, plasma transglutaminase). The fibrin gel acts as a haemostatic plug.

fibrinogen Soluble plasma protein (340kD; 46nm long), composed of 6 peptide chains (2 each of Aα, Bβ, and γ) and present at about 2–3mg/ml.

fibrinolysis Solubilisation of fibrin, chiefly by the proteolytic action of *plasmin*.

fibrinopeptides Very negatively-charged peptide fragments cleaved from fibrinogen by thrombin. Two peptides (A and B) are produced from each fibrinogen.

fibroblast Resident cell of connective tissue, mesodermally derived, that secretes fibrillar *procollagen, fibronectin* and *collagenase*.

fibroblastic Many types of cultured cell become fibroblastic in appearance — this does not mean that they *are fibroblasts*.

fibroin Structural protein of silk, one of the first to be studied with X-ray diffraction. It has a repeat sequence (gly-ser-gly-ala-gly-ala) and is unusual in that it consists almost entirely of stacked antiparallel *beta-pleated sheets*.

fibronectin Glycoprotein of high molecular weight (2 chains each of 250kD linked by disulphide bonds) which occurs as an insoluble fibrillar form in extracellular matrix of animal tissues, and as a soluble form in plasma, the latter previously known as cold-insoluble globulin. The various slightly different forms of fibronectin appear to be generated by tissue-specific differential splicing of fibronectin mRNA, transcribed from a single gene. Fibronectins have multiple domains which confer the ability to interact with many extracellular substances such as collagen, fibrin and heparin, and also with specific membrane receptors on responsive cells. Interaction of a cell's fibronectin receptors (members of the *integrin* family) with fibronectin adsorbed to a surface results in adhesion and spreading of the cell.

fibrosarcoma Malignant tumour derived from connective tissue fibroblast.

fibrosis Scar formation: deposition of avascular collagen-rich matrix (*fibrous tissue*), usually as a consequence of slow fibrinolysis or extensive tissue damage as in sites of chronic inflammation.

fibrous lamina Alternative name for the *nuclear lamina*, the region lying just inside the inner nuclear membrane.

fibrous plaque Thickened area of arterial *intima* with accumulation of smooth muscle cells and fibrous tissue (collagen etc.) produced by the fat-laden smooth muscle cells. Below the thickening may be free extracellular lipid and debris which, if much necrosis is also present, is referred to as an *atheroma*.

fibrous tissue Although most connective tissue has fibrillar elements, the term usually refers to tissue laid down at a wound site — well-vascularised at first (*granulation tissue*) but later avascular and dominated by collagen-rich extracellular matrix, forming a scar. Excessive contraction and hyperplasia leads to formation of a *keloid*.

filaggrin Basic protein component of *keratohyalin granules* of the suprabasal cells of the skin.

filaments See *thick filaments*, *thin filaments*, *intermediate filaments* and *microfilaments*.

filamin A protein which binds to F-*actin*, cross-linking it to form an isotropic network; the binding does not require calcium ions. It was originally isolated from smooth muscle and is a homodimer $2 \times 250kD$. Similar to actin-binding protein (ABP) from leucocytes.

filiform papillae Curved tapering cone-shaped bodies on the tongue of rodents, of which the epithelial cell columns have been investigated in detail.

filopodium (*pl.* filopodia) 1. A thin protrusion from a metazoan cell, usually supported by microfilaments; may be functionally the linear equivalent of the leading lamella. 2. Fine, pointed pseudopodium of an amoeba.

fimbriae See *pili*.

fimbrin Actin-binding protein (68kD) from the core of epithelial brush-border *microvilli*.

fixation Any chemical or physical treatment of cellular material which tends to result in its insolubilisation, thus making it suitable for various types of processing for microscopy, such as *embedding* or staining. Typically, fixation involves protein denaturation.

flagellin Subunit protein (40kD) of the bacterial flagellum.

flagellum (*pl.* flagella) Long thin projection from a cell used in movement. In eukaryotes flagella (like *cilia*) have a characteristic axial $9 + 2$ microtubular array (*axoneme*) and bends are generated along the length of the flagellum by restricted sliding of the nine outer doublets. In prokaryotes the flagellum is made of polymerised *flagellin* and is rotated by the basal motor.

flanking sequence Short *DNA* sequences bordering a *transcription unit*. Often these do not code for proteins.

flare streaming Phenomenon described in isolated cytoplasm of giant amoeba when the medium contains calcium ions and ATP. A loop of cytoplasm flows outward and then returns to the main mass — the appearance is reminiscent of flares around the eclipsed sun.

flat revertant Variant of a morphologically-transformed animal tissue-culture cell in which the characteristic high *saturation density* and piled-up morphology have reverted to the flatter morphology associated with non-transformed cells.

flavine adenine dinucleotide See *flavine nucleotides*.

flavine nucleotides Flavine derivatives which act as prosthetic groups (covalently-linked co-factors) for flavine enzymes.

flip-flop A term used to describe the coordinated transfer of two phospholipid molecules from opposite sides of a lipid bilayer membrane. Now used to mean the passage of a phospholipid species from one lamella of a lipid bilayer membrane to the other.

florigen Hypothetical plant growth substance postulated to induce flowering. Existence not proven: recently suggested that it might be an *oligosaccharin*.

flow cytometry Slightly imprecise but common term for the use of the Fluorescence Activated Cell Sorter (FACS). Cells are labelled with fluorescent dye and then passed, in suspending medium, through a narrow dropping nozzle so that each cell is in a small droplet. A laser-based detector system is used to excite fluorescence, and droplets with positively-fluorescent cells are given an electric charge. Charged and uncharged droplets are separated as they fall between charged plates, and so collect in different tubes. The machine can be used either as an analytical tool, counting the number of labelled cells in a population, or to separate the cells for subsequent growth of the selected population. Further sophistication can be built into the system by using a second laser system at right angles to the first to look at a second fluorescent label, or to gauge cell size on the basis of light-scatter. The great strength of the system is that it looks at large numbers of individual cells, and makes possible the separation of populations with, for example, particular surface properties.

fluctuation analysis Method used to determine (for example) how many ion channels contribute to the transmembrane current. On the assumption that each channel is either open or shut, the noise in the recorded current can be considered to arise from the statistical fluctuation in the number of channels open, and the magnitude of the fluctuation gives an estimate of the conductance of a single channel.

fluctuation test Test devised by Luria and Delbruck to determine whether genetic variation in a bacterial population arises spontaneously or adaptively. In the original version the statistical variance in the number of bacteriophage-resistant cells in separate cultures of bacteriophage-sensitive cells was compared with variance in replicate samples from bulk culture. The greater variance in the isolated populations indicates that mutation occurs spontaneously before challenge with phage. (The proportion of resistant cells depends upon when after isolation the mutation arises — which will be very different in separate populations).

fluid bilayer model Generally accepted model for membranes in cells. In its original form, the model held that proteins "floated" in a "sea" of phospholipids arranged as a bilayer with a central hydrophobic domain. Although it is now recognised that some proteins are restrained by interactions with cytoskeletal elements, and that the phospholipid annulus around a protein may contain only specific types of lipid, the model is still considered broadly correct.

fluorescein Fluorophore commonly used in microscopy. Fluorescein di-acetate can be used as a vital stain, or can be conjugated to proteins (particularly antibodies) using isothiocyanate. Excitation is at 365nm, and the emitted light is green-yellow (450–490nm). The emission spectrum is pH-sensitive and fluorescein can therefore be used to measure pH in intracellular compartments.

fluorescence The emission of one or more photons by a molecule or atom activated by the absorption of a quantum of electromagnetic radiation. Typically the emission, which is of longer wavelength than the excitatory radiation, occurs within 10^{-8}s: phosphorescence is a phenomenon with a longer or much longer delay in re-radiation. Note that gamma rays, X-rays, ultraviolet radiation and visible light may all stimulate fluorescence.

fluorescence activated cell sorter See *flow cytometry*.

fluorescence microscopy Any type of microscopy in which intrinsic or applied reagents are visualised. Intrinsic fluorescence is often referred to as autofluorescence. The applied reagents typically include fluorescently-labelled proteins which are reactive with sites in the specimen. In particular, fluorescently-labelled antibodies are widely used to detect particular antigens in biological specimens.

fluorescence recovery after photobleaching (FRAP) Many fluorochromes are bleached by exposure to exciting light. If, for example, the cell surface is labelled with a fluorescent probe and an area bleached by laser illumination, then the bleached patch which starts off as a dark area will recover fluorescence, due to the re-population of the area by unbleached molecules and diffusion of bleached molecules to other areas. The rate and extent of recovery are a measure of the fluidity of the membrane and the proportion of labelled molecules which are free to exchange with adjacent areas. The technique is usually applied to cell surface fluidity or viscosity measurements, but is also applicable to other structures.

fluoride The fluoride ion F⁻. Low levels of fluoride in drinking water markedly decrease the incidence of dental caries, probably because bacterial metabolism is much more sensitive than eukaryotic metabolism to low fluoride levels.

fluorite objective Microscope objective corrected for *spherical* and *chromatic aberration* at two wavelengths. Better than an ordinary objective corrected at one wavelength but inferior to (and much cheaper than) a planapochromatic objective.

fluorochromes Those molecules which are fluorescent when appropriately excited; fluorochromes such as fluorescein or tetramethyl rhodamine are usually used in their isothiocyanate forms (FITC, TRITC).

fMLP See *formyl peptides*.

FMN (flavine adenine nucleotide) See *flavine nucleotides*.

foam cells Lipid-laden smooth muscle cells found in *fatty streaks* on the arterial wall.

focal adhesions Areas of close apposition, and thus presumably anchorage points, of the plasma membrane of a fibroblast (for example) to the substratum over which it is moving. Usually $1 \times 0.2\mu m$ with the long axis parallel to the direction of movement; always associated with a cytoplasmic microfilament bundle which is attached *via* several proteins to the plasma membrane at an area of high protein concentration (this is noticeably electron-dense in electron-micrographs). Focal adhesions tend to be characteristic of slow-moving cells.

fodrin Tetrameric protein (α 240kD, β 235kD) found in brain: an isoform of *spectrin*.

foetal calf serum (USA fetal calf serum) Expensive component of standard culture media for many types of animal tissue cells.

folate (tetrahydrofolate; pteroylglutamate) Molecule which acts as a carrier of one-carbon units in intermediary metabolism. It contains residues of p-aminobenzoate, glutamate and a substituted pteridine. The latter cannot be synthesised by mammals, which must obtain tetrahydrofolate as a vitamin or from intestinal micro-organisms. One-carbon units are carried at three different levels of oxidation, as methyl-, methylene- or formimino- groups. Important biosyntheses dependent on tetrahydrofolate include those of *methionine*, *thymine* and *purines*. Analogues of dihydrofolate, such as *aminopterin* and *methotrexate* block the action of tetrahydrofolate by inhibiting its regeneration from dihydrofolate.

follicle Generally a small sac or vesicle. *Bot.* A kind of fruit formed from a single carpel, which splits to release its seeds. *Zool.* Its use includes: hair follicle, an invagination of the epidermis into the dermis surrounding the hair root; ovarian follicle, an oocyte surrounded by one or more layers of granulosa cells. As the ovarian follicle develops a cavity forms and it is then termed a Graafian follicle.

follicle-stimulating hormone (FSH; follitropin) Pituitary hormone which is an acidic glycoprotein. It induces development of ovarian follicles and stimulates the release of oestrogens.

foreign body giant cell Syncytium formed by the fusion of macrophages in response to an indigestible particle too large to be phagocytosed (eg. talc, silica or asbestos fibres). There may be as many as 100 nuclei randomly distributed: similar cells but with the nuclei more peripherally located (*Langhans' giant cells*) are found at the centre of tuberculous lesions.

formaldehyde Commonly used fixative and antibacterial agent. As a fixative it is cheap and more rapidly penetrating than glutaraldehyde. It is, however, a less effective cross-linking agent and tends to cause less denaturation of proteins than does glutaraldehyde, particularly if used in a well-buffered solution ("buffered formalin", "formal saline"). Old formaldehyde solutions usually contain cross-linking contaminants, and it is therefore often preferable to used a formaldehyde-generating agent such as paraformaldehyde. Formalin fumes, particularly in conjunction with HCl vapour, are potently carcinogenic.

formyl peptides Informal term for small peptides with a formylated N-terminal methionine and usually a hydrophobic amino acid at the carboxy-terminal end (fMetLeuPhe is the most commonly used). These peptides stimulate the motor and secretory activities of leucocytes, particularly neutrophils and monocytes, which have a specific receptor (about 60kD) of high affinity (K_d approximately $10^{-8}M$). Leucocytes show chemotaxis towards formyl peptides but the term "chemotactic peptides" understates the range of activities the molecules will trigger. Thought to be synthetic analogues of bacterial signal sequences — though this is unproven. The leucocytes of many animals (eg. pig, cow, chicken) do not respond.

Forssman antigen A glycolipid heterophile antigen present on tissue cells of many species. It was first described for sheep red cells, and is not present on human, rabbit, rat, porcine or bovine cells.

founder cell Cell which gives rise to tissue by clonal expansion. For most mammalian tissues there are considerably more than two founder cells, as can be determined by forming chimeras from genetically distinguishable embryos, but single founder cells have been found for the intestine and germ line in **Caenorhabditis elegans**.

fraction I protein Synonym for *ribulose bisphosphate carboxylase/oxidase* (RUBISCO).

fractionation A term used to describe any method for separating and purifying materials. See also *cell fractionation*.

fragmin An actin-binding protein (42kD) from *Physarum polycephalum* that has calcium-sensitive severing and capping properties.

frame-shift mutation Insertion or deletion of a number of bases not divisible by three in an open reading frame in a DNA sequence. Such mutations usually result in the generation, downstream, of nonsense, chain-termination codons.

Frankia Genus of actinomycete bacteria capable of nitrogen fixation, both independently and in symbiotic association with roots of certain non-leguminous plants.

free-energy (Gibbs free energy, G) A thermodynamic term used to describe the energy that may be extracted from a system at constant temperature and pressure. In biological systems the most important relationship is:- $\Delta G = -RT.\ln(K_{eq})$, where K_{eq} is an equilibrium constant.

freeze cleavage See *freeze fracture*.

freeze drying Method commonly adopted to produce a dry and stable form of biological material which has not been seriously denatured. By freezing the specimen, often with liquid nitrogen, and then subliming water from the specimen under vacuum, proteins are left in reasonably native form, and can usually be rehydrated to an active state. Since the freeze-dried material will store without refrigeration for long periods, it is a convenient method for holding back-up or reference material, or for the distribution of antibiotics, vaccines etc.

freeze etching If a *freeze fractured* specimen is left for any length of time before shadowing, then water will sublime off from the specimen etching (lowering) those surfaces which are not protected by a lipid bilayer. Some etching will take place following any freeze cleavage process; in deep etching the ice surface is substantially lowered to reveal considerable detail of, for example, cytoplasmic filament systems.

freeze fracture Method of specimen preparation for the electron microscope in which rapidly frozen tissue is "cracked" so as to produce a fracture plane through the specimen. The surface of the fracture plane is then shadowed by heavy metal vapour, strengthened by a carbon film, and the underlying specimen is digested away, leaving a replica which can be picked up on a grid and examined in the transmission electron microscope. The great advantage of the method is that the fracture plane tends to pass along the centre of lipid bilayers, and it is therefore possible to get *en face* views of membranes which reveal the pattern of integral membrane proteins. The *E-face* is the outer lamella of the plasma membrane viewed as if from within the cell, the P-face the inner lamella viewed from outside the cell. Fracture planes also often pass along lines of weakness such as the interface between cytoplasm and membrane, so that outer and inner membrane surfaces can be viewed. Further information about the structure can be revealed by *freeze etching*. Extremely rapid freezing followed by deep etching has allowed the structure of the cytoplasm to be studied without the artefacts that might be introduced by fixation.

French flag problem The French flag (tricolor) is used to illustrate a problem in the determination of pattern in a tissue, that of specifying three sharp bands of cells with discrete properties which do not have blurred edges using, for example, a gradient of a diffusible morphogen.

Freund's adjuvant A water-in-oil emulsion used experimentally for stimulating a vigorous immune response to an antigen (which is in the aqueous phase). Complete Freund's adjuvant contains heat-killed tubercle bacilli; these are omitted from Freund's incomplete adjuvant. Unsuitable for use in humans because it elicits a severe granulomatous reaction.

Friend helper virus Mouse (lymphoid) leukaemia virus present in stocks of Friend virus, which was believed at one time to assist its replication. Molecular cloning of Friend virus has since shown that it is non-defective.

Friend murine erythroleukaemia cells Lines of mouse erythroblasts transformed by the Friend virus, that can be induced to differentiate terminally, producing haemoglobin, by various agents such as *DMSO*.

Friend murine leukaemia virus Murine leukaemia virus isolated by Charlotte Friend in 1956 whilst attempting to transmit the *Ehrlich ascites* tumour by cell-free extracts. Causes an unusual erythroblastosis-like leukaemia, in which anaemia is accompanied by large numbers of nucleated red cells in blood. Does not carry a host-derived oncogene, but seems to induce tumours by proviral insertion into specific regions of host genome.

Friend spleen focus-forming virus Defective virus found in certain strains of *Friend murine leukaemia virus*, detected by its ability to form foci in spleens of mice, and believed to be responsible in those strains for the production of a leukaemia associated with polycythemia rather than anaemia.

frontal zone contraction theory Model proposed to account for the movement of giant amoebae in which cytoplasmic contraction at the front of the leading pseudopod (fountain zone) pulls viscoelastic cytoplasm forward in the centre of the cell and forms a tube of more rigid cytoplasm immediately below the plasma membrane behind the active region. The peripheral contracted cytoplasm relaxes into a weaker gel at the rear and is pulled forward in its turn. Contrasts with the *ectoplasmic tube contraction* model.

frozen stocks Because cell lines tend to change their properties with continuous rounds of subculturing, it is common practice to keep stocks of cells frozen (either in liquid nitrogen or at -70°C) and to keep returning to this stock so that experiments are all carried out on cells of comparable passage number. The method also allows strains to be stored for long periods. Cells are usually frozen down in the presence of a cryoprotectant such as *DMSO* or glycerol. The method is also extensively used for storing semen for artificial insemination.

fructose A 6-carbon sugar (hexose) abundant in plants. Fructose has its reducing group (carbonyl) at C2, and thus is a ketose, in contrast to glucose which has its carbonyl at C1 and thus is an aldose. Sucrose, common table sugar, is the non-reducing disaccharide formed by an α-linkage from C1 of glucose to C2 of fructose (latter in furanose form). Fructose is a component of polysaccharides such as inulin, levan.

FSH See *follicle-stimulating hormone*.

fucose L-fucose (6-deoxy-L-galactose) is found as a constituent of N-glycan chains of glycoproteins; it is the only common L-form of sugar involved. D-fucose is usually encountered as a synthetic galactose analogue.

fucosyl transferase An enzyme catalyzing the transfer of fucosyl residues from the nucleotide sugar GDP-fucose.

fucoxanthin *Carotenoid* pigment of certain brown algae (*Phaeophyta*) and bacteria: absorbs at 500–580nm.

fumarate A dicarboxylic acid intermediate in the tricarboxylic acid cycle (TCA cycle). Can be derived from aspartate, phynylalanine and tyrosine for input into the TCA cycle.

fura-2 A fluorescent dye, used in measurement of intracellular free calcium levels.

furosemide (frusemide) Potent diuretic which increases the excretion of sodium, potassium and chloride ions, by inhibiting their resorption in the proximal and distal renal tubules.

G

G0 Used to describe the phase of the eukaryotic *cell cycle* in quiescent (non-dividing) cells. Either a distinct phase from *G1* or a prolonged G1, depending on opinion.

G1 Phase of the eukaryotic *cell cycle* between the end of cell division and the start of DNA synthesis, *S phase*. G stands for gap.

G2 Phase of the eukaryotic *cell cycle* between the end of DNA synthesis and the start of cell division.

G banding (Giemsa banding) Spreads of metaphase chromosomes, treated briefly with protease then stained with *Giemsa*, produce characteristic patterns of staining that allow identification of the separate chromosomes. The deeply-staining G bands do not coincide with the pattern of *Q bands*.

GABA (γ-amino butyric acid) A *neurotransmitter* (MW 103) widespread in *inhibitory chemical synapses*. GABA released from the presynaptic cell, in response to depoloarisation, opens channels on the *postsynaptic cell* that reduce its tendency to fire. Used medically in the treatment of epilepsy.

GAG See *glycosaminoglycan*.

G-actin Globular actin: see *actin*.

G-proteins See *GTP-binding proteins*.

gag-protein (group specific antigen) The protein of the nucleocapsid shell around the RNA of a retrovirus.

galactose Hexose identical to glucose except that orientation of –H and –OH on carbon 4 are exchanged. A component of *cerebrosides* and *gangliosides*, glycoproteins. *Lactose*, the disaccharide of milk, consists of galactose joined to glucose by a β(1–4)-glycosidic link.

galactose binding protein A bacterial periplasmic protein, most studied in *E. coli*, which acts both as a sensory element in the detection of galactose as a chemotactic signal, and in the uptake of the sugar.

galactosemia Inborn disorder (autosomal recessive in humans) in which the enzyme (galactose-1-phosphate uridyl transferase) that converts galactose-1-phosphate into glucose-1-phosphate is absent. Excess galactose-1-phosphate accumulates in the blood and a variety of problems result.

β-galactosidase (EC 3.2.1.23) Enzyme which catalyzes hydrolysis of β-glycosidic link between galactose and various hydroxyl donors. The enzyme from *E. coli* enables the bacterium to utilise lactose as its sole source of carbon. Studies of the induction of this enzyme by galactosides led Jacob and Monod to the operon model for regulation of protein synthesis.

galactosyl transferase Enzyme catalyzing the transfer of galactose units from the sugar-nucleotide, uridine diphosphogalactose (UDP-galactose) to an acceptor, commonly N-acetyl glucosamine in a glycan chain, forming a glycosidic bond involving C1 of galactose.

galanin Neuropeptide (29 amino acids) isolated from the upper small intestine of pig but subsequently found throughout the central and peripheral nervous system. Regulates gut motility and the activity of endocrine pancreas.

galvanotaxis The directed movement of cells induced by an applied voltage. This movement is almost always directed toward the cathode, occurs at fields around 1mV/mm, and is argued to be involved in cell guidance during *morphogenesis*, and in the repair of wounds. The term galvanotropism is used for neurons, since the cell body remains stationary and the neurites grow toward the cathode. Note that these processes involve *cell locomotion*, and are distinct from *cell electrophoresis*.

galvanotropism See *galvanotaxis*.

gametes Specialised haploid cells produced by meiosis and involved in sexual reproduction. Male gametes are usually small and motile (spermatozoa), whereas female gametes (oocytes) are larger and non-motile.

gametogenesis Process leading to the production of *gametes*.

gamma-toxin (γ-toxin) Toxin (33.4kD) produced by *Staphylococcus aureus*, having high haemolytic activity against rabbit erythrocytes and sensitive to inhibition by heparin, *Trypan blue* and sulphated polysaccharides in agar. On addition of a second component "A" (35kD), the toxin displays high haemolytic activity towards sheep and human erythrocytes.

ganglion A physical cluster of *neurons*. In vertebrates, the ganglia are appendages to the central nervous system; in invertebrates, the majority of neurons are organised as separate ganglia.

ganglion cell A cell within a *ganglion* or, in the retina, a type of *interneuron* which conveys information from the retinal *bipolar*, *horizontal* and *amacrine cells* to the brain.

ganglioside A *glycosphingolipid* which contains one or more residues of N-acetyl or other neuraminic acid derivatives. Gangliosides are found in highest concentration in cells of the nervous system, where they can constitute as much as 5% of the lipid.

gangliosidoses Diseases, such as *Tay–Sachs*, caused by inherited deficiency in enzymes necessary for the breakdown of gangliosides. Cause gross pathological changes in the nervous system, with devastating neurological symptoms.

gap junction A junction between two cells characterised by the presence of many pores that allow the passage of molecules up to about 900D. Each pore is formed by an hexagonal array of six transmembrane proteins (connexon) in each plasma membrane: when mated together the pores open allowing communication and the interchange of metabolites between cells. *Electrical synapses* are gap junctions, and metabolic cooperation depends upon the formation of gap junctions.

gargoyle cells Fibroblasts with large deposits of mucopolysaccharide, commonly found in storage diseases such as *Hurler's disease*.

gastric inhibitory polypeptide Peptide hormone (43 amino acids) that stimulates insulin release and inhibits the release of gastric acid and pepsin.

gastrin A group of peptide *hormones* secreted by the mucosal gut lining of some mammals in response to mechanical stress or high pH. They stimulate secretion of protons and pancreatic enzymes. Several different gastrins have been identified; human gastrin I has 16 amino acids (MW 2116). Gastrin is competitively inhibited by *cholecystokinin*.

gastrin-releasing peptide (GRP) A regulatory peptide (27 amino acids) thought to be the mammalian equivalent of *bombesin*. It elicits gastrin release, causes bronchoconstriction and vasodilation in the respiratory tract, and stimulates the growth and mitogenesis of cells in culture.

gastrula Embryonic stage of an animal when *gastrulation* occurs; follows *blastula* stage.

gastrulation During embryonic development of most animals a complex and co-ordinated series of cellular movements occurs at the end of *cleavage*. The details of these movements, gastrulation, vary from species to species, but usually result in the formation of the three primary germ layers, *ectoderm, mesoderm* and *endoderm*.

gated ion channel *Transmembrane proteins* of excitable cells, which allow a flux of ions to pass only under defined circumstances. Channels may be either *voltage-gated*, such as the *sodium channel* of neurons, or *ligand-gated* such as the acetylcholine receptor of *cholinergic neurons*. Channels tend to be relatively ion-specific and allow fluxes of typically 1000 ions to pass in around 1ms; they are thus much faster at moving ions across a membrane than transport *ATPases*.

Gaucher's disease Familial defect of glucocerebrosidase (β-glucosidase), most common in Ashkenazi Jews. Associated with hepatosplenomegaly and, in severe early onset forms of the disease, with neurological dysfunction. Autosomal, recessive.

gel Jelly-like material formed by the coagulation of a colloidal liquid. Many gels have a fibrous matrix and fluid-filled interstices: gels are viscoelastic rather than simply viscous and can resist some mechanical stress without deformation. Examples are the gels formed by large molecules such as collagen (and gelatin), agarose, polyacrylamide and starch.

gel electrophoresis *Electrophoresis* using a gel supporting-phase. Usually applied to systems where the gel is based on polyacrylamide.

gel filtration An important method for separating molecules according to molecular size by percolating the solution through beads of solvent-permeated polymer that has pores of similar size to the solute molecules. Unlike a continous filter which retards flow according to molecular size, separation is achieved because molecules which can enter the beads take a longer path (ie. are retarded) than those which cannot. Typical gels for protein separation are made from polyacrylamide, or from flexible (Sephadex) or rigid (agarose, Sepharose) sugar polymers. The size separation range is determined by the degree of cross-linking of the gel.

gelatin Heat-denatured collagen.

gelatinous lesion A small area of oedema in the arterial intima, possibly a precursor of a *fibrous plaque*.

gelsolin Actin-binding protein (90kD) which nucleates actin polymerisation, but at high calcium ion concentrations (10^{-6}M) causes severing of filaments.

gemfibrozil Drug which lowers plasma lipoprotein levels.

gene Originally defined as the physical unit of heredity but the meaning has changed with increasing knowledge. A term best used as an inclusive, non-specific way of referring to any genetic determinant. Exactly what (at the DNA level) the term refers to depends upon the particular experimental approach through which the genetic determinant is identified. See *allele*, *cistron*, *exon*, *intron*, *locus*.

gene amplification Selective replication of DNA sequence within a cell, producing multiple extra copies of that sequence. The best-known example occurs during the maturation of the oocyte of *Xenopus*, where the set (normally 500 copies) of ribosomal RNA genes is replicated some 4,000 times to give about 2,000,000 copies.

gene cloning The insertion of a DNA sequence into a *vector* which can then be propagated in a host organism, generating a large number of copies of the sequence.

gene conversion A phenomenon in which alleles are segregated in a 3:1 not 2:2 ratio in meiosis. May be a result of *DNA polymerase* switching templates and copying from the other homologous sequence, or a result of mismatch repair (nucleotides being removed from one strand and replaced by repair synthesis using the other strand as template).

gene dosage Number of copies of a particular *gene locus* in the genome; in most cases either one or two.

gene expression The full use of the information in a gene *via transcription* and *translation* leading to production of a protein and hence the appearance of the *phenotype* determined by that gene. Gene expression is assumed to be controlled at various points in the sequence leading to protein synthesis and this control is thought to be the major determinant of cellular *differentiation* in eukaryotes.

gene regulatory protein Any protein which interacts with DNA sequences of a *gene* and controls its transcription.

generation time Time taken for a cell population to double in numbers, and thus equivalent to the average length of the *cell cycle*.

genetic code The matrix of interactions between nucleotide base triplets in messenger RNA (*codons*) and triplet anticodons in transfer RNA, which enable a sequence of bases in nucleic acid to be translated during protein synthesis into a specific sequence of amino acids in a protein. See Table C3.

genetic engineering General term covering the use of various experimental techniques to produce molecules of DNA containing new *genes* or novel combinations of genes, usually for insertion into a host cell for cloning.

genetic linkage The term refers to the fact that certain genes tend to be inherited together, because they are on the same chromosome. Thus parental combinations of characters are found more frequently in offspring than non-parental. Linkage is measured by the percentage recombination between *loci*, unlinked genes showing 50% recombination. See *linkage equilibrium*, *linkage disequilibrium*.

genetic load (genetic burden) The decrease in fitness of a population (as a result of selection acting on phenotypes) due to deleterious mutations in the population gene pool.

genetic locus The position of a *gene* in a linkage map or on a chromosome.

genetic recombination Formation of new combinations of alleles in offspring (viruses, cells or organisms) as a result of exchange of DNA sequences between molecules. It occurs naturally in meiosis as a result of independent *segregation* of chromosome pairs and of *crossing over* between homologous chromosomes. Other forms of genetic recombination include *gene conversion*, somatic crossing over, sister-strand exchange and site-specific exchange.

genetic transformation Genetic change brought about by the introduction of exogenous DNA into a cell. See *transformation*.

genome The total set of genes carried by an individual or cell.

genomic library A collection of DNA molecules, derived from *restriction fragments* that have been cloned in *vectors*, which includes all or part of the genetic material of an organism.

genotype The genetic constitution of an organism or cell, as distinct from its expressed features or *phenotype*.

gentamycin A group of aminoglycoside antibiotics produced by *Micromonospora* spp. Members include the closely related gentamycins C1, C2 and C1a, together with gentamycin A. They inhibit protein synthesis on 70S ribosomes by binding to the 23S core protein of the small subunit, which is responsible for binding mRNA. Mode of action similar to that of kanamycin, neomycin, paromomycin, spectinomycin and streptomycin. Active against some strains of the bacterium *Pseudomonas aeruginosa*.

geotaxis See *gravitaxis*. The prefix gravi- is preferable since the gravitational fields used as cues need not necessarily be the Earth's.

geotropism See *gravitropism*.

GERL Old name for *trans-Golgi network*. The Golgi-Endoplasmic Reticulum-Lysosome system. See also individual entries for each of these membranous compartments.

germ cell Cell specialised to produce *haploid gametes*. The germ cell line is often formed very early in embryonic development.

germ layers The main divisions of tissue types in multicellular organisms. Diploblastic organisms (eg. coelenterates) have two layers, ectoderm and endoderm; triploblastic organisms (all higher animal groups) have mesoderm between these two layers. Germ layers become distinguishable during late blastula/early gastrula stages of embryogenesis, and each gives rise to a characteristic set of tissues, the ectoderm to external epithelia and to the nervous system for example, although some tissues contain elements derived from two layers.

GFAP Glial fibrillary acidic protein, a member of the family of *intermediate filament* proteins, characteristic of glial cells.

ghosts See *erythrocyte ghosts*

GI Common abbreviation for gastrointestinal.

giant axons Extraordinarily large unmyelinated axons found in invertebrates. Some, like the squid giant axon, can approach 1mm diameter. Large axons have high conduction speeds; the giant axons are invariably involved in panic or escape responses, and may (eg. crayfish) have *electrical synapses* to further increase speed. Vertebrate axons with high conduction velocities are much narrower: they are myelinated, allowing *saltatory conduction*.

gibberellic acids Diterpenoid compounds with *gibberellin* activity in plants. At least 70 related gibberellic acids have been described and designated as a series GA1, GA2 etc. Not all the C20 or C19 diterpenoids which have the appropriate structural features to be considered as gibberellic acids have biological activity.

gibberellin *Plant growth substance* involved in promotion of stem elongation, mobilisation of food reserves in seeds, and other processes. Its absence results in the dwarfism of some plant varieties. Chemically all known gibberellins are *gibberellic acids*, and may be converted to the acid before acting.

Giemsa A Romanovsky-type stain that is often used to stain blood films which are believed to contain protozoan parasites. Contains both basic and acidic dyes and will therefore differentiate acid and basic granules in granulocytes.

GIP See *gastric inhibitory polypeptide*.

glands Organs specialised for secretion by the infolding of an epithelial sheet. The secretory epithelial cells may either be arranged as an acinus with a duct or as a tubule. Glands from which release occurs to a free epithelial surface are exocrine; those which release product to the circulatory system are endocrine glands.

glandular fever Self-limiting disorder of lymphoid tissue usually caused by infection with *Epstein Barr virus* (infectious mononucleosis). Characterised by the appearance of many large lymphoblasts in the circulation.

Glanzmann's thrombasthenia Platelet

dysfunction in which aggregation is deficient. A specific glycoprotein complex (IIb/IIIa) is absent from the plasma membrane: this seems to be the fibronectin/fibrinogen receptor and is an *integrin* similar to the LFA-1 class of membrane glycoproteins found on leucocytes.

glial cells Specialised cells which surround *neurons*, providing mechanical and physical support, and electrical insulation between neurons. See *astrocyte*, *oligodendrocyte* and *microglia*.

glial fibrillary acidic protein (GFAP) Protein (50kD) of the intermediate filaments of glial cells.

glial filaments Intermediate filaments of glial cells, made of *glial fibrillary acidic protein*.

gliomas Neuroectodermal tumours of neuroglial origin: include astrocytomas, oligodendroglioma and ependymoma, derived from astrocytes, oligodendrocytes and ependymal cells respectively. All infiltrate the adjacent brain tissue, but they do not metastasise.

globin The polypeptide moiety of haemoglobin. In the adult human the haemoglobin molecule has two α (141 residues) and two β (146 residues) globin chains.

globular protein Any protein which adopts a compact morphology is termed globular. Generally applied to proteins in free solution, but may also be used for compact folded proteins within membranes.

glomerulonephritis Inflammatory disease of the kidney glomerulus, often characterised by the presence of immune complexes which cannot pass through the basement membrane of the fenestrated epithelium where plasma filtration occurs. Circulating neutrophils are trapped on the accumulated adhesive immune complexes (which also activate complement). Immune complex tends to be irregularly distributed in contrast to the picture in *Goodpasture's syndrome*.

glucagon A polypeptide hormone (MW 3485) secreted by the *A cells* of the *Islets of Langerhans* in response to a fall in blood sugar levels. Induces *hyperglycaemia*. Physiological antagonist of insulin.

glucans Glucose-containing polysaccharides, including cellulose, *callose*, *laminarin*, *starch*, and *glycogen*.

glucocorticoids Steroid hormones (both natural and synthetic) which promote gluconeogenesis and the formation of glycogen at the expense of lipid and protein synthesis. They also have important anti-inflammatory activity. Type compound is hydrocortisone (cortisol), other common examples are cortisone, prednisone, prednisolone, dexamethasone, betamethasone.

glucomannan *Hemicellulosic* plant cell-wall polysaccharide containing glucose and mannose linked by β(1–4)-glycosidic bonds. May contain some side-chains of galactose, in which case it may be termed galactoglucomannan. A major polysaccharide of gymnosperm cell walls (softwood).

gluconeogenesis Synthesis of glucose from non-carbohydrate precursors, such as pyruvate, amino acids and glycerol. Takes place largely in liver, and serves to maintain blood glucose under conditions of starvation or intense exercise.

glucosamine Amino-sugar (2-amino-2-deoxyglucose); component of *chitin*, heparan sulphate, chondroitin sulphate, and many complex polysaccharides. See also *N-acetyl glucosamine*.

glucose (dextrose) Six-carbon sugar (aldohexose) widely distributed in plants and animals. Breakdown of glucose (*glycolysis*) is a major energy source for metabolic processes. In green plants, glucose is a major product of photosynthesis, and is stored as the polymer *starch*. In animals it is obtained chiefly from dietary di- and polysaccharides, but also by *gluconeogenesis*, and is stored as *glycogen*. Storage polymer in microorganisms is *dextran*.

glucose-1-phosphate Product of glycogen breakdown by phosphorylase. Converted to glucose-6-phosphate by phosphoglucomutase.

glucose-6-phosphate Phosphomonoester of glucose which is formed by transfer of phosphate from ATP, catalyzed by the enzyme *hexokinase*. It is an intermediate both of the glycolytic pathway (next converted to fructose-6-phosphate), and of the NADPH-generating *pentose phosphate pathway*.

glucose-related protein (GRP) One of the stress-related proteins: identical to *endoplasmin*.

β-glucosidase (EC 3.2.1.21) Enzyme

catalysing the release of glucose by hydrolysis of the glycosidic link in various β-D-glucosides, R-β-D-glucose, where the group R may be alkyl, aryl, mono- or oligosaccharide. Favoured source: almonds, from which the enzyme is known as emulsin.

glucosylation Transfer of glucose residues, usually from the nucleotide-sugar derivative UDPG. Enzymic glucosylation to generate the glucosyl-galactosyl disaccharide on the hydroxylysine of collagen is a normal process as is the glucosylation of hydroxymethyl cytosine in phage T2. A recent theory suggests that glucosylation of certain long-lived proteins by a non-enzymic reaction with free glucose may contribute to ageing.

glucuronoxylan *Hemicellulosic* plant cellwall polysaccharide containing glucuronic acid and xylose as its main constituents. Has a β(1–4)-xylan backbone, with 4–0-methylglucuronic acid side-chains. Arabinose and acetyl side-chains may also be present. Major polysaccharide of angiosperm cell walls (hardwood).

glutamic acid (Glu; E; MW 147) One of the 20 α-amino acids commonly found in proteins. Plays a central role in amino acid metabolism, acting as precursor of *glutamine*, *proline* and *arginine*. Also acts as amino group donor in synthesis by transamination of alanine from pyruvate, and aspartic acid from oxaloacetate. Glutamate is also a neurotransmitter; the product of its decarboxylation is the inhibitory neurotransmitter *GABA*. See also Table A2.

glutamine (Gln; Q; MW 146) One of the 20 amino acids commonly found (and directly coded for) in proteins. It is the amide at the γ-carboxyl of the amino acid *glutamate*. Glutamine can participate in covalent cross-linking reactions between proteins, by forming peptide-like bonds by a transamidation reaction with lysine residues. This reaction, catalyzed by clotting Factor XIII stabilises the aggregates of fibrin formed during blood-clotting. Media for culture of animal cells contain some 10 times more glutamine than other amino acids, the excess presumably acting as a carbon source. One problem in media is that glutamine is unstable and is rapidly depleted; it is usual to add glutamine separately from frozen stock. See also Table A2.

glutaraldehyde A dialdehyde used as a fixative, especially for electron microscopy.

By its interaction with amino groups (and others) it forms cross-links between proteins.

glutathione Naturally occurring tri-peptide (gamma-glutamyl-cysteinyl-glycine) which acts as a biological redox agent and sulphydryl buffer within the cytoplasm and also as a cofactor, coenzyme or substrate in some reactions. It protects against the toxicity of hydrogen peroxide and organic peroxides.

glyceraldehyde-3-phosphate Three-carbon intermediate of the glycolytic pathway formed by the cleavage of fructose 1,6-bisphosphate, catalyzed by the enzyme aldolase. Also involved in reversible interchange between *glycolysis* and the *pentose phosphate pathway*.

glycerination Permeabilisation of the plasma membrane of cells by incubating in aqueous glycerol at low temperature. The technique was first applied to muscle which, once glycerinated, can be made to contract by adding exogenous ATP and calcium.

glycerol A metabolic intermediate, but primarily of interest as the central structural component of the major classes of biological lipids, triglycerides and phosphatidyl phospholipids. Also used as a *cryoprotectant*.

glycine (Gly; G; MW 75.1) The simplest amino acid. It is a common residue in proteins, especially collagen and elastin, and is not optically active. See Table A2.

glycocalyx The region, seen by electron microscopy, outwith the outer dense line of the plasmalemma rich in glycosidic compounds such as proteoglycans and glycoproteins. Since these molecules are often integral membrane proteins and may be denatured by the processes of fixation for electron microscopy, it might be better to avoid the term or to refer to membrane glycoproteins or to proteoglycans associated with the cell surface.

glycogen Branched polymer of D-glucose (mostly α(1–4)-linked, but some α(1–6) at branch points). Size range very variable, up to 10^5 glucose units. Major short-term storage polymer of animal cells, and is particularly abundant in the liver and to a lesser extent in muscle. In the electron microscope glycogen has a characteristic "asterisk/star" appearance.

glycolic acid Hydroxyacetic acid; found in young plants and green fruits. Glycolate is formed from ribulose-1,5-bisphosphate in a seemingly wasteful side reaction of photosynthesis, known as photorespiration.

glycolipid Oligosaccharides covalently attached to lipid as in the glycosphingolipids (GSL) found in plasma membranes of all animal and some plant cells. The lipid part of GSLs is sphingosine in which the amino group is acylated by a fatty chain, forming a *ceramide*. Most of the oligosaccharide chains belong to one of four series, the ganglio-, globo-, lacto-type 1 and lacto-type 2 series. Blood group antigens are GSLs.

glycolysis The conversion of a monosaccharide (generally glucose) to pyruvate *via* the glycolytic pathway (ie. the *Embden–Meyerhof pathway*) in the cytosol. Generates ATP without consuming oxygen and is thus anaerobic.

glycophorins A class of abundant transmembrane glycoproteins of the human erythrocyte. The major component is a 131 residue peptide chain which is highly O-glycosylated and is rich in terminal sialic acid. The peptide chain carries the *MN blood group antigens* at its N-terminus.

glycoprotein Proteins with covalently attached sugar units, either bonded via the OH group of serine or threonine (O-glycosylated) or through the amide NH_2 of asparagine (N-glycosylated). Includes most secreted proteins (serum albumin is the major exception) and proteins exposed at the outer surface of the plasma membrane. Sugar residues found include:- mannose, N-acetyl glucosamine, N-acetyl galactosamine, galactose, fucose and sialic acid.

glycosaminoglycans (old name = mucopolysaccharides) Polysaccharide side-chains of *proteoglycans* made up of repeating disaccharide units (more than 100) of amino sugars, at least one of which has a negatively charged side-group (carboxylate or sulphate). Commonest are hyaluronate (D-glucuronic acid-N-acetyl D-glucosamine: MW up to 10 million), chondroitin sulphate (D-glucuronic acid-N-acetyl D-galactosamine -4- or -6-sulphate), dermatan sulphate (D-glucuronic acid- or L-iduronic acid-N-acetyl D-galactosamine), keratan sulphate (D-galactose-N-acetyl D-glucosamine-sulphate) and heparan sulphate (D-glucuronic acid- or L-iduronic acid-N-acetyl D-glucosamine). Glycosaminoglycan side chains (with the exception of hyaluronate) are

covalently attached to a core protein at about every 12 amino acid residues to produce a proteoglycan; these proteoglycans are then non-covalently attached by link proteins to hyaluronate, forming an enormous hydrated space-filling polymer found in *extracellular matrix*. The extent of sulphation is variable and the structure allows tremendous diversity.

glycosidase (glycosylase) General and imprecise term for an enzyme which degrades the linkage between a sugar and another molecule. The sugars may be subunits of a polysaccharide or may be linked to nucleic acid bases, for example. The enzymes may distinguish between α- and β-links but they are not very substrate-specific.

glycosome Microbody containing glycolytic enzymes, found in protozoa of the Kinetoplastida (eg. trypanosomes).

glycosphingolipids *Ceramide* derivatives containing more than one sugar residue. If sialic acid is present these are called *gangliosides*.

glycosyl transferase Enzyme that catalyzes the transfer of a sugar (monosaccharide) unit from a sugar nucleotide derivative to a sugar or amino acid acceptor.

glycosylase Common term for what should, more correctly, be a *glycosidase*.

glycosylation The process of adding sugar units such as in the addition of glycan chains to proteins.

glyoxisome Organelle found in plant cells and some protozoa, containing the enzymes of the *glyoxylate cycle*. Also contains catalase and enzymes for β-*oxidation* of fatty acids. Together with the *peroxisome* and *glycosome* makes up the class of organelles known as *microbodies*.

glyoxylate cycle Metabolic pathway present in bacteria and in the *glyoxysome* of plants, in which two acetyl-CoA molecules are converted to a 4-carbon dicarboxylic acid, initially succinate. Includes two enzymes not found elsewhere, isocitrate lyase and malate synthase. Permits net synthesis of carbohydrates from lipid, and hence is prominent in those seeds in which lipid is the principal food reserve.

glypiation A common modification of the C-terminus of membrane-attached proteins in which a phosphatidyl inositol moiety is linked through glucosamine and mannose to a phosphoryl ethanolamine residue which is linked to the C-terminal amino acid of the protein by its amino group. Glypiation is the sole means of attachment of such proteins to the membrane. The name derives from the glycosyl phosphatidyl inositol (PI) tail which is added. May be the extracellular equivalent of myristoylation.

GM-CSF (granulocyte-macrophage colony stimulating factor) A cytokine which stimulates the formation of granulocyte or macrophage colonies from myeloid stem cells isolated from bone marrow.

Gm-types Genetically determined allotypic antigens found on the γ chain of human IgG. See *allotype*.

goblet cell 1. Cell of the epithelial lining of the intestine which secretes mucus and has a very well-developed Golgi apparatus. 2. Cell type characteristic of larval lepidopteran midgut, containing a potent K$^+$-*ATPase*, and thought to be involved in maintenance of ion and pH gradients.

Golgi apparatus Intracellular stack of membrane-bounded vesicles in which glycosylation and packaging of secreted proteins takes place. Also known as the Golgi body, Golgi vesicles; in plants, the *dictyosome*; in flagellate protozoa, the parabasal body.

Gomori procedure Cytochemical staining procedure used to localise acid phosphatases. Depends upon the production of phosphate ions from organic phospho-esters such as β-glycerophosphate. The phosphate in the presence of lead ions causes the formation of a precipitate of lead salt which is converted to the brown sulphide of lead by the action of yellow ammonium sulphide.

gonadotrophins A group of glycoprotein hormones from the anterior lobe of the pituitary gland. They stimulate growth of the gonads and the secretion of sex hormones. Examples: *follicle stimulating hormone*, *luteinising hormone*, *chorionic gonadotrophin*.

Goodpasture's syndrome A disease of lungs and kidneys in which there is accumulation of a very uniform layer of autoantibodies to basement membrane on the kidney glomerular basement membrane.

gout Recurrent acute arthritis of peripheral joints caused by the accumulation of monosodium urate crystals. Usually due to overproduction of uric acid but may be a result of under-excretion. The problems partly arise because neutrophils release lysosomal enzymes as a result of damage to the phagosome membrane by ingested crystals: colchicine acts to reduce the attack by inhibiting lysosome-phagosome fusion.

Graafian follicle Final stage in the differentiation of follicles in the mammalian ovary. Consists of a spherical fluid-filled blister on the surface of the ovary which bursts at ovulation to release the oocyte.

gradient perception Problem faced by a cell which is to respond directionally to a gradient of, for example, a diffusible attractant chemical. In a spatial mechanism the cell would compare receptor occupancy at different sites on the cell surface; a temporal mechanism would involve comparison of concentrations at different times, the cell moving randomly between readings. In *pseudospatial sensing*, the cell would detect the gradient as a consequence of positive feedback to protrusive activity if receptor occupancy increased with time as the protrusion moved up-gradient. Few cell types have been unambiguously shown to detect gradients.

graft-versus-host response (GVH) Immunological reaction of grafted lymphocytes against host antigens.

Gram negative bacteria Bacteria which fail to retain the *Gram stain*. They have thin *peptidoglycan* walls bounded by an outer membrane containing *endotoxin* (lipopolysaccharide).

Gram positive bacteria Bacteria which retain the *Gram stain*. They have thick cell walls containing *teichoic* and *lipoteichoic acid* complexed to the *peptidoglycan*.

Gram stain Procedure developed by Christian Gram in the 19th century. A heat-fixed bacterial smear is stained with crystal violet (methyl violet), treated with 3% iodine/potassium iodide solution, washed with alcohol and counterstained. The method differentiates bacteria into two main classes, *Gram positive* and *Gram negative*. Certain bacteria, notably mycobacteria, which have walls with high lipid content, show a different stain reaction referred to as acid-fast staining — the stain (which is different from Gram stain) resists decolouration in strong acid.

gramicidin A Antibiotic from *Bacillus brevis* consisting of a linear peptide of alternate D- and L-amino acids which acts as a cation *ionophore* in lipid bilayer membranes. It is proposed that two molecules form a membrane-spanning helix containing a pore lined with polar residues.

granular component of nucleolus Area of nucleolus which appears granular in the electron microscope and contains 15nm diameter particles which are maturing ribosomes. Contrasts with the pale-staining and *fibrillar regions*.

granulation tissue Highly vascularised tissue which replaces the initial fibrin clot in a wound. Vascularisation is by ingrowth of capillary endothelium from the surrounding vasculature. The tissue is also rich in fibroblasts (which will eventually produce *fibrous tissue*) and leucocytes.

granule cell Type of neuron found in the cerebellum.

granulocyte Leucocyte with conspicuous cytoplasmic granules. In humans the granulocytes are also classified as polymorphonuclear leucocytes and are subdivided according to the staining properties of the granules into *eosinophils*, *basophils* and *neutrophils* (using a Romanovsky type stain); some invertebrate blood cells (granular haemocytes) are also referred to as granulocytes.

granulocyte/macrophage colony stimulating factor See *GM-CSF*.

granulocytopenia Low granulocyte number in circulating blood.

granuloma Chronic inflammatory lesion characterised by large numbers of cells of various types (macrophages, lymphocytes, fibroblasts, giant cells), some degrading and some repairing the tissues.

granulopoiesis The production of granulocytes in the bone marrow.

granum (*pl.* grana) Stack of *thylakoids* in the chloroplast, containing the pigments of the *light-harvesting system* and the enzymes responsible for the *light dependent reactions* of photosynthesis.

gravitaxis Directed locomotory response to gravity.

gravitropism Directional growth of a plant organ in response to a gravitational field — roots grow downwards, shoots grow upwards. Achieved by differential growth on the sides of the root or shoot. A gravitational field is thought to be sensed by sedimentation of statoliths (starch grains) in root caps.

gray See *grey*.

Green algae See *Chlorophyta*.

gregarine movement Peculiar gliding movement shown by gregarines (Protozoa), the mechanism of which is poorly understood.

grex The multicellular aggregate formed by cellular slime moulds (*Acrasidae*): the slug-like grex migrates, showing positive phototaxis and negative gravitaxis, until culmination (the formation of a fruiting body) takes place. Coordination of the activities of the hundreds of thousands of individual amoebae that compose the grex may involve pulses of cyclic AMP in *Dictyostelium discoideum*, a species in which cAMP is the chemotactic factor for aggregation.

grey crescent A region near the equator of the surface in the fertilised egg of various amphibia, often of greyish colour, which appears to contain special morphogenetic properties.

ground tissues Plant tissues other than those of the vascular system and the *dermal tissues*. Undifferentiated cells which are not involved in specialised functions.

growth cone A specialised region at the tip of a growing *neurite*, which is responsible for sensing the local environment and moving toward the neuron's target cell. Growth cones are hand-shaped, with several long *filopodia* which differentially adhere to surfaces in the embryo. Growth cones can be sensitive to several guidance cues, for example, surface adhesiveness, *growth factors*, *neurotransmitters* and electric fields (*galvanotropism*).

growth control When applied to cells usually means control of growth of the population, ie. of the rate of division rather than of the size of an individual cell.

growth factors Polypeptide hormones which regulate the division of cells, for example EGF (epidermal growth factor),

PDGF (platelet-derived GF), FGF (fibroblast GF). Insulin and somatomedin are also growth factors; the status of NGF (*nerve growth factor*) as a growth factor, as opposed to a trophic factor, is more uncertain. Perturbation of growth factor production or of the response to growth factor is important in neoplastic transformation. See Table H2.

growth hormone (somatotropin) Polypeptide (191 amino acids) produced by anterior pituitary which stimulates liver to produce *somatomedin* 1 (insulin-like growth factor 1) but not somatomedin 2. Release is controlled by the hypothalamus.

growth substances See *plant growth substances*.

GRP See *glucose-related protein* or *gastrin-releasing peptide*.

GTP (guanosine 5'-triphosphate) Like *ATP* a source of phosphorylating potential, but is separately synthesised and takes part in a limited, distinct set of energy-requiring processes. GTP is required in RNA and protein synthesis, the assembly of *microtubules*, and for the activation of regulatory *GTP-binding proteins*.

GTP-binding proteins (G-proteins) Important intermediaries in signal transduction systems. Specifically a class of trimeric αβγ proteins in which the Gα unit has a GTP-binding site with slow GTPase activity. The α subunit is triggered to release GDP and bind GTP as a result of the primary signal (eg. hormone receptor interaction). GTP binding causes release of the Gα unit from the complex and allows it to modulate the rate at which an effector enzyme generates a *second messenger*. The GTP is slowly hydrolysed by intrinsic GTPase to give GDP and reassociation of the G protein subunits. Gα subunits have a covalently-linked myristoyl residue. G proteins are subdivided into groups Gs (stimulatory), Gi (inhibitory), Go and Gx classes which differ in structure and function of the α subunit. See Table G1.

guanine (2-amino-6-hydroxy purine) One of the constituent bases of nucleic acids, nucleosides and nucleotides.

guanosine (9-β-D-ribofuranosyl guanine) The nucleoside formed by linking ribose to guanine.

guanylate cyclase Enzyme catalyzing the synthesis of guanosine 3',5'- cyclic monophosphate from guanosine 5'-triphosphate.

guard cells Plant cells occurring in pairs enclosing a pore in the *epidermis*, the whole (guard cells and pore) constituting a *stoma*. Changes in *turgor* pressure in the guard cells cause the stoma to open and close.

Guarnieri body Acidophilic inclusion body found in cells infected with *vaccinia* virus.

guidance See *contact guidance*.

Guillain–Barre syndrome (Landry–G–B syndrome) Acute infective polyneuritis, associated with *cytomegalovirus* infection, in which there is cell-mediated immunity to a component of myelin; the disease may be autoimmune in origin. Causes a temporary paralysis, particularly of the extremities.

gut-associated lymphoid tissue (GALT) Peripheral lymphoid organ consisting of lymphoid tissue associated with the gut (Peyer's patches, tonsils, mesenteric lymph nodes and the appendix).

Table G1. The properties of G protein subunits

Subunit G	MW (kD)	Signal detector	Effector	Toxin	Comments
sα	46	β-adrenergic, glucagon and many other receptors	Activates adenylate cyclase	Cholera	Molecular masses determined from cDNA. Two forms run on SDS-PAGE gels with M_s of 52,000 and 45,000 respectively. Relative concentrations of two $G_{s\alpha}$ forms varies from one cell type to another. Thought to be alternative splicing products of one gene.
sα	44.5				
iα1	40.4	α-adrenergic, muscarinic cholinergic, opiate and many other receptors	Inhibits adenylate cyclase (Others?)	Pertussis	The sequence for $G_{i\alpha1}$ and $G_{i\alpha2}$ were derived from two different cDNA clones. The relative importance of the two forms, in the cell, is not yet clear
iα2	40.5	?	?	Pertussis	
oα	39	Unknown	Unknown	Pertussis	"o" stands for other G protein. It was first detected in large amounts in brain and its function is unknown
tα1	40	Rhodopsin	Retinal cGMP:Phosphodiesterase	Cholera Pertussis	Subunit of Transducin, found in the rod cells of the retina
tα2	40.4	—	cGMP:Phosphodiesterase?	Cholera Pertussis	Found in the cone cells of the retina: probably the α subunit of the cone's analogue of Transducin
β36	36				β subunits are functionally interchangeable in vitro. $G_{\beta36}$ and $G_{\beta35}$ are distinguishable on SDS-PAGE gels and may be found in one cell
β35	35				
t β	37.4				Single form of Transducin β subunit. Molecular mass determined from cDNA
γ	8.4				Although γ subunits are functionally interchangeable in vitro, more than one form may exist. Molecular mass given was determined from cDNA for the Transducin γ subunit

H

H-chain Heavy chain of immunoglobulin; see *IgG*, *IgM*, etc.

H2 antigen An antigen of the H2 region of the major histocompatibility complex of mice. Divided into Class I and Class II antigens. See *histocompatibility antigen*.

H2 complex Mouse equivalent of the human MHC (*major histocompatibility complex*) system, a set of genetic loci coding for Class I and Class II MHC antigens and for comple-ment components.

haem (USA heme) Compounds of iron complexed in a porphyrin (tetrapyrrole) ring which differ in side chain composition. Haems are the prosthetic groups of *cytochromes* and are found in most oxygen carrier proteins.

haemagglutination (USA hemagglutination) Agglutination of red blood cells, often used to test for the presence of antibodies directed against red-cell surface antigens, carbohydrate-binding proteins or viruses in a solution. Requires that the agglutinin has at least two binding sites.

haemagglutinin (USA hemagglutinin) Substance that will bring about the *agglutination* of erythrocytes.

haemangioblast (USA hemangioblast) Earliest mesodermal precursor of both blood and vascular endothelial cells. Described in embryonic yolk-sac blood-islands of birds.

haematopoiesis (USA hematopoesis) Production of blood cells involving both proliferation and differentiation from stem cells. In adult mammals usually occurs in bone marrow.

haematopoietic stem cell (USA hematopoetic stem cell) Cell which gives rise to distinct daughter cells, one a replica of the stem cell, one a cell which will further proliferate and differentiate into a mature blood cell. Pluripotent stem cells can give rise to all lineages, committed stem cells (derived from the pluripotent stem cell) only to some.

haematoxylin (USA hematoxylin) Basic stain which gives a blue colour (to the nucleus of a cell for example), commonly used in conjunction with eosin which stains the cytoplasm pink/red. Various modifications of haematoxylin have been developed. The histopathologist's "H & E" is haematoxylin and eosin.

haemocyanins (USA hemocyanins) Blue, oxygen-transporting, copper-containing protein found in the blood of molluscs and crustacea. A very large protein with 20–40 subunits and molecular weight of 2–8 million, and having a characteristic cuboidal appearance under the electron microscope. Prior to the introduction of immunogold techniques, it was used for electron-microscopic localisation by coupling to antibody.

haemocytes (USA hemocytes) Blood cells, associated with a haemocoel, particularly those of insects and crustacea. Despite the name they are more leucocyte-like, being phagocytic and involved in defence and clotting of haemolymph, and not involved in transport of oxygen.

haemoglobin (USA hemoglobin) Four-subunit globular oxygen-carrying protein of vertebrates and some invertebrates. There are two α and two β chains (very similar to myoglobin) in adult humans; the haem moiety (an iron-containing substituted porphyrin) is firmly held in a non-polar crevice in each peptide chain.

haemoglobinopathies (USA hemoglobinopathies) Disorders due to abnormalities in the haemoglobin molecule, the best known being sickle-cell anaemia in which there is a single amino acid substitution (valine for glutamate) in position 6 of the β chain. In other cases one of the globin chains is synthesised at a slower rate, despite being normal in structure. See also *thassalemia*.

haemolysins (USA hemolysins) Bacterial *exotoxins* which lyse erythrocytes.

haemolysis (USA hemolysis) Leakage of haemoglobin from erythrocytes due to membrane damage.

haemolytic anaemia (USA hemolytic anemia) *Anaemia* resulting from reduced red-cell survival time, either because of an intrinsic defect in the erythrocyte (hereditary spherocytosis or ellipsocytosis, enzyme defects, *haemoglobinopathy*), or an extrinsic damaging agent, for example autoantibody (autoimmune haemolytic anaemia), iso-antibody, parasitic invasion of the cells (malaria), bacterial or chemical haemoly-sins, mechanical damage to erythrocytes.

haemopexin (USA hemopexin) Single-chain haem-binding plasma β_1-glycoprotein (57kD) — unlike *haptoglobin* does not bind haemoglobin. Present at around 1mg/ml in plasma.

haemophilia (USA hemophilia) Sex-linked recessive deficiency of blood-clotting system, usually of Factor VIII.

Haemophilus influenzae Bacterium sometimes associated with influenza virus infections, causes pneumonia and men-ingitis.

haemosiderin (USA hemosiderin) A mam-malian iron-storage protein related to *ferri-tin* but less abundant.

haemostasis (USA hemostasis) Arrest of bleeding through blood clotting and contrac-tion of the blood vessels.

Hagemann factor Plasma β-globulin (110kD), blood-clotting Factor XII, which is activated by contact with surfaces to form Factor XIIa, which in turn activates Factor XI. Factor XIIa also generates *plasmin* from plasminogen and *kallikrein* from prekallik-rein. Both plasmin and kallikrein activate the complement cascade. Hagemann factor is important both in clotting and activation of the inflammatory process.

hair cells 1. Cells found in the epithelial lining of the labyrinth of the inner ear. The hairs are *stereocilia* (stereovilli) up to 25μm long that restrict the plane in which defor-mation of the apical membrane of the cell can be brought about by movement of fluid or by sound. Movement of the single ste-reovillus transduces mechanical movements into electrical receptor potentials. 2. *Bot.* Many plant surfaces are covered with fine hairs (*Tradescantia* stamens are a common source); the hairs are made up of thin-walled cells which are convenient for studying cyto-plasmic streaming and for observing mitosis. Hairs may be unicellular or multicellular, simple or glandular.

half-life ($t_{1/2}$) The period over which the activ-ity or concentration of a specified chemical or element falls to half its original activity or concentration. Typically applied to the half-life of radioactive atoms but also applicable

to any other situation where the population is of molecules of diminishing concentration or activity.

halobacteria Bacteria which live in condi-tions of high salinity.

Halobacterium halobium Photosynthetic bacterium which has patches of purple membrane containing the pigment *bacterio-rhodopsin*.

halophilic Literally, salt-loving: able to survive in environments with high ionic strength such as salt lakes.

halorhodopsin Light-driven chloride ion pump of halobacteria, a retinylidene protein very similar to *bacteriorhodopsin*.

hamartoma Tumour-like but non-neo-plastic overgrowth of tissue which is dis-ordered in structure. Examples are hae-mangiomas (which include the vascular naevus or birthmark) and the pigmented naevus (mole).

haploid Describes a nucleus, cell or organism possessing a single set of unpaired *chromo-somes*. *Gametes* are haploid.

haplotype The set, made up of one *allele* of each gene, comprising the *genotype*. Also used to refer to the set of alleles on one *chromosome* or a part of a chromosome, ie. one set of alleles of linked genes. Its main current usage is in connection with the linked genes of the *major histocompatibility complex*.

hapten A chemical substance capable of binding to antibody but which cannot stimu-late an immune response unless present on a carrier molecule. Could be considered an isolated *epitope*. The hapten constitutes a single antigenic determinant; perhaps the best known example is dinitrophenol (DNP) which can be conjugated to bovine serum albumin (BSA) and against which anti-DNP antibodies are produced (antibodies to the BSA can be adsorbed out). Because the hapten is monovalent, immune complex formation will be blocked if the soluble hap-ten is present as well as the hapten-carrier conjugate. Such competitive inhibition by the soluble small molecule is sometimes referred to as haptenic inhibition, and this term has carried over into other systems.

haptenic inhibition See *hapten*.

haptoglobin Acid α_2-plasma glycoprotein that binds to oxyhaemoglobin which is free in the plasma, and the complex is then removed in the liver. Tetrameric (2 α, 2 β subunits): the existence of two different α chains in humans means that haptoglobins can exist in three variants in heterozygotes.

haptotaxis Strictly speaking, a directed response of cells in a gradient of adhesion, but nearly always loosely applied to situations where an adhesion gradient is thought to exist and local trapping of cells seems to occur.

Hartnup disease Amino acid transport defect that leads to excessive loss of mono-amino monocarboxylic acids (cystine, lysine, ornithine, arginine) in the urine, and poor absorption in the gut. See *iminoglycinuria*.

Harvey sarcoma virus See *ras* *gene*.

HAT medium A selective growth medium for animal tissue cells that contains hypoxanthine, the folate antagonist aminopterin (or amethopterin) and thymidine. Used for selection of hybrid somatic cell lines, as in the production of monoclonal antibodies. In HAT medium, cells are forced to use these exogenous bases, *via* the salvage pathways, as their sole source of purines and pyrimidines. Parental cells lacking enzymes such as HGPRT or TK can be eliminated whilst hybrids grow.

Hatch-Slack-Kortshak pathway (Hatch-Slack pathway; HSK pathway) Metabolic pathway responsible for primary CO_2 fixation in *C4 plant* photosynthesis. The enzymes which are found in *mesophyll* chloroplasts include *PEP carboxylase*, which adds CO_2 to phosphoenolpyruvate to give the 4-carbon compound, oxaloacetate. Four-carbon compounds are transferred to bundle-sheath chloroplasts, where the CO_2 is liberated and re-fixed by the *Calvin–Benson cycle*. The HSK pathway permits efficient photosynthesis under conditions of high light intensity and low CO_2 concentration; oxygen is prevented from association with *RUBISCO*, so avoiding the non-productive effects of photorespiration.

haustorium A projection from a cell or tissue of a fungus or higher plant, which penetrates another plant and absorbs nutrients from it. In fungi, it is a *hyphal* projection which penetrates the cell wall of a host plant cell, but is separated from the cytoplasm by a characteristic membrane; in parasitic angiosperms, it is a modified root.

Hayflick limit See *cell death*.

Heaf test A commonly-used tuberculin test in which tuberculin is injected intradermally with a multiple puncture apparatus. A positive reaction indicates the presence of T-cell reactivity to mycobacterial products.

heart muscle A striated but involuntary muscle responsible for the pumping activity of the vertebrate heart. The individual muscle cells are joined through a junctional complex known as the *intercalated disc*, and are not fused together into multinucleate structures as they are in skeletal muscle.

heat shock proteins A specific group of proteins whose synthesis is induced on exposure to an abnormally high temperature and certain other stresses, and which are presumed to enable cells to adapt to these stresses. The phenomenon is found in both pro- and eukaryotic cells. Similar proteins can be induced in vertebrates and invertebrates.

heavy chain In general, the larger polypeptide in a multimeric protein. Thus the immunoglobulin heavy chain is of 50kD, the light chain of 22kD, whereas in myosin the heavy chain is very much larger (220kD) than the light chains (approximately 20kD).

heavy water (deuterium oxide; D_2O) Most commonly used by cell biologists to stabilise microtubules.

HeLa cells An established line of human epithelial cells derived from a cervical carcinoma (said to be from Helen Lane or Lange).

helicase General term for DNA replication enzymes which unwind the helix. They are ATPases which move at the replication fork, disrupting hydrogen bonds. Also known as unwindases.

helicoidal cell wall Type of plant cell wall in which each wall layer contains parallel *microfibrils*, but in which the orientation of the microfibrils changes by a fixed angle from one layer to the next. Gives a characteristic "herringbone" pattern in transmission electron microscopy.

Heliozoa Order Heliozoida. A group of amoeboid *Protozoa*. They are generally free-floating, spherical cells with many straight, slender microtubule-supported *pseudopods* radiating from the cell body like a sunburst. These modified pseudopods are termed *axopodia*. Genera include *Actinophrys* and *Echinosphaerium*.

helix-destabilising proteins (single-stranded DNA-binding proteins) Proteins involved in DNA metabolism. They bind cooperatively to single-stranded DNA, stabilising it and extending the DNA backbone. Their use makes it possible to visualise single-stranded DNA in electron micrographs.

helper factor A group of factors apparently produced by helper T-lymphocytes which act specifically or non-specifically to transfer T-cell help to other classes of lymphocytes. The existence of specific T-cell helper factor is uncertain.

hema-, hemo- See *haema-, haemo-*.

heme See *haem*.

hemicellulose Class of plant cell-wall polysaccharide which cannot be extracted from the wall by hot water or chelating agents, but can be extracted by aqueous alkali. Chemically distinct from cellulose because of the presence of pentose, xylose and/or the hexoses arabinose, galactose and mannose, as well as glucose. May also be branched. Some are well-characterised, eg. *xylan, glucuronoxylan, arabinoxylan, arabinogalactan* II, *glucomannan, xyloglucan*, and galactomannan. Part of the cell-wall matrix.

hemidesmosome Specialised junction between an epithelial cell in a simple epithelium and its basal lamina. Although morphologically similar to half a desmosome (into which intermediate cytokeratin filaments are also inserted) different proteins are involved.

hemizygote Nucleus, cell or organism which has only one of a normally diploid set of genes. In mammals the male is hemizygous for the *X chromosome*.

heparan sulphate A *glycosaminoglycan*.

heparin Sulphated mucopolysaccharide, found in granules of mast cells, which inhibits the action of thrombin on fibrinogen by potentiating anti-thrombins, thereby interfering with the blood-clotting cascade. Platelet Factor IV will neutralise heparin.

hepatitis A Small (27nm diameter) single-stranded RNA virus with some resemblance to *enteroviruses* such as polio. Causes "infectious hepatitis".

hepatitis B Virion (*Dane particle*), 42nm diameter, with an outer shell enclosing inner 27nm core particle containing the circular viral DNA. Aggregates of the envelope proteins are found in plasma and are referred to as hepatitis B surface antigen (HBsAg; previously called Australia antigen). Causes "serum hepatitis"; virus can persist for long periods (and in asymptomatic carriers); association of integrated virus with hepatocellular carcinoma is now well established.

hepatitis non-A, non-B A virus somewhat similar in size to hepatitis A but has no antigenic cross-reaction with either A or B.

hepato-carcinoma (hepatocellular carcinoma) Malignant tumour derived from hepatocytes. Associated with *hepatitis B* in 80–90% of cases.

hepato-pancreas Digestive gland of crustaceans with functions approximately analogous to liver and pancreas of vertebrates — enzyme secretion, food absorption and storage.

hepatocyte Epithelial cell of liver. Often considered the paradigm for an unspecialised animal cell. Blood is directly exposed to hepatocytes through fenestrated endothelium, and hepatocytes have receptors for sub-terminal N-acetyl galactosamine residues on asialo-glycoproteins of plasma.

hepatoma Carcinoma derived from liver cells: better term is hepato-carcinoma.

hereditary angio-oedema (USA hereditary angioedema) Condition in which there seems to be uncontrolled production of C2-*kinin* because of a deficiency in C1-inhibitor levels.

Herpetoviridae A family of large Class I DNA viruses: herpes simplex causes cold-sores and genital herpes; Varicella-zoster causes chicken-pox and shingles; cytomegalovirus causes congenital abnormalities and is an opportunistic pathogen; *Epstein Barr virus* (EBV) causes glandular fever and is associated with *Burkitt's lymphoma* and nasopharyngeal carcinoma. All establish persistent or latent infections with a tendency to reactivation if the immune system is suppressed.

Hers disease Glycogen-storage disease in which there is a deficiency of liver phosphorylase.

HETEs A family of hydroxytetraeicosenoic acid (20-carbon) derivatives of arachidonic acid produced by the action of lipoxygenase. Potent pharmacological agents with diverse actions.

heterochromatin The chromosomal regions which are condensed during interphase and at the time of nuclear division. They show what is to be considered an abnormal pattern of staining as opposed to *euchromatin*. Can be subdivided into constitutive regions (present in all cells) and facultative heterochromatin (present in some cells only); the inactive X chromosome of female mammals is an example of facultative hetrochromatin (see *Lyon hypothesis*).

heterocyst Specialised cell type found at regular intervals along the filaments of certain *Cyanobacteria*; site of nitrogen fixation.

heterodimer A dimer in which the two subunits are different. One of the best known examples is tubulin which is found as an α-tubulin/β-tubulin dimer. Heterodimers are relatively common, and it may be that the arrangement has the advantage that, for example, several different binding subunits may interact with a conserved signalling subunit.

heterogenous nuclear RNA (hnRNA) Originally identified as a class of RNA, found in the nucleus but not the nucleolus, which is rapidly labelled and with a very wide range of sizes, 2–40 kilobases. It represents the primary transcripts of *RNA polymerase* II and includes precursors of all *messenger RNAs*.

heterokaryon Cell which contains two or more genetically different nuclei. Found naturally in many fungi and produced experimentally by cell fusion techniques, eg. *hybridoma*.

heterophile antibody An antibody raised against an antigen from one species which also reacts against antigens from other species. Also used of *Forssman*-type systems where antibody against a variety of species antigens is present without immunisation.

heteroplasmy (heteroplasia) The occurrence of a tissue in the wrong place in an organism, as a result of inappropriate cellular differentiation.

heterospory Condition in vascular plants where the spores are of two different sizes, the smaller producing male prothalli, the larger female prothalli.

heterotypic Of different types. Thus heterotypic adhesion would be between dissimilar cells, in contrast to homotypic adhesion between cells of the same type.

heterozygosity index Measure of the number of *loci* for which an individual is heterozygous.

heterozygote Nucleus, cell or organism with different *alleles* of one or more specific genes. A heterozygous organism will produce unlike gametes and thus will not breed true.

hexitol Sugar alcohol with six carbon atoms. Natural examples are sorbitol, mannitol.

hexokinase Enzyme which catalyzes the transfer of phosphate from ATP to glucose to form glucose-6-phosphate, the first reaction in the metabolism of glucose *via* the glycolytic pathway.

hexon Subunit of a hexameric structure or with hexameric symmetry, in particular the arrangement of most of the capsomers of *Adenoviridae* — one capsomer surrounded by six others to form the hexon.

hexose Monosaccharide containing six carbon atoms, eg. *glucose*, *galactose*, *mannose*.

hexose monophosphate shunt See *pentose phosphate pathway*.

HGPRT (hypoxanthine–guanine phosphoribosyl transferase) Enzyme which catalyzes the first step in the pathway for salvage of the purines hypoxanthine and guanine. The phosphoribosyl moiety is transferred from an activated precursor, 5-phosphoribosyl-1-pyrophosphate. Since animal cells can synthesise purines *de novo*, HGPRT⁻ mutants can be selected by their resistance to toxic purine analogues. A genetic lesion in HGPRT in humans underlies the *Lesch–Nyhan syndrome*.

high mannose oligosaccharide A subset of the N-glycan chains which are added post-translationally to certain asparagine residues of secreted or membrane proteins in eukaryotic cells; contain 5–9 mannose residues, but lack the sialic acid-terminated antennae of the so-called complex type.

high density lipoproteins (HDL) Involved in cholesterol transport in serum. See *lipoproteins*.

high voltage electron microscope (HVEM) The high voltage electron microscope has two advantages, the increased voltage shortens the wavelength of the electrons (and therefore increases resolving power) but, more importantly for the biologist, the penetrating power of the beam is increased, and it becomes possible to look at thicker specimens. Thus it is possible, by using stereoscopic views (obtained with a tilting stage) to get a three-dimensional picture of the interior of a cell.

high-energy bond Chemical bonds which release more than 25kJ/mol on hydrolysis: their importance is that the energy can be used to transfer the hydrolysed residue to another compound. The risk in using the term is that students may think the bond itself is different in some way, whereas it is the compound that matters. Hydrolysis of creatine phosphate yields 42.7kJ/mol; of phosphoenolpyruvate, 53.2; ATP to ADP, 30.5: the latter is important because it shows that energetically the hydrolysis of creatine phosphate will suffice to reconstitute ATP — hence the use of creatine phosphate in muscle.

Hill coefficient A measure of cooperativity in a binding process. A Hill coefficient of 1 indicates independent binding, a value of greater than 1 shows positive cooperativity — binding of one ligand facilitates binding of subsequent ligands at other sites on the multimeric receptor complex. Worked out originally for the binding of oxygen to haemoglobin (Hill coefficient of 2.8).

Hill reaction Reaction, first demonstrated by Hill in 1939, in which illuminated chloroplasts evolve oxygen when incubated in the presence of an artificial electron acceptor (eg. ferricyanide). The reaction is a property of *photosystem II*.

Hind II First type II *restriction endonuclease* identified, by Hamilton Smith in 1970. Isolated from *Haemophilus influenzae* Rd, it cleaves the sequence GTPyPuAC between the unspecified pyrimidine and purine generating blunt ends.

Hind III Commonly used type II *restriction endonuclease* isolated from *Haemophilus influenzae* Rd, it cleaves the sequence AAGCTT between the two As generating *sticky ends*.

hinge region Flexible region of a polypeptide chain — for example, in immunoglobulins between Fab and Fc regions, and in myosin the S2 portion of heavy meromyosin.

hispid flagella Eukaryotic flagella with two rows of stiff protrusions (*mastigonemes*) at right-angles to the long axis of the shaft. In hispid (pentonematic) flagella, the normal relationship between the direction of flagellar wave-propagation and the direction of movement is reversed; a proximal-to-distal wave pulls the organism forward.

histamine Formed by decarboxylation of histidine. Potent pharmacological agent acting through receptors in smooth muscle and in secretory systems. Stored in mast cells and released by antigen (see *hypersensitivity*). Responsible for the early symptoms of anaphylaxis. Also present in some venoms.

histidine (His; H; MW 155) An amino acid with an imidazole side chain with a pK_a of 6–7. Acts as a proton donor or acceptor and has high potential reactivity and diversity of chemical function. Forms part of the catalytic site of many enzymes. See Table A2.

histiocytes Long-lived resident macrophages found within tissues.

histoblasts Population of small diploid epithelial cells in Dipteran larvae which do not form typical imaginal discs, yet resemble them in some ways.

histochemistry Study of the chemical composition of tissues by means of specific staining reactions.

histocompatibility If tissues of two organisms are histocompatible, then grafts between the organisms will not be rejected. If, however, major histocompatibility antigens are different then an immune response will be mounted against the foreign tissue.

histocompatibility antigen (major histocompatibility antigen) A set of plasmalemmal glycoproteins which are crucial for T-cell recognition of antigens; particularly the HLA system in humans and the H2 system in mice. There are two classes of histocompatibility antigens: 1. Class I; histocompatibility antigens composed of two glycosylated subunits, a heavy chain of 44kD and β_2-microglobulin (12kD). The heavy chain may be coded by K, D or L genes of mouse H2 and A, B or C genes of human HLA complex. Class I antigens are important in T-cell killing (particularly of virus-infected cells) and are recognised in conjunction with the foreign cell surface antigens (*MHC restriction*). 2. Class II antigens; heterodimeric histocompatibility antigens composed of α (32kD) and β (28kD) chains. Found constitutively on dendritic cells, and inducibly on mononuclear phagocytes and B-cells. Can be induced on many other cell types, eg. vascular endothelium by lipopolysaccharides, interferon-γ, etc. The response of T-helper cells requires that the foreign antigen is presented in conjunction with the appropriate Class II antigens. (Murine H2 Ia antigens and human HLA-DR antigens are Class II).

histogenesis The process of formation of a tissue, involving differentiation, morphogenesis, and other processes such as angiogenesis, growth control, cellular infiltration etc.

histones Proteins found in the nuclei of all eukaryotic cells where they are complexed to DNA in *chromatin* and *chromosomes*. They are of relatively low molecular weight and are basic, having a very high arginine/lysine content. Highly conserved and can be grouped into five major classes. Two copies of H2A, H2B, H3 and H4 bind to about 200 base pairs of DNA to form the repeating structure of chromatin, the *nucleosome*, with H1 binding to the linker sequence. May act as non-specific repressors of gene transcription. See Table H1.

histoplasmin Filtrate of a culture of *Histoplasma capsulatum*, the causative agent of histoplasmosis. Histoplasmin is intradermally injected in a skin test for the disease.

histotope A site on an MHC Class I or Class II antigen (see *histocompatibility antigen*) recognised by a T-cell.

HIV (human immunodeficiency virus) Previously known as HTLV-III, human lymphotrophic virus type III, and also referred to as LAV, lymphadenopathy-associated virus; the retrovirus that causes acquired immunodeficiency syndrome (AIDS) in humans, by killing T4-lymphocytes (T-helper cells). Now clear that there are multiple forms, HIV-1 and HIV-2 being recognised at present.

HLA (human leucocyte antigen) Refers to the *histocompatibility antigens* found in humans.

HMG proteins (high mobility group) Group of non-histone chromosomal proteins of high electrophoretic mobility, some of which are apparently associated with actively transcribed genes.

HMM (heavy meromyosin) Soluble tryptic fragment of myosin which retains the ATPase activity and which will bind to F-actin to produce a characteristic arrowhead pattern in electron micrographs (unless ATP is present, in which case it detaches). Papain cleavage of HMM yields S1 and S2 subfragments, the former having the ATPase activity.

hnRNA (heterogenous nuclear RNA) Primary transcripts from DNA from which *introns* are removed by splicing.

Hodgkin's disease A human lymphoma that appears to originate in a particular lymph node and later spreads to the spleen, liver and bone marrow. Giant cells, the Sternberg–Reed cells, with mirror-image nuclei are diagnostic. Immunological depletion, caused perhaps by the excessive growth of neoplastic histiocytes, occurs. Four types of the disease are recognised depending on the relative predominance of various neoplastic derivatives of the lymphoid series. Pyrexia is often a feature of the disease and death often results from generalised immunological inability to respond to infections.

Hoechst 33258 dye A fluorescent dye which is a specific stain for DNA, and can therefore be used to visualise chromosomes,

Table H1. Classes of histones

Class	Average MW (kD)	% Arginine	% Lysine
H1	23	1.5	29
H2A	14	8	11
H2B	14	5	16
H3	15	13.5	10
H4	11	14	11

and to monitor animal cell cultures for contamination by micro-organisms such as mycoplasma.

holandry Inheritance of characters borne on the male chromosome and therefore only expressed in the male.

holocrine Form of secretion in which the whole cell is shed from the gland — usually after becoming packed with the main secretory substance. In mammals, sebaceous glands are one of the few examples.

homeobox (homoeobox) Conserved DNA sequence originally detected by DNA–DNA *hybridisation* in many of the genes which give rise to homeotic and segmentation mutants in *Drosophila*. The homeobox consists of about 180 nucleotides coding for a sequence of 60 amino acids in a protein, sometimes termed the homeodomain, of which about 80–90% are identical in the various homeodomains identified from *Drosophila*. Homeoboxes have also been detected in the genomes of vertebrates, with about 75% amino acid homology, and a similar sequence has been found in the MAT gene of yeast. The homeobox codes for a protein domain which is involved in binding to DNA.

homeotic mutant (homoeotic mutant) A mutant in which one body part, organ or tissue, is transformed into another part normally associated with another segment. Examples are the *antennapedia* and *bithorax* mutants of *Drosophila*.

homocystinuria Recessive condition in which the enzyme (cystathione synthetase) which converts homocysteine and serine into cystathione, a precursor of cysteine, is missing. Deficiency of this enzyme has widespread consequences in connective tissue, circulation and nervous system.

homograft Outmoded term for a graft from one individual of a species to another. Includes *allogeneic* grafts (allografts) between genetically dissimilar individuals, and syngeneic grafts between identical individuals (eg. twins).

homologous chromosomes Chromosomes that are homologous with respect to genetic *loci*, and that tend to pair or *synapse* during *meiosis*. The two chromosomes (maternal and paternal) of each of the pairs occurring in the nuclei of diploid cells are homologous.

homologous recombination *Genetic recombination* involving exchange of homologous loci.

homozygote Nucleus, cell or organism with identical *alleles* of one or more specific genes.

hook Basal portion of bacterial flagellum, to which is distally attached the *flagellin* filament. Proximally the hook is attached to the rotating spindle of the motor. In some bacteria (Myxobacteria) the rotation of the hook itself (without an attached flagellum) may directly cause forward gliding movement.

horizontal cell A type of *non-spiking interneuron* found in the retina, named for its morphology. Horizontal cells process the information from a large number of *photoreceptors* and synapse onto *ganglion cells* and other cell types in the retina.

hormone A substance secreted by specialised cells which affects the metabolism or behaviour of other ("target") cells possessing functional receptors for the hormone. Hormones may be hydrophilic, like *insulin*, in which case the receptors are on the cell surface, or lipophilic, like the *steroid hormones*, where the receptor can be intracellular. See Tables H2 and H3.

horseradish peroxidase A large enzyme, frequently used in conjunction with diaminobenzidine as an intracellular marker to identify cells both at light- and electron-microscopic levels.

host-versus-graft reaction The normal lymphocyte-mediated reactions of a host against allogeneic or xenogeneic cells acquired as a graft or otherwise, which lead to damage or/and destruction of the grafted cells. The opposite of *graft-versus-host reaction*. More commonly just referred to as graft rejection.

host-range mutant A mutant of phage or animal virus which grows normally in one of its host cells, but has lost the ability to grow in cells of a second host type.

housekeeping proteins Those sets of proteins involved in the basic functioning of a cell or the set of cells in an organism, eg. enzymes involved in synthesis and processing of DNA, RNA and proteins or in the major metabolic pathways. As opposed to *luxury proteins*.

Table H2. Polypeptide hormones and growth factors

Name	MW	Residues	Source	Actions
Parathyroid hormone (PTH)	9500D	84	Parathyroid glands	Raises kidney cAMP and blood calcium by stimulating bone release
Calcitonin[a]	4500D	32	Thyroid, parathyroid	Opposes parathyroid hormone, hypocalcaemic, hypophosphataemic
Growth factors				
Insulin[a]	6kD	21 + 30	B cells of pancreas	Hypoglycaemic, growth factor
Insulin-like growth factor I (IGF I, somatomedin C)		70	Liver, kidney	Growth factor, released into plasma
Insulin-like growth factor II (IGF II, multiplication stimulating activity, MSA)		67	Cultured hepatocytes	Mitogen for several cell types
Somatomedin A	3000D	50–80	Liver, kidney	Stimulates growth of peripheral nervous system
Epidermal growth factor (urogastrone, EGF)		53	Mouse submaxillary gland, urine	Stimulates epidermal growth, formation of GI tract
Platelet-derived growth factor (PDGF)	3000D		Platelets	Stimulates fibroblast proliferation, wound healing
Fibroblast growth factor (FGF)	130kD		Brain, pituitary	Stimulates proliferation of fibroblasts, adrenal cells, chondrocytes, endothelia
β-Nerve growth factor (NGF)	130kD	118	Salivary gland, snake venom	Tropic and trophic effects, mainly on sensory and sympathetic neurons. Peripheral nerve targets
Transforming growth factor (TGFα)	5–20kD		Carcinomata	Acts via EGF receptor
TGFβ	2 × 25kD	112	Tumour cells	Multifunctional role in tissue damage. Promotes or inhibits proliferation depending on cell type
GM-CSF (MG1, CSFα)	23kD		Many tissues	Promotes granulocyte or macrophage colony formation
G-CSF (murine; CSFβ in human)	25 or 30kD		Many tissues	Promotes differentiation in myeloleukaemic cells
M-CSF (CSF-1)	40–70kD		Mouse tissues	Promotes macrophage differentiation
MultiCSF (IL3)	23–30kD		Mouse primed T cells	Proliferation of various haemopoietic stem cells
Hypothalamic releasing hormones				
Luteinising hormone releasing factor[a] (luliberin, LHRH)		10	Hypothalamus	Stimulates release of LH, FSH
Hypothalamic thyrotropic hormone releasing factor (TRF, TRH)[a]	362D	3	Hypothalamus	Stimulates release of thyrotropin from anterior pituitary
Corticotrophin releasing factor (CRH)[a]		41	Hypothalamus	Stimulates release of corticotrophin
Growth hormone release factor (somatoliberin)		44	Hypothalamus	Stimulates release of growth hormone
Somatostatin (SRIF)[a]		14	D-cells of pancreas	Inhibits release of several hormones, including growth hormone

(Continued)

Table H2. (Continued)

Name	MW	Residues	Source	Actions
Pituitary gland				
Oxytocin[a]	1007D	8	Posterior pituitary	Uterine contraction, lactation
Vasopressin (antidiuretic hormone, ADH)[a]		8	Posterior pituitary	Antidiuretic, vasopressor
ACTH-MSH family				
Adrenocorticotrophin (ACTH)[a]	4500D	39	Anterior pituitary	Stimulates glucocortoid production from adrenals
Melanotropin (melanocyte stimulating hormone, MSH)[a]		11–12	Anterior pituitary	Stimulates darkening of skin
Lipotropins[a]		50–90	Anterior pituitary	Stimulate lipid breakdown
Endorphins[a]		15–31	Pituitary, brain	Opiates
Enkephalins[a]		5	Adrenal medulla, brain, gut	Opiates
Growth hormone (somatotropin, GH)[a]	21.5kD	191	Pituitary	Regulates organismal growth
Prolactin	23kD	199	Pituitary	Promotes mammary growth, lactation
Gonadotropin/thyrotrophin family (glycoproteins of the anterior pituitary with α + β subunits)				
Luteinising hormone (LH)	30kD		Anterior pituitary	Acts on gonads
Follicle stimulating hormone (FSH)	30kD		Anterior pituitary	Acts on gonads
Thyroid stimulating hormone (TSH)	30kD		Anterior pituitary	Stimulates release of thyroid hormones
Human chorionic gonadotrophin (hCG)	30kD		Anterior pituitary, placenta	
Gastrointestinal tract				
Vasoactive intestinal peptide (VIP)[a]		28	Lung, gut H-cells, nervous tissue	Vasodilation, bronchodilation, stimulates insulin, glucagon, prolactin secretion
Gastrin[a]		17 or 34	Gut G-cells	Stimulates acid secretion, muscle contraction
Secretin[a]		27	Gut S-cells	Stimulates alkali secretion from pancreas, decreases gastric acid secretion
Bombesin[a]		14	Skin, gut P-cells, nerves	Acts on CNS, gut smooth muscle, pancreas, pituitary, kidney, heart; mitogenic *in vitro*
Motilin[a]		22	Gut enterochromaffin-2 cells	Increases contractile response of stomach muscle
Pancreatic polypeptide[a]		36	PP-cells of pancreas; gut mucosa	Released on feeding; alters gut muscle tone

(Continued)

Table H2. (Continued)

Name	MW	Residues	Source	Actions
Gastrointestinal tract (continued)				
Pancreozymin-cholecystokinin (PZ-CCK)[a]		8, 33 or 39	Gall bladder, brain, gut I-cells	Secretion of enzymes & electrolytes, secretion of insulin, glucagon, pancreatic polypeptide, contraction of gall bladder
Gastric inhibitory peptide (GIP)		43	Gut	Inhibits gastric acid secretion, stimulates intestinal secretion, stimulates insulin & glucagon secretion
Substance P[a]		11	Gut enterochromaffin-1 cells, nerves	Contracts gut musculature, decreases blood pressure
Neurotensin[a]		13	Gut, hypothalamus	Effects on smooth muscle tone; may also be a neurotransmitter
Plasma kinins				
Bradykinin[a]		9	Formed in plasma	Dilates blood vessels, increases capillary permeability
Kallidin		10	Formed in plasma	Dilates blood vessels, increases capillary permeability
Other				
Renin	40kD	gp	Kidney	Cleaves angiotensinogen to angiotensin I in plasma
Angiotensin II[a]		8	Formed from angiotensinogen, *via* angiotensin I	Acts on adrenal gland to stimulate aldosterone release, elevates blood pressure, mitogen for vascular smooth muscle
Atrial natriuretic peptide (ANP, ANF)		21–28	Atria of heart, brain, adrenal glands	Acts on kidney to produce natriuresis & diuresis, relaxes vascular smooth muscle, inhibits catecholamine release from adrenal medulla
Glucagon (HGF)[a]	3550D	29	A cells of pancreas	Hyperglycaemic
Placental lactogen		191	Placenta	Promotes lactation
Caerulein	1352D	10	Amphibian skin	Hypotensive, stimulates gastric secretion

[a]Neuropeptide. Other neuropeptides include calcitonin gene related peptide, neurokinin, neuromedin, neuropeptide Y, proctolin, and carnosine.

HPLC (high pressure liquid chromatography) Chromatographic method in which the sample is forced at high pressure through a tightly-packed column of finely-divided particles which present a very large surface area. Because HPLC gives good separation very rapidly (but is expensive), manufacturers tend to speak of "high performance liquid chromatography" as an encouragement to purchasers.

HPV Human *Papillomavirus*; causes warts.

HRGP (hydroxyproline-rich glycoprotein) Class of plant glycoproteins and proteoglycans rich in hydroxyproline, that includes *AGP*, *extensin* and certain *lectins*. Found in the cell wall and are produced in response to injury.

HSK pathway See *Hatch–Slack–Kortshak pathway*.

HTLV-I (human T-cell leukaemia/lymphoma virus type I) A retrovirus causing leukaemia and sometimes a mild immunodeficiency. In addition to *gag*, *pol* and *env*, the virus causes a coding sequence *pX* which does not seem to have a normal genomic homologue and is not a conventional oncogene. The protein product of the *pX* region is a short-lived nuclear protein (around 40kD). T-cells transformed with HTLV-I continue to proliferate and are independent of *interleukin-2*.

HTLV-II (human T-cell leukaemia/lymphoma virus type II) Originally isolated from a T-cell line from a patient with hairy-cell leukaemia. It has partial homology with *HTLV-I*, and its pathological potential is uncertain.

HTLV-III See *HIV*.

humoral immune responses Those immune responses mediated by antibody.

Hunter syndrome Recessive mucopolysaccharidosis, X-linked, in which dermatan and heparan sulphates are not degraded because of a deficiency of heparan sulphate sulphatase. Two lysosomal enzymes (heparan sulphate sulphatase and α-iduronidase) are involved in the breakdown of these glycosaminoglycans, fibroblasts from Hunter's syndrome will complement the fibroblasts

Table H3. Steroid hormones

Name	MW (D)	Distribution	Actions
Female sex hormones			
β-Oestradiol	272	Ovary, placenta, testis	Oestrogen
Progesterone	314	Corpus luteum, adrenal, testis, placenta	Regulates pregnancy, menstrual cycle
Male sex hormones			
Testosterone	288	Testes	Male secondary sexual characteristics, anabolic effects
Dihydrotestosterone			
Adrenal cortex: glucocorticoids			
Cortisol	362	Adrenal cortex	Gluconeogenesis, anti-inflammatory
Cortisone	360	Adrenal cortex	Glucocorticoid, anti-inflammatory
Corticosterone	346	Adrenal cortex	
Adrenal cortex: Mineralocorticoids			
Aldosterone	360	Adrenal cortex	Stimulates resorption of sodium from kidneys
Thyroid hormones			
Thyroxine (T4)	777	Thyroid	Control basal metabolic rate
Triiodothyronine (T3)	651	Thyroid	
Other			
Vitamin D (calciferol)	384	Skin (and sunlight)	Calcium and phosphate metabolism
Ecdysone	481	—	Moulting hormone of insects and nematodes

from Hurler's patients in culture; by recapture of lysosomal enzymes from the medium, both types of cells in mixed culture become competent to digest glycosaminoglycans.

Hurler's disease Autosomal recessive storage disease in which α-iduronidase is absent, leading to accumulation of heparan and dermatan sulphates. Extensive deposits of mucopolysaccharide are found in *gargoyle cells*, and in neurons. See *Hunter syndrome*.

Hurler–Scheie syndrome Although clinically distinct diseases, fibroblasts from patients with Hurler's and with Scheie syndrome do not cross-complement in culture, suggesting that the enzyme defect is the same.

HVEM See *high voltage electron microscope*.

hyaline Clear, transparent, granule-free; as, for example, hyaline cartilage and the hyaline zone at the front of a moving amoeba.

hyaluronic acid Polymer composed of repeating dimeric units of glucuronic acid and N-acetyl glucosamine. May be of extremely high molecular weight (up to several million) and forms the core of complex proteoglycan aggregates found in extracellular matrix.

hyaluronidase Enzyme which degrades *hyaluronic acid*; found in lysosomes.

hybrid cell Any cell type containing components from one or more genomes, other than zygotes and their derivatives. Hybrid cells may be formed by *cell fusion* or by *transfection*. See *heterokaryons*.

hybridisation (of nucleic acids) Technique in which single-stranded nucleic acids are allowed to interact so that complexes, or hybrids, are formed by molecules with sufficiently similar, complementary sequences. By this means the degree of sequence identity can be assessed and specific sequences detected. The hybridisation can be carried out in solution or with one component immobilised on a gel or, most commonly, nitrocellulose or nylon membranes. Hybrids are detected by various means: visualisation in the electron microscope; by radioactively labelling one component and removing non-complexed DNA; or by washing or digestion with an enzyme that attacks single-stranded nucleic acids and then estimating the radioactivity

bound. Hybridisations are done in all combinations: DNA–DNA (DNA can be rendered single-stranded by heat denaturation), DNA–RNA or RNA–RNA. *In situ* hybridisations involve hybridising a labelled nucleic acid (often labelled with a fluorescent dye) to suitably prepared cells or histological sections. This is used particularly to look for specific *transcription*, and localisation of genes to particular chromosomes.

hybridoma A cell hybrid in which a tumour cell forms one of the original source cells. In practice, confined to hybrids between T- or B-lymphocytes and appropriate *myeloma cell* lines.

Hydra Genus of freshwater coelenterates (cnidarians). Hydras are small, solitary and only exist in the polyp form, which is a radially-symmetrical cylinder that is attached to the substratum at one end and has a mouth surrounded by tentacles at the other. They have considerable powers of *regeneration* and have been used in studies on positional information in morphogenesis.

hydraulic motor By altering the internal osmotic pressure within a cell, water will enter and a considerable expansion of the compartment will occur. This has been used as a motor device in plants (turgor pressure), in eversion of *nematocysts*, and possibly in the production of other cellular protrusions.

hydrogen peroxide (H_2O_2) Hydrogen peroxide is produced by vertebrate phagocytes and is used in bacterial killing (the *myeloperoxidase*–halide system).

hydrogenosome Organelle found in certain anaerobic trichomonad and some ciliate protozoa: contains hydrogenase and produces hydrogen from glycolysis.

hydrolase One of a class of enzymes (EC Class 3) catalyzing hydrolysis of a variety of bonds, such as esters, glycosides, peptides.

hydrolytic enzymes See *hydrolase*.

hydrophilic group A polar group or one which can take part in hydrogen bond formation, eg. OH, COOH, NH₂. Confers water solubility, or in lipids and macromolecules causes part of the structure to make close contact with the aqueous phase.

hydrophobic bonding Interaction driven by the exclusion of non-polar residues from water. It is an important determinant of protein conformation and of lipid structures, and is considered to be a consequence of maximising polar interactions rather than a positive interaction between apolar residues.

hydroxyapatite The calcium phosphate mineral found both in rocks of non-organic origin and as a component of bone and dentine. Used as column packing for chromatography, particularly for separating double-stranded DNA from mixtures containing single-stranded DNA.

hydroxylysine Post-translationally hydroxylated lysine is found in *collagen* and commonly has galactose and then glucose added sequentially by glycosyl transferases. The extent of glycosylation varies with the collagen type.

hydroxyproline Specific proline residues on the amino side of a glycine residue in collagen become hydroxylated at C4, before the polypeptides become helical, by the activity of prolyl hydroxylase. This enzyme has a ferrous ion at the active site, and a reducing agent such as ascorbate is necessary to maintain the iron in the ferrous state. The presence of hydroxyproline is essential to produce stable triple-helical tropocollagen, hence the problems caused by ascorbate deficiency in scurvy. This unusual amino acid is also present in considerable amounts in the major glycoprotein of primary plant cell walls; see *HRGP*.

hydroxyproline rich glycoprotein See *HRGP*.

5-hydroxytryptamine See *serotonin*.

hydroxyurea Inhibitor of DNA synthesis (but not repair).

hypercholesterolaemia (USA hypercholesterolemia) High serum levels of cholesterol. Can in some cases be caused by a defect in lipoprotein metabolism or, for example, defects in the *low-density lipoprotein receptor* (familial hypercholesterolaemia).

hyperglycaemia (USA hyperglycemia) An excess of plasma glucose which can arise through a deficiency in insulin production. See *diabetes mellitus, diabetes insipidus*.

hyperimmune serum Serum prepared from animals that have recently received repeated injections or applications of a chosen antigen; thus the serum should contain a very high concentration of polyclonal antibodies against that antigen.

hyperlipidaemia (USA hyperlipidemia) Elevated levels of serum low-density lipoprotein, correlated with increased risk of cardiovascular disease. See *hypercholesterolaemia*.

hyperlipoproteinaemia (USA hyperlipoproteinemia) The same as *hyperlipidaemia*.

hypermastigote Large multi-flagellate symbiotic protozoa found in the gut of termites and wood-eating cockroaches. Most bizarre example of the group is *Mixotricha paradoxica* which actually has few flagella and is propelled by spirochaetes (bacteria) which are attached to special bracket-like regions of the cell wall.

hyperosmotic Of a liquid, having a higher osmotic pressure (usually than the physiological level).

hyperplasia Increase in the size of a tissue as a result of enhanced cell division. Once the stimulus (wound healing, mechanical stress, hormonal overproduction) is removed the division rate returns to normal (whereas in neoplasia proliferation continues in the absence of a stimulus).

hyperpolarisation A negative shift in a cell's *resting potential* (which is normally negative), thus making it numerically larger ie. more polarised. The opposite of *depolarisation*.

hypersensitive response *Bot.* An active response of plant cells to pathogenic attack in which the cell undergoes rapid necrosis and dies. Associated with the production of *phytoalexins*, *lignin*, and sometimes *callose*. The response is thought to prevent a potential pathogen from spreading through the tissues.

hypersensitivity In immunology, a state of excessive and potentially damaging immune responsiveness as a result of previous exposure to antigen. If the hypersensitivity is of the immediate type (antibody-mediated), then the response occurs in minutes; in delayed hypersensitivity the response takes much longer (about 24hr) and is mediated by primed T-cells. Hypersensitivity responses are not simply divisible into the two types, and it is now more common to subdivide immediate responses into types I, II, and III, the delayed response being of type IV. Type

I responses involve antigen reacting with IgE fixed to cells (usually mast cells) and are characterised by histamine release; anaphylactic responses and urticaria are of this type. In type II responses circulating antibody reacts with cell-surface or cell-bound antigen, and if complement fixation occurs, cytolysis may follow. In type III reactions immune complexes are formed in solution and lead to damage (serum sickness, glomerulonephritis, *Arthus reaction*). Delayed-type responses of Type IV involve primed lymphocytes reacting with antigen and lead to formation of a lymphocyte-macrophage granuloma without involvement of circulating antibody.

hyperthyroidism (Graves's disease; thyrotoxicosis) Excess production of thyroid hormone which may be caused by autoantibodies that bind to the *thyroid stimulating hormone* receptor and induce secretion of *thyroxine* by raising cyclic AMP levels in the thyroid cells.

hypertonic Of a fluid, sufficiently concentrated to cause osmotic shrinkage of cells immersed in it. Note that a mildly *hyperosmotic* solution is not necessarily hypertonic for viable cells, which are capable of regulating their volumes by *active transport*. See also *isotonic*.

hypervariable region Those regions within the variable regions of the heavy or light chains of immunoglobulins in which there is high sequence diversity within that set of immunoglobulins in a single individual. These regions are believed to specify the antigen specificity of each antibody.

hypha Filament of fungal tissue which may or may not be separated into a file of cells by cross-walls (septa). It is the main growth form of filamentous fungi, and is characterised by growth at the tip followed by lateral branching.

hypocotyl Part of the axis of a plant embryo or seedling between the point of insertion of the *cotyledon(s)* and the top of the radicle (root). In some *etiolated* seedlings, the hypocotyl is greatly extended. Includes the transition zone where the vascular system changes from the root configuration to that of the stem.

hypogammaglobulinaemia (USA hypogammaglobulinemia) Syndromes in humans and other vertebrates in which the immunoglobulin level is depressed below the normal range. Congenital, chronic and transient types are known.

hypoxanthine Purine base present in inosine monophosphate (IMP) from which adenosine monophosphate (AMP) and guanosine monophosphate (GMP) are made. The product of deamination of adenine, 6-hydroxy-purine.

hypoxanthine guanine phosphoribosyl transferase See *HGPRT*.

I

IAA See *indole acetic acid*.

iC3b Inactivated C3b (C3bi). See *complement*.

I-band The isotropic band of the sarcomere of striated muscle, where only thin filaments are found. Unlike the A-band, the I band can vary in width depending upon the state of contraction of the muscle when fixed.

I-cell disease A human disease in which the lysosomes lack hydrolases but high concentrations of these enzymes are found in the extracellular fluids. The gene defect responsible probably prevents the addition of the lysosome recognition marker (mannose-6-phosphate) to these enzymes so that they are not directed into the lysosomes but are released.

I-region 1. The gene in *E. coli* coding for the repressor of the lactose operon. 2. Region of the murine genome coding for products involved in many aspects of the immune response (hence Immune-region). Most of the proteins are Class II *histocompatibility antigens*.

I-region associated antigens Class II major *histocompatibility antigens*.

Ia antigen Antigens coded for by the I-region of the major histocompatibility complex; now more generally known as the Class II major *histocompatibility antigens*. The majority, if not all, such antigens are Class II molecules composed of α and β polypeptide chains.

ibotenate (ibotenic acid) An *excitotoxin* from *Amanita* sp., which acts on the *NMDA* receptor.

ICAM (intercellular cell adhesion molecule) ICAM-1 (90kD) is the ligand for the leucocyte *integrin LFA-1*. It is expressed on fibroblasts and endothelial cells, and its levels can be up-regulated by lymphokine treatment. One of the immunoglobulin superfamily.

idioblast See *sclereid*.

idiotope An antigenic determinant (*epitope*) unique to the immunoglobulin product of a single clone of cells. Found in the variable region and usually forms the antigen-binding site. Any single immunoglobulin may have more than one idiotope.

idiotype Collective term for the *idiotopes* on a V region. Usually a feature of the antigen-binding site.

idoxuridine Analogue of thymidine which inhibits the replication of DNA viruses. Has been used in the treatment of herpes simplex and *varicella-zoster*.

α-L-iduronic acid A uronic acid, derived from the sugar idose, and bearing one terminal carboxyl group. With N-acetyl galactosamine-4-sulphate, a component of dermatan sulphate.

α-1-iduronidase Lyses the bonds between iduronic acid and N-acetyl galactosamine-4-sulphate, a lysosomal enzyme absent in *Hurler's disease*.

IgA Major class of immunoglobulin of external secretions in mammals, also found in serum. In secretions, found as a dimer (400kD) joined by a short J-chain and linked to a secretory piece or transport piece. In serum found as a monomer (170kD) or a polymer (but without the secretory piece). IgAs are believed to be the main means of providing local immunity against infections in the gut or respiratory tract and may act by reducing the binding between an IgA-coated micro-organism and a host epithelial cell. Present in human colostrum but not transferred across the placenta. Have α heavy chains.

IgD This class of immunoglobulin (184kD) is present at a low level (3–400μg/ml) in serum but is a major immunoglobulin on the surface of B-lymphocytes where it may play a role in antigen recognition. Its structure resembles that of IgG but the heavy chains are of the δ type.

IgE This class of immunoglobulin (188kD) is associated with immediate-type *hypersensitivity* reactions and helminth infections. Present in very low amounts in serum and mostly bound to mast cells and basophils which have an IgE-specific Fc-receptor. IgE has a high carbohydrate content and is also present in external secretions. Heavy chains are of the ε type.

IgG The classic immunoglobulin class, also known as 7S IgG (150kD). Composed of two identical light and two identical heavy chains, the constant region sequence of the heavy chains being of the γ type. The molecule can be described in another way as being composed of two *Fab* and an *Fc* fragment. The Fabs include the antigen combining sites; the Fc region consists of the remaining constant sequence domains of the heavy chains and contains cell-binding and complement-binding sites. IgGs act on pathogens by opsonising them, by activating complement-mediated reactions against cellular pathogens and by neutralising toxins. They can pass across the placenta to the foetus as maternal antibodies, unlike other Ig classes. In humans four main subclasses are known, IgG2 differs from the rest in not being transferred across the placenta and IgG4 does not fix complement. IgG is present at 8–16mg/ml in serum.

IgM Also known as macroglobulin or 19S antibody (970kD). The IgM molecule is built up from five IgG-type monomers joined together with the assistance of J chains to form a cyclic pentamer. IgM binds complement, and a single IgM molecule bound to a cell surface can lyse that cell. IgM is usually produced first in an immune response, before IgG. The human red cell isoantibodies are IgM antibodies. Heavy chain (μ chain) is rather larger than heavy chains of other immunoglobulins.

IgX An immunoglobulin class found in Amphibia.

IL-1, IL-2, etc See *interleukins*.

imaginal disc Epithelial infoldings in the larvae of holometabolous insects (eg. Lepidoptera, *Diptera*) which rapidly develop into adult appendages (legs, antennae, wings etc.) during metamorphosis from larval to adult form. By implanting discs into the haemocoele of an adult insect their differentiation can be blocked, though their *determination* remains unchanged except occasionally (*transdetermination*). The hierarchy of transdetermination has been studied in great detail in **Drosophila**.

iminoglycinuria A defect in amino acid transport leading to abnormal excretion of glycine, proline and hydroxyproline in the urine: more seriously, absorption in the intestine may be inadequate. See *Hartnup disease*.

immaturin Soluble protein produced by

Paramecium caudatum which represses its mating activity.

immortalisation Escape from the normal limitation on growth of a finite number of division cycles (Hayflick limit), by variants in animal cell cultures, and cells in some tumours. Immortalisation in culture may be spontaneous, as happens particularly readily in mouse cells, or induced by mutagens or by *transfection* of certain oncogenes.

immotile cilia syndrome Congenital defect in dynein (either absent or inactive) which leads to male sterility and poor bronchial function. Interestingly, non-ciliated cells show altered locomotion and 50% of patients have *Kartagener's syndrome*.

immune complex Multimolecular antibody–antigen complexes which may be soluble or insoluble depending upon their size and whether or not complement is present. Immune complexes can in some cases be filtered from plasma in the kidney, and the deposition of the complexes gives rise to glomerulonephritis, probably because of the trapping of neutrophils *via* their Fc receptors.

immune complex diseases Diseases characterised by the presence of immune complexes in body fluids. Hypersensitivity of the *Arthus reaction* type and serum sickness are examples.

immune deficiency diseases Those diseases in which immune reactions are suppressed or reduced. Reasons may include congenital absence of B- and/or T-lymphocytes, or viral killing of helper lymphocytes (see *HIV*).

immune response Alteration in the reactivity of an organism's immune system in response to an antigen; in vertebrates, this may involve antibody production, induction of cell-mediated immunity, or of immunological tolerance.

immune response gene See *Ir gene*.

immune surveillance See *immunological surveillance*.

immunity A state in which the body responds specifically to antigen and/or in which a protective response is mounted against a pathogenic agent. May be innate or may be induced by infection or vaccination, or by the passive transfer of antibodies or immunocompetent cells.

immunoadsorbent Any insoluble material, eg. cellulose, with either an antigen or an antibody bound to it and which will bind its corresponding antibody or antigen thus removing it from a solution.

immunoblotting Techniques, such as *Western blotting*, in which very small amounts of protein are transferred from gels to nitrocellulose sheets by electrophoresis and then detected by their antibody binding, usually in combination with peroxidase- or radioactively-labelled IgG. A widely-used technique for the specific recognition of very small amounts of protein.

immunoconglutinin Antibodies which react with fixed complement components. Usually directed against C3b or C4, and found in high titre in patients with rheumatoid arthritis.

immunocytochemistry Techniques for staining cells or tissues using antibodies against the appropriate antigen. Although in principle the first antibody could be labelled, it is more common (and improves the visualisation) to use a second antibody directed against the first (an anti-IgG). This second antibody is conjugated either with fluorochromes, or appropriate enzymes (for colorimetric reactions), or gold beads (for electron microscopy), or with the biotin–avidin system, so that the location of the primary antibody, and thus the antigen, can be recognised.

immunodeficient See *immune deficiency diseases*.

immunoelectrophoresis Any form of *electrophoresis* in which the molecules separated by electrophoresis are recognised by precipitation with an antibody.

immunofluorescence A test or technique in which one or other component of an immunological reaction is made fluorescent by coupling with a *fluorochrome* such as fluorescein, phycoerythrin or rhodamine so that the occurrence of the reaction can be detected as a fluorescing antigen–antibody complex. Used in microscopy to localise small amounts of antigen or specific antibody.

immunogenicity The property of being able to evoke an immune response within an organism. Immunogenicity depends partly upon the size of the substance in question, and partly upon how unlike host molecules it is. Highly conserved proteins tend to have rather low immunogenicity.

immunoglobulins See *IgA, IgD, IgE, IgG*, and *IgM*.

immunological memory The systems responsible for the situation where reactions to a second or subsequent exposure to an antigen are more extensive than those seen on first exposure (but see also *immunological tolerance*). The memory is best explained by clonal expansion and persistence of such clones following the first exposure to antigen.

immunological network The concept due to Jerne that the entire specific immune system within an animal is made up of a series of interacting molecules and cell surface receptors, based on the idea that every antibody combining-site carries its own marker antigens or *idiotypes* and that these in turn may be recognised by another set of antibody combining-sites and so on.

immunological surveillance The hypothesis, proposed by Burnet, that lymphocyte traffic ensures that all or nearly all parts of the vertebrate body are surveyed by visiting lymphocytes in order to detect any altered self material, eg. mutant cells, tumours.

immunological tolerance Specific unresponsiveness to antigen. Self-tolerance is a process occurring normally early in life due to suppression of self-reactive lymphocyte clones. Tolerance to foreign antigens can be induced in adult life by exposure to antigens under conditions in which specific clones are suppressed. Note that tolerance is not the same as immunological unresponsiveness, since the latter may be very non-specific as in immunodeficiency states.

immunoprecipitation The precipitation of a multivalent antigen by a bivalent antibody, resulting in the formation of a large complex. The antibody and antigen must be soluble. Precipitation usually occurs optimally when there is near equivalence between antibody and antigen concentrations.

immunoregulation An imprecise term which might be applied to the various processes by which antibodies may regulate immune responses. At a simple level, secreted antibody neutralises the antigen with which it reacts thus preventing further antigenic stimulation of the antibody-producing clone. At a more complex level, anti-idiotype antibodies can be shown to develop against the first antibodies in some cases, and perhaps further anti-idiotype antibodies against them. This is the major concept of the *immunological network* theory.

immunosuppression This occurs when T- and/or B-clones of lymphocytes are depleted in size or suppressed in their reactivity, expansion or differentiation. It may arise from activation of specific or non-specific T-suppressor lymphocytes of either T- or B-clones, or by drugs that have generalised effects on most or all T- or B-lymphocytes. Cyclosporin A acts on T-cells, as does anti-lymphocyte serum; alkylating agents such as cyclophosphamide are less specific in their action and damage DNA replication, while base analogues interfering with guanine metabolism act in a similar way.

immunotoxins Any toxin that is conjugated to either an immunoglobulin or *Fab* fragment directed against a specified antigen. Thus if the antigen is borne by a particular type of cell, such as a tumour cell, the toxin may be targetted at the specified cell by the immunological reaction.

impermeable cell junction See *zonula occludens*.

inbred strain Any strain of animal or plant obtained by a breeding strategy which tends to lead to homozygosity. Such breeding strategies include brother–sister mating and back-crossing of offspring with parents. See also *congenic*.

inclusion bodies Nuclear or cytoplasmic structures with characteristic staining properties, especially those found at the site of virus multiplication. Semi-crystalline arrays of *virions*, *capsids*, or other viral components.

indirect immunofluorescence A method of fluorescence staining in which the first antibody, which is directed against the antigen to be localised, is used unlabelled, and the location of the first antibody is then detected by use of a fluorescently labelled anti-IgG (against IgGs of the species in which the first antibody was raised). The advantage is that there is some amplification, and a well characterised goat anti-rabbit IgG antibody can, for example, be used against a scarce specific antibody raised

in rabbits. The same technique can be used for ultrastructural localisation of the first antibody by substituting peroxidase- or gold-labelled second antibody.

indole acetic acid (IAA) The most common naturally occurring *auxin*. promotes growth in excised plant organs, induces adventitious root formation, prevents axillary bud growth, regulates *gravitropism* and controls latex flow.

indomethacin Non-steroidal anti-inflammatory drug which blocks the production of arachidonic acid metabolites by inhibiting *cyclo-oxygenase*.

inducer cells Cells which induce other nearby cells to differentiate in specified pathways. Perhaps the distinction should be made, as of old, between those cells that evoke a predetermined pathway of differentiation in the target cells and those cells that can actually induce new and unexpected differentiations.

induction See *embryonic induction* or *enzyme induction*.

infarction Death of tissue as a result of loss of blood supply, often as a result of thrombotic occlusion of vessels.

infectious hepatitis See *hepatitis A*.

infectious mononucleosis See *glandular fever*.

inflammation Response to injury. *Acute inflammation* is characterised by vascular changes and, in some but not all cases, by a predominance of *neutrophil* leucocytes in the early stages, mononuclear phagocytes later on. The cellular part of the response involves local adhesion of leucocytes and their emigration into the tissue through gaps in the endothelial lining of the post-capillary venules. Plasma exudation from vessels may lead to tissue swelling, but the early vascular changes are independent of and not essential for the later cellular response. In type I immediate *hypersensitivity*, for example, the vascular changes occur without cellular infiltration. In chronic inflammation, where the stimulus is persistent and usually antigenic, the characteristic cells are *macrophages* and *lymphocytes*.

influenza virus Member of the *Orthomyxoviridae* which causes influenza in humans. There are three types of influenza virus; Type A causes the world-wide epidemics (pandemics) of influenza and can infect other mammals and birds; Type B only affects humans; Type C causes only a mild infection in humans. Types A and B virus evolve continuously, resulting in changes in the antigenicity of their spike proteins, preventing the development of prolonged immunity to infection. The spike proteins, external haemagglutinin (HA) and *neuraminidase* have been studied as models of membrane glycoproteins.

inhibin Polypeptide hormone secreted by the Graafian follicle and possibly seminiferous tubules, which selectively suppresses the secretion of pituitary FSH (*follicle stimulating hormone*). The molecule has two subunits (14kD and 18kD), and is a product of the gene family which includes *transforming growth factor*-β.

inhibitory synapse A synapse in which an action potential in the presynaptic cell reduces the probability of an action potential occurring in the postsynaptic cell. The most common inhibitory neurotransmitter is *GABA*; this opens channels in the postsynaptic cell which tend to stabilise its *resting potential*, thus rendering it less likely to fire. See also *excitatory synapse*.

initial cell Actively dividing plant cell in a *meristem*. At each division one daughter cell remains in the meristem as a new initial cell, and the other is added to the growing plant body. Animal equivalent is a stem cell (terminology which would be confusing in a plant).

initiation codon (start codon) The *codon* 5'-AUG in messenger RNA, at which polypeptide synthesis is started. It is recognised by formyl-methionyl-tRNA in bacteria and by methionyl-tRNA in eukaryotes.

initiation complex Complex between messenger RNA, 30S ribosomal subunit and formyl-methionyl-tRNA which requires *GTP* and *initiation factors*.

initiation factors (IFs) The set of catalytic proteins required, in addition to mRNA and ribosomes, for protein synthesis to begin. In bacteria three distinct proteins have been identified: IF-1 (8kD), IF-2 (75kD) and IF-3 (30kD). At least 6–8 proteins have been identified in eukaryotes. IFs 1 and 2 enhance the binding of initiator tRNA to the *initiation complex*.

inner cell mass A group of cells found in the mammalian blastocyst which give rise to the embryo and are potentially capable of forming all tissues, embryonic and extra-embryonic, except the trophoblast.

inner sheath The material which encases the two central microtubules of the ciliary axoneme.

insertion sequence (IS elements) Mobile nucleotide sequences which occur naturally in the genomes of bacterial populations. When inserted into bacterial DNA, they inactivate the gene concerned; when they are removed the gene regains its activity. Closely related to transposons and range in size from a few hundred to a few thousand bases, but are usually less than 1500. Characteristically have repeat sequences at both ends.

insertional mutagenesis Mutation or alteration of a DNA sequence as a result of the insertion of DNA.

inside-out patch A variant of the *patch-clamping* technique, in which a disc of plasma membrane covers the tip of the electrode, with the inner face of the plasma membrane facing outward, to the bath.

inside-out vesicle (IOV) Mechanical disruption of cell membranes gives rise to small closed vesicles surrounded by a bilayer membrane. These may be right-side out (ROV), or IOV if the topography is inverted.

insulin A polypeptide hormone (bovine insulin MW 5780) found in both vertebrates and invertebrates. Secreted by the B cells of endocrine pancreas in response to high blood-sugar levels, it induces hypoglycaemia. Defective secretion of insulin is the cause of *diabetes mellitus*. Insulin is also a mitogen, has sequence homologies with other *growth factors*, and is a frequent addition to cell culture media for demanding cell types.

insulin-like growth factor (IGF) IGFs I and II are polypeptides with considerable sequence similarity to insulin. They are capable of eliciting some of the same biological responses, including mitogenesis in cell culture. On the cell surface, there are two types of IGF receptor, one of which (IGF-1 receptor) closely resembles the insulin receptor (which is also present). IGF I = somatomedin A = somatomedin C; IGF II

= MSA (Multiplication-stimulating activity). IGF-1 is released from the liver in response to *growth hormone*.

integral membrane protein A protein that is firmly anchored in a membrane (unlike a peripheral membrane protein). Most is known about the integral proteins of the plasma membrane, where important examples include hormone receptors, ion channels, and transport proteins. An integral protein need not cross the entire membrane; those that do are referred to as transmembrane proteins.

integrase protein Enzyme of the bacteriophage λ which catalyses the integration of λ DNA into the host DNA.

integrins Superfamily of cell surface proteins which are involved in binding to extracellular matrix components in some cases. Most are heterodimeric with a β subunit of 95kD which is conserved through the superfamily, and a more variable α subunit of 150–170kD. The first examples described were *fibronectin* and *vitronectin* receptors of fibroblasts, which bind to an RGD (Arg-Gly-Asp) sequence in the ligand protein, though the "context" of the RGD seems important and there is also a divalent cation-dependence. Subsequently the platelet IIb/IIIa surface glycoprotein (fibronectin and fibrinogen receptor) and the *LFA-1* class of leucocyte surface protein were recognised as integrins, together with the *VLA* surface protein. The requirement for the RGD sequence in the ligand does not seem to be invariable, and the LFA-1 class will bind *lipopolysaccharide* which does not have RGD, as well as *C3bi* which does. See Table I1.

intercalated disc An electron-dense junctional complex, at the end-to-end contacts of cardiac muscle cells, that contains *gap junctions* and *desmosomes*. Most of the disc is formed of a convoluted fascia adhaerens type of junction into which the actin filaments of the terminal sarcomeres insert (they are therefore equivalent to half Z-discs); desmosomes are also present. The lateral portion of the stepped disc contains gap junctions which couple the cells electrically and thus coordinate the contraction.

intercalation Insertion into a pre-existing structure; eg. (a) nucleotide sequences into DNA (or RNA), (b) molecules into structures such as membranes.

intercellular Between cells: can be used either in the sense of connections between cells (as in intercellular junctions), or as an antonym for intra-cellular.

interdigitating cells Cells found particularly in thymus-dependent regions of lymph nodes; they have dendritic morphology and *accessory cell* function.

interference diffraction patterns The patterns arising from the recombination of beams of light or other waves after they have been split and one set of rays have undergone a phase retardation relative to the other. Such patterns formed by simple objects give information on the correctness of the focus and the presence or absence of optical defects.

interference microscopy Although all image formation depends on interference, the term is generally restricted to systems in which contrast comes from the recombi-

Table I1. Vertebrate integrins

Family	Designations		Ligands
$\alpha_1 \beta_0$	—	(Chick integrin 1)	—
$\alpha_1 \beta_1$	VLA1	—	
$\alpha_1 \beta_2$	VLA2	Platelet Ia/IIa	Collagens
$\alpha_1 \beta_3$	VLA3	Chick integrin	Collagen, fibronectin, laminin
$\alpha_1 \beta_4$	VLA4	—	—
$\alpha_1 \beta_{5(F)}$	VLA5	"Fibronectin receptor" Ic/IIa (disputed)	Fibronectin
$\alpha_1 \beta_6$	VLA6	Platelet Ic/IIa	Laminin
$\alpha_2 \beta_L$	LFA-1, CD11a		ICAM-1
$\alpha_2 \beta_M$	Mac-1, Mo-1, CR3, CD11b		Fibrinogen, C3bi, Factor X
$\alpha_2 \beta_X$	gp150,95, CR4, CD11c		C3bi
$\alpha_3 \beta_{IIb}$	Platelet IIb/IIIa		Fibrinogen, vitronectin, fibronectin, von Willebrand factor
$\alpha_3 \beta_V$	"Vitronectin receptor"		Vitronectin

nation of a reference beam with light that has been retarded by passing through the object. Because the phase retardation is a consequence of the difference in refractive index between specimen and medium, and because the refractive increment is almost the same for all biological molecules, it is possible to measure the amount of dry mass per unit area of the specimen by measuring the phase retardation. Quantification of the phase retardation is usually done by using a compensator to reduce the bright object to darkness (see *Sénarmont* and *Ehrlinghaus compensators*). Two major optical systems have been used — the *Jamin–Lebedeff system* and the *Mach–Zehnder system*. These instruments are often referred to as interferometers, since they are designed for measuring phase retardation. Although their use has passed out of fashion, it may be that they will be employed more frequently in future in conjunction with image analysing systems.

interference reflection microscopy An optical technique for detecting the topography of the side of a cell in contact with a planar substrate, and for providing information on the separation of the plasmalemma from the substrate. Interference between the reflections from the substrate–medium interface and the reflections from the plasmalemma–medium interface generate the image.

interferons (IFN) A family of glycoproteins produced in mammals which prevent virus multiplication in cells. IFN-α is made by leucocytes and IFN-β by fibroblasts after viral infection. IFN-γ is produced by immune cells after antigen stimulation; it is the major macrophage-activating factor and also induces expression of Class II major *histocompatibility antigens* on many cell types. IFN-γ is an important lymphokine

and immunostimulatory molecule. IFN-α and -β are also known as Type I and IFN-γ as Type II interferons.

interleukin A variety of substances produced by leucocytes (not necessarily exclusively) and which function during inflammatory responses. (This is the definition based on a recommendation by the IUIS–WHO Nomenclature Committee). Interleukins are of the larger class of *lymphokines*.

interleukin-1 (IL-1) (lymphocyte activating factor; LAF) Protein (17kD: 152 amino acids) secreted by *macrophages* or *accessory cells* involved in the activation of both T- and B-lymphocytes in response to antigens or mitogens, as well as affecting a wide range of other cell types, including central nervous system (*endogenous pyrogen*) and vascular endothelium. At least two IL-1 genes are active and α and β forms of IL-1 are recognised. See also *catabolin*.

interleukin-2 (IL-2) (T-cell growth factor) Released by antigen- or mitogen-stimulated T-lymphocytes in *G1*, causes activation and differentiation of other T-lymphocytes independently of antigen. T-cells must make both IL-1 and IL-2 receptors to proceed to *S phase*. IL-2 also acts on B-cells.

interleukin-3 (IL-3) Product of mitogen-activated T-cells: colony stimulating factor for bone-marrow stem cells and *mast cells*.

interleukin-4 (IL-4) Also known as: B-cell stimulating factor (BSF-1), B-cell differentiation factor-γ (BCDF-γ), B-cell growth factor-1 (BCGF-1), and IgG1-induction factor. Acts early in B-cell activation (*G0–G1* transition).

interleukin-5 (IL-5) A B-cell growth and differentiation factor; also stimulates eosinophil precursor proliferation and differentiation.

Table I2. Intermediate filaments and sequence-related proteins

Name of protein	Sequence homology group	MW (kD)	Cell type
Cytokeratins (19 in humans)	I (acidic)	40–60	Epithelia
	II (neutral-basic)	50–70	
Vimentin	III	53	Many, especially mesenchymal
Desmin	III	52	Muscle
Glial fibrillary acidic protein	III	51	Glial cells. Astrocytes
Neurofilament polypeptides	IV	57–150	Neurons (vertebrates)
		60–120	Neurons (invertebrates)
Nuclear Lamins	V	60–70	All eukaryotic cells

interleukin-6 (IL-6) (interferon β₂) Recently identified cytokine that is described as being: co-induced with interferon from fibroblasts, a B-cell differentiation factor, a hybridoma growth factor, an inducer of acute phase proteins, and a colony stimulating factor acting on mouse bone marrow. Some of these proposed functions may disappear as its role is consolidated.

interleukin-7 Lymphopoietin 1.

intermediate filaments A class of cytoplasmic filaments of animal cells so named originally because their diameter (nominally 10nm) in muscle cells was intermediate between thick and thin filaments. Unlike microfilaments and microtubules, the protein subunits of intermediate filaments show considerable diversity and tissue specificity. See *cytokeratins*, *desmin*, *glial fibrillary acidic protein*, *vimentin* and Table I2.

intermembrane space Region between the two membranes of mitochondria and chloroplasts. On the *endosymbiont hypothesis*, this space would represent the original phagosome.

internal bias Applied to the motile behaviour of crawling cells which, in the short term, show *persistence* and do not behave as true random walkers. Any intrinsic regulation of the random motile behaviour of the cell could be considered as internal bias.

internal membranes General term for intracellular membrane systems such as endoplasmic reticulum. Not particularly helpful, but has the advantage of being non-committal.

interneurons Neurons which connect only with other neurons, and not with either sense organs or muscles. They are thus involved in the intermediate processing of signals.

interphase The stage of the cell or nucleus when it is not in mitosis, hence comprising most of the cell cycle.

interstitial cells 1. Cells lying between but distinct from other cells in a tissue, a good example being the interstitial cells in *Hydra* which serve as stem cells. 2. Cells lying between the testis tubules of vertebrates and which are responsible for the secretion of testosterone.

intervening sequence Alternative but uncommon name for an *intron*.

intestinal epithelium The endodermally-derived epithelium of the intestine varies considerably, but the absorptive epithelium of small intestine is usually implied. The apical surfaces of these cells have *microvilli* (possibly to increase the absorptive surface, but probably also to provide a larger surface area for enzyme activity). The lateral subapical regions have well-developed *junctions*.

intima (tunica intima) Inner layer of blood vessel, comprising an endothelial monolayer on the luminal face with a subcellular elastic extracellular matrix containing a few smooth muscle cells. Below the intima is the *media*, then the *adventitia*. The term may be applied to other organs.

intine Inner layer of the wall of a pollen grain, resembling a *primary cell wall* in structure and composition. Also used for the inner wall layer of a *spore*.

intramembranous particles (IMP) Particles (or complementary pits) seen in *freeze fractured* membranes. The cleavage plane is through the centre of the bilayer, and the particles are usually assumed to represent *integral membrane proteins* (or polymers of such proteins).

intrinsic factor A mucoprotein normally secreted by the epithelium of the stomach and which binds vitamin B₁₂; the intrinsic factor/B₁₂ complex is selectively absorbed by the distal ileum, though only the vitamin is taken into the cell.

intron A non-coding sequence of DNA within a gene (cf *exon*), which is transcribed into *hnRNA* but is then removed by RNA splicing in the nucleus, leaving a mature *mRNA* which is then translated in the cytoplasm. Introns are poorly conserved and of variable length, but the regions at the ends are self-complementary, allowing a hairpin structure to form naturally in the hnRNA; this is the cue for removal by RNA splicing. Introns are thought to play an important role in allowing rapid evolution of proteins by exon shuffling. Genes may contain as many as 80 introns.

inulin A polysaccharide of variable molecular weight (around 5kD), that is a polymer of fructofuranose. Widely used as a marker of extracellular space, an indicator of blood volume in insects (by measuring the dilution of the radio-label), and in food for diabetics.

invasion A term which should be used with caution; although most cell biologists would follow Abercrombie in meaning the movement of one cell type into a territory normally occupied by a different cell type, some pathologists might not agree.

invasion index An index devised by Abercrombie and Heaysman as a means to estimate the invasiveness of cells *in vitro*. The index is derived from measurements on confronted explants of the cells and embryonic chick heart fibroblasts: it is the ratio of the estimated movement, had the cells not been hindered, and the actual movement in the zone in which collision occurs.

inversion heterozygote Individual in which one chromosome contains an inversion whereas the homologous chromosome does not.

invertase ("sucrase") Enzyme catalysing the hydrolysis of sucrose to glucose and fructose, so-called because the sugar solution changes from dextro-rotatory to laevo-rotatory during the course of the reaction.

ion channel A transmembrane pore which presents a hydrophilic channel for ions to cross a lipid bilayer down their electrochemical gradients. Some degree of ion specificity is usually observed, and typically a million ions per second may flow (see *transport proteins*). Channels may be permanently open, like the potassium leak channel; or they may be *voltage-gated*, like the *sodium channel*; or *ligand-gated* like the acetylcholine receptor.

ion-exchange chromatography Separation of molecules by absorption and desorption from charged polymers. An important technique for protein purification. For small molecules the support is usually polystyrene, but for macromolecules, cellulose, polyacrylamide or agarose supports give less non-specific absorption and denaturation. Typical charged residues are CM (carboxymethyl) or DEAE (diethylaminoethyl).

ion-selective electrode An electrode half-cell, with a *semipermeable membrane*, which is permeable only to a single ion. The electrical potential measured between this and a reference half-cell (eg. a calomel electrode) is thus the *Nernst* potential for the ion. Given that the solution filling the ion-selective electrode is known, the activity (rather than concentration) of the ion in the unknown solution can be measured. Commercial ion-selective electrodes frequently use a hydrophobic membrane containing an *ionophore*, such as *valinomycin* (for potassium) or *monensin* (for sodium). A pH electrode is made with a thin membrane of pH-sensitive (ie. proton-permeable) glass.

ionic coupling The same as *electrical coupling*.

ionising radiation Radiation capable of ionising, either directly or indirectly, the substances it passes through. α and β radiation are far more effective at producing ionisation (and therefore are more likely to cause tissue or cell damage) than γ radiation or neutrons.

ionophore A molecule which allows ions to cross lipid bilayers. There are two classes: carriers and channels. Carriers, like *valinomycin*, form cage-like structures around specific ions, diffusing freely through the hydrophobic regions of the bilayer. Channels, like *gramicidin*, form continuous aqueous pores through the bilayer, allowing ions to diffuse through. See *ion channels* and Table I3.

Ir associated antigens Antigens coded for by IR (immune response) genes or antigens coded for by the genome close to the *Ir genes*. See also *Ia antigens*.

Ir genes Immune response genes, located within the major histocompatibility complex of vertebrates. Originally recognised as controlling the level of immune response to various synthetic polypeptides, they are now also recognised as mapping within the regions controlling T-cell help and suppression.

iridoviruses A group of non-occluded viruses of insects; the crystalline array of virus particles in the cytoplasm of epidermal cells gives infected insects an iridescent appearance.

Is element See *insertion sequence*.

ischaemia (USA ischemia) Inadequate blood flow leading to hypoxia in the tissue.

ISCOMS (immunostimulatory complexes) Small cage-like structures which make it possible to present viral proteins to the immune system in an array, much as they would appear on the virus. Produced by mixing the viral protein with Quill A, a substance isolated from the Amazonian oak, in the presence of detergent. ISCOMS are being used successfully in vaccines.

islet cells Cells of the *Islets of Langerhans* within the pancreas. See *A cells, B cells, D cells*.

Islets of Langerhans Groups of cells found within the pancreas: *A cells* and *B cells* secrete *insulin* and *glucagon*. See also *D cells*.

isocitrate An intermediate in the *tricarboxylic acid cycle*.

isoelectric focusing *Electrophoresis* in a stabilised pH gradient. High-resolution method for separating molecules, especially proteins, which carry both positive and negative charges. Molecules migrate to the pH corresponding to their *isoelectric point*. The gradient is produced by electrophoresis of amphiphiles, which are heterogenous molecules giving a continuum of isoelectric points. Resolution is determined by the number of amphiphile species and the evenness of distribution of their isoelectric points.

isoelectric point The pH at which a protein carries no net charge. Below the isoelectric point proteins carry a net positive charge; above it a net negative charge. Due to a preponderance of weakly acid residues in almost all proteins, they are nearly all negatively charged at neutral pH. The isoelectric point is of significance in protein purification because it is the pH at which solubility is often minimal, and at which mobility in an electrofocusing system is zero (and therefore the point at which the protein will accumulate).

isoenzymes Variants of enzymes which catalyze the same reaction, but owing to differences in amino acid sequence can be distinguished by techniques such as electrophoresis or isoelectric focusing. Different tissues often have different isoenzymes. The sequence differences generally confer different enzyme-kinetic parameters which can sometimes be interpreted as fine-tuning to the specific requirements of the cell types in which a particular isoenzyme is found.

Table I3. Ionophores[a]

	MW (D)	Ion selectivity	Comments
Neutral:			
Valinomycin	1110	Rb > K > Cs > Ag >> NH$_4$ > Na > Li	Depsipeptide, uncoupler
Enniatin A	681	K > Rb ≈ Na > Cs >> Li	Cyclic hexadepsipeptides
Enniatin B	639	Rb > K > Cs > Na >> Li; Ca > Ba > Sr > Mg	
Nonactin	736	NH$_4$ > K ≈ Rb > Cs > Na	Macrotetralide, product of *Actinomyces* strains
Monactin	750	NH$_4$ > K > Rb > Cs > Na > Ba	Macrotetralide
Cryptate 211	288	Li > Na > K ≈ Rb ≈ Cs; Ca > Sr ≈ Ba	Amino ether; one of a substantial family
Carboxylic:			
Monensin	670	Na >> K > Rb > Li > Cs	Blocks transport through Golgi
Nigericin	724	K > Rb > Na > Cs >> Li	
X-537A (lasalocid)	590	Cs > Rb ≈ K > Na > Li; Ba > Sr > Ca > Mg	Macrotetralide
A23187	523	Li > Na > K; Mn > Ca > Mg > Sr > Ba	Predominantly selective for divalent cations
Channel-forming			
Gramicidin A	~1700	H > Cs ≈ Rb > NH$_4$ > K > Na > Li	Peptide; works as a dimer
Alamethicin	—	K > Rb > Cs > Na	Peptide; voltage dependent
Monazomycin	~1422	Cs > Rb > K > Na > Li	Polyene-like; voltage sensitive

[a]Many uncouplers such as FCCP (carbonyl cyanide-trifluoro-methoxyphenylhydrazone) also act as ionophores. Amphotericin, filipin and nystatin may be anion-specific ionophores.

isoform A protein having the same function and similar (or identical) sequence, but the product of a different gene and (usually) tissue-specific. Rather stronger in implication than "homologous".

isohaemagglutinins Natural antibodies that react against normal antigens of other members of the same species, eg. ABO blood group antibodies.

isoleucine (Ilu; I; MW 131) Hydrophobic amino acid. See Table A2.

isomers Alternative forms of molecules containing the same atoms.

isometric tension Tension generated in a muscle without contraction occurring: cross-bridges are being re-formed with the same site on the thin filament, and the tension (in striated muscle) is proportional to the overlap between thick and thin filaments.

isoproterenol (isoprenaline: isopropyl-noradrenaline) Synthetic β-adrenergic agonist; causes peripheral vasodilation, bronchodilation, and increased cardiac output.

isotonic Of a fluid; of a concentration which will not cause osmotic volume changes of cells immersed in it. Note that an isotonic solution is not necessarily isoosmotic. See *hypertonic*.

isotonic contraction Contraction of a muscle, the tension remaining constant. Since the contractile force is proportional to the overlap of the filaments, and the overlap is varying, the numbers of active cross-bridges must be changing.

isotropic environments Environments in which the properties are the same at all points, and there are no vectorial or axial cues.

isotype 1. Applied to a set of macromolecules sharing some features in common. In immunology, isotype describes the class, subclass, light chain type and subtype of an immunoglobulin. 2. Antigenic determinant that is uniquely present in individuals of a single species. 3. A conventionalised method for the graphical display of statistical data.

isotype switching The switch of immunoglobulin isotype which occurs, for example, as the immune response progresses (IgM to IgG). The switch from IgM to IgG involves only the constant region of the heavy chains (from μ to γ), the light chain and variable regions of the heavy chain remaining the same, and involves the switch regions, upstream (on the 5′ side) of the constant region genes, at which recombination occurs. Similarly, IgM and IgD with the same variable region of the heavy chain, but with different heavy chain constant regions (μ and δ), seem to coexist on the surface of some lymphocytes.

isotypic variation Variability of antigens common to all members of a species, for example the five classes of immunoglobulins found in humans. See *idiotype* and *allotype*.

isozyme See *isoenzyme*.

J

J chain Also known as J-piece. A polypeptide chain (15kD) found in IgA and in IgM joining heavy chains (H chains) to each other to form dimers of IgA and pentamers of IgM. Disulphide bonds are formed between the J chain and H chains near the Fc ends of the heavy chains. Despite the similar name, it is not identical with the *J region* or coded for by the *J gene*.

J gene Gene(s) coding for the Joining segment of polypeptide chain which links the V (variable regions) to the C (constant) regions of both light and heavy chains of immunoglobulins. During lymphoid development the DNA is rearranged so that the V genes are linked to the J region sequences.

J region The polypeptide chains coded for by *J gene*.

Jamin–Lebedeff system *Interference microscopy* in which object and reference beams are split and later recombined by birefringent calcite plates, but pass through the same optical components (in contrast to the *Mach–Zehnder system*).

jaundice Yellowing of the skin (and whites of eyes) by bilirubin, a bile pigment.

JC virus Member of the human *Papovaviridae* similar to *polyoma virus*, but which has not been found associated with any human cancer.

Job's syndrome Rather poorly-defined syndrome, thought to be due to a defect in neutrophil chemotaxis which predisposes to infection by staphylococci, often without the normal signs of inflammation ("cold abscesses"). All patients described have been female, with red hair, and elevated plasma IgE levels.

jumping gene Populist term for *transposon*.

junctional basal lamina Specialised region of the extracellular matrix surrounding a muscle cell, at the neuromuscular junction. May be responsible for localisation of *acetylcholine* receptors in the synaptic region, and also binds *acetylcholine esterase* to this region.

junctions See *adhaerens junction, desmosome, gap junction, zonula occludens*.

juvenile hormone (JH) A hormone found in insects which affects the balance between mature and juvenile attributes of certain tissues at each moult. In particular, the *imaginal discs* of many larval insects only develop into adult wings, sexual organs or limbs when haemolymph juvenile hormone levels fall below a threshold level. There is a complex interaction between juvenile hormone and *ecdysone*. Synthetic analogues of JH, which include farnesol and methoprene, have been tested for insecticide potential (known, with diflubenzuron, as Insect Growth Regulators, IGRs; see also *chitin*).

K

K cells See *killer cells*.

K_a 1. Acid *dissociation constant*. Often encountered as pK_a (ie. $-\log_{10} K_a$). 2. Association constant (K_{ass}). The equilibrium constant for association, the reciprocal of K_d, with dimensions of litres/mole. Better to use K_d, thereby removing any ambiguity.

kainate An agonist for the K-type *excitatory amino acid* receptor. It can act as an *excitotoxin* producing symptoms similar to those of Huntingdon's chorea, and is also used as an anthelminthic drug. Originally isolated from the alga *Digenea simplex*.

kainic acid See *kainate*.

kallidin Decapeptide (lysyl-bradykinin; amino acid sequence KRPPGFSPFR) produced in kidney. Like *bradykinin*, an inflammatory mediator (a *kinin*); causes dilation of renal blood vessels and increased water excretion.

kallikrein Plasma proteases normally present as inactive prekallikreins which are activated by *Hagemann factor*. Act on *kininogens* to produce *kinins*.

Kanagawa haemolysin See Table E1.

kanamycin Aminoglycoside antibiotic. See Table A3.

Kaposi's sarcoma A sarcoma of spindle cells mixed with angiomatous tissue, found on the skin. Usually classed as an angioblastic tumour. A fairly frequent concomitant to *HIV* infection or long-term immunosuppression.

kappa (κ) class of light chain Light chains (*L chains*) of mammalian immunoglobulins are of two types, kappa and lambda, referring to sequence differences in their constant domain. In a mammal heterozygous for the kappa/lambda isotypes both types of chain are found but each immunoglobulin is either kappa or lambda.

kappa particle Gram negative bacterial endosymbiont of *Paramecium* spp., (*Caedobacter taeniospiralis*), that confers the "killer" trait; infected *Paramecium* are resistant to the toxin liberated by infected forms. Killing activity is associated with the *induction* of defective phage in the endosymbiont, leading to the release of R-bodies, coded for by the phage genome and apparently composed of mis-assembled phage-coat protein.

Kartagener's syndrome (situs inversus) Condition in which the normal asymmetry of the viscera is reversed. Associated with a *dynein* defect (dynein is absent or dysfunctional in some cases) and with *immotile cilia syndrome*.

karyoplast A nucleus isolated from a eukaryotic cell surrounded by a very thin layer of cytoplasm and a plasma membrane. The remainder of the cell is a cytoplast.

karyorrhexis Degeneration of the nucleus of a cell. There is contraction of the chromatin into small pieces, with obliteration of the nuclear boundary.

karyotype The complete set of chromosomes of a cell or organism. Used especially for the display prepared from photographs of mitotic chromosomes arranged in homologous pairs.

K_d (K_{diss}) An equilibrium constant for dissociation. Thus for the reaction $A + B \rightleftharpoons C$, at equilibrium $K_d = [A][B]/[C]$ and has dimensions of moles/litre. K_d is the reciprocal of K_{eq}. The concept of K_d is perhaps more readily understood than that of K_{eq}; for example, in considering the conversion of A to C by ligand B, the $K_d = [B]$ when $[A] = [C]$ ie. is the concentration of ligand which produces half-maximal conversion (response).

keloid A large bulging scar, the result of excess collagen production. Tendency to produce keloids seems to be heritable (particularly in Negroes) and is associated in some cases with low plasma fibronectin levels.

K_{eq} The equilibrium constant for a reversible reaction. $K_{eq} = [AB]/[A][B]$.

keratan sulphate See *glycosaminoglycans*.

keratinising epithelium An epithelium such as vertebrate epidermis in which a keratin-rich layer is formed from intracellular *cytokeratins* as the outermost cells die.

keratins Group of highly insoluble fibrous proteins (of high α-helical content) which are found as constituents of the outer layer of vertebrate skin and of skin-related structures such as hair, wool, hoof and horn, claws, beaks and feathers. Extracellular keratins are derived from cytokeratins, a large and diverse group of intermediate filament proteins.

keratitis Inflammation of the cornea, often associated with herpes virus I infection and with congenital syphilis.

keratohyalin granules Granules found in living cells of keratinising epithelia and which contribute to the keratin content of the dead cornified cells. Some, but not all, contain sulphur-rich keratin.

keratoses (actinic keratoses) Benign but precancerous lesions of skin associated with ultraviolet irradiation.

α-ketoglutarate Intermediate of the *tricarboxylic acid cycle*, also formed by deamination of glutamate.

ketosis Metabolic production of abnormal amounts of ketones. A consequence of *diabetes melittus*.

killer cells 1. (K cells) Mammalian cells which can lyse antibody-coated target cells. They have a receptor for the Fc portion of IgG, and are often of the mononuclear phagocyte lineage, though some may be lymphocytes. Not to be confused with *cytotoxic T-cells* (CTL) which recognise targets by other means and are clearly a sub-set of T-lymphocytes: this confusion exists in the early literature. See also *ADCC*. 2. (NK cells) Natural killer cells are CD3-negative large granular lymphocytes, mediating cytolytic reactions that do not require expression of Class I or II major *histocompatibility antigens* on the target cell. 3. (LAK cells) Lymphokine-activated killer cells are NK cells activated by *interleukin-2*.

kinase Widely used abbreviation for phosphokinase, an enzyme catalyzing transfer of phosphate from ATP to a second substrate usually specified in less abbreviated name, eg. creatine phosphokinase (*creatine kinase*), *protein kinase*.

kinesin Cytoplasmic protein (110kD) which is responsible for moving vesicles and particles towards the distal (+) end of microtubules. Differs from cytoplasmic dynein (*MAP1C*) in the direction in which it moves

particles and its relative insensitivity to vanadate.

kinesis Alteration in the movement of a cell, without any directional bias. Thus speed may increase or decrease (*orthokinesis*) or there may be an alteration in turning behaviour (*klinokinesis*). See *chemokinesis*.

kinetin (6-furfurylaminopurine) A *cytokinin* used as a component of plant tissue culture media. Obtained by heat-treatment of DNA, and does not occur naturally in plants.

kinetochore Multilayered structure, a pair of which develop on the mitotic chromosome, adjacent to the *centromere*, and to which spindle microtubules attach — but not at the end normally associated with a *microtubule organising centre*.

kinetodesma Longitudinally-oriented cytoplasmic fibrils associated with, and always on the right of, the *kinetosomes* of ciliate protozoa.

kinetoplast Stainable mass of mitochondrial DNA in flagellate protozoa, usually adjacent to flagellar basal body.

kinetosome *Basal body* of cilium: term used mostly of ciliates.

kinety A row of *kinetosomes* and associated *kinetodesmata* in a ciliate protozoan.

kininogen Inactive precursor in plasma from which *kinin* is produced by proteolytic cleavage.

kinins Inflammatory mediators which cause dilation of blood vessels and altered vascular permeability. Kinins are small peptides produced from *kininogen* by *kallikrein*, and are broken down by kininases. Act on phospholipase and increase arachidonic acid release and thus prostaglandin (PGE2) production. See *bradykinin*, *kallidin*.

Kirsten sarcoma virus A murine sarcoma-inducing retrovirus, generated by passaging a murine erythoblastosis virus in newborn rats. Source of the *Ki-ras* oncogene.

Klebsiella Genus of Gram negative bacteria, non-motile and rod-like, associated with respiratory, intestinal and urinogenital tracts of mammals; *K. pneumoniae* associated with pneumonia in humans. Also a genus of euglenid flagellate protozoa.

Klenow fragment Larger part of the bacterial DNA polymerase I (76kD) which remains after treatment with subtilisin; retains some but not all exonuclease and polymerase activity.

Klinefelter's syndrome Human genetic abnormality in which the individual, phenotypically apparently male, has three sex chromosomes (XXY).

klinokinesis *Kinesis* in which the frequency or magnitude of turning behaviour is altered. Bacterial chemotaxis can be considered as an adaptive klinokinesis; the probability of turning is a function of the change in concentration of the substance eliciting the response.

K$_m$ (Michaelis constant) A kinetic parameter used to characterise an enzyme, defined as the concentration of substrate which permits half-maximal rate of reaction.

km-fibres Bundles of microtubules running longitudinally below and to one side of the bases of cilia in a *kinety*.

Krebs cycle *Tricarboxylic acid cycle* or citric acid cycle.

Kupffer cell Specialised macrophage of the liver sinusoids; responsible for the removal of particulate matter from the circulating blood (particularly old erythrocytes).

Kurloff cells Cells found in the blood and organs of guinea pigs that contain large secretory granules but are of unknown function.

Kuru Degenerative disease of the central nervous system found in members of the Fore tribe of New Guinea: a *slow virus* infection similar to *scrapie* and *Creutzfeld–Jacob disease*, and thought to be transmitted by cannibalism.

L

L-cell Mouse (methylcholanthrene-induced) sarcoma cells maintained as a line in culture since the 1950s.

L-chain (light chain) Although *light chains* are found in many multimeric proteins, L-chain usually refers to the light chain of an immunoglobulin. These are of 22kD and of one of two types, *kappa* (κ) or lambda (λ). A single immunoglobulin has identical light chains (2κ or 2λ). Light chains have one variable and one constant region. There are isotype variants of both κ and λ.

L-forms Bacteria lacking cell walls, a phenomenon usually induced by inhibition of cell-wall synthesis, sometimes by mutation.

L-ring Outermost ring of the basal part of the bacterial flagellum in Gram negative bacteria. It may serve as a bush to anchor the flagellum relative to the lipopolysaccharide layer.

La protein Protein (45kD) transiently bound to unprocessed cellular precursor RNAs which have been produced by polymerase III. Mainly located in the nucleus.

***lac* operon** See *lactose operon*.

β-lactamase Enzyme secreted by many species of bacteria which inactivates penicillins by opening the lactam ring.

lactate (2-hydroxypropionic acid) Important as the terminal product of anaerobic glycolysis. Accumulation of lactate in tissues is responsible for the so-called oxygen debt.

lactic dehydrogenase (LDH) The enzyme which catalyses the formation and removal of lactate according to the equation: pyruvate + NADH = lactate + NAD. The appearance of LDH in the medium is often used as an indication of cell death and the release of cytoplasmic constituents.

lactoferrin Iron-binding protein of very high affinity (K_d 10^{-19} at pH 6.4; 26-fold greater than that of *transferrin*) found in milk and in the specific granules of neutrophil leucocytes.

lactoperoxidase Peroxidase enzyme from milk which finds an important use in generating active iodine as a non-permeant radiolabel for membrane proteins.

lactose (4-O-β-ᴅ-galactopyranosyl β-ᴅ-glucose) The major sugar in human and bovine milk. Conversion of lactose to lactic acid by *Lactobacilli* etc. is important in the production of yoghurt and cheese.

lactose carrier protein The best known example is the product of the *lacY* gene, coded for in the *lactose operon* and responsible for the uptake of lactose by *E. coli*.

lactose operon Group of adjacent and coordinately controlled genes concerned with the metabolism of lactose in *E. coli*. The lac operon was the first example of a group of genes under the control of an *operator* region to which a *lactose repressor* binds. When the bacteria are transferred to lactose-containing medium, allolactose (which forms by transglycosylation when lactose is present in the cell) binds to the repressor, inhibits the binding of the repressor to the operator, and allows transcription of mRNA for enzymes involved in lactose transport across the membrane and metabolism (β-galactosidase, galactoside permease and thiogalactoside transacetylase).

lactose repressor Protein (tetramer of 37kD subunits) which normally binds with very high affinity to the *operator* region of the *lactose operon* and inhibits transcription of the downstream genes by blocking access of the polymerase to the promoter region. When the lactose repressor binds allolactose, its binding to the operator is reduced and the gene set is derepressed.

lacuna Space or gap; for example, the space in bone where an *osteoblast* is found.

λ bacteriophage Bacterial DNA virus, first isolated from *E. coli*, with structure similar to that of the T-even phages. It has cycles of lysis and of *lysogeny*; studies on the control of these alternative cycles have been very important in understanding the regulation of gene *transcription*. The phage is used as a cloning *vector*, accommodating fragments of DNA up to 15 kilobase pairs long. For larger pieces, the cosmid vector was constructed from its ends.

lambda chain See *L-chain*.

lamellar phase See *phospholipid bilayer*.

lamellipodium Flattened projection from the surface of a cell, often associated with locomotion of fibroblasts.

lamina Flat sheet; as in *basal lamina*.

laminarin Storage polysaccharide of *Laminaria* and other brown algae; made up of β(1–3)-glucan with some β(1–6)-linkages.

laminin Glycoprotein (α subunits 200kD, β subunits 400kD) of the extracellular matrix, which is a major component of basement membranes. Induces adhesion and spreading of many cell-types in culture and promotes neurite outgrowth. Originally isolated from line of mouse *EHS cells*.

lamins Proteins which form the nuclear lamina, a polymeric structure intercalated between chromatin and the inner nuclear envelope. Lamins A and C (70 and 60kD respectively) have C-terminal sequences homologous to the head and tail domains of *keratins*; their peptide maps are similar, and significantly different from that of lamin B (67kD), although there are some common epitopes.

lampbrush chromosomes Large chromosomes (as long as 1mm), actually meiotic *bivalents*, seen during prophase of the extended meiosis in the oocytes of some Amphibia. Segments of DNA form loops in pairs along the sides of the sister chromosomes, giving them a brush-like appearance. These loops are not permanent structures but are formed by the unwinding of *chromomeres* and represent sites of very active RNA synthesis.

Langerhans cells 1. See *Islets of Langerhans*. 2. *Dendritic cells* of epidermis. It is thought that they may transport antigen to lymph nodes where they act as *accessory cells*.

Langhans' giant cells Multinucleate cells formed by fusion of epithelioid macrophages and associated with the central part of early tuberculous lesions. Similar to *foreign body giant cells*, but with the nuclei peripherally located.

Langmuir trough A device for studying the properties of lipid monolayers at an air/water interface.

lanthanum Lanthanum salts are used as a negative stain in electron microscopy, and as calcium-channel blockers.

Lassa fever See *Lassa virus*.

Lassa virus Virulent and highly transmissible arenavirus whose normal host is a rodent (*Mastomys natalensis*); first recorded from Nigeria.

latent virus Virus genome maintained within the host cell, usually in the absence of infective particles; may be reactivated to produce infective particles by stress such as ultraviolet irradiation.

lateral diffusion Diffusion in two dimensions, usually referring to movement in the plane of the membrane, such as the motion of fluorescently-labelled lipids or proteins measured by the technique of *fluorescent recovery after photobleaching* (FRAP).

lateral inhibition A simple form of information processing. The classic example is found in the eye, whereby ganglion cells are stimulated if photoreceptors in a well-defined field are illuminated, but their response is inhibited if neighbouring photoreceptors are excited (an "on field/off surround" cell), or *vice versa* an "off field/on surround" cell. The effect of lateral inhibition is to produce edge- or boundary-sensitive cells, and to reduce the amount of information which is sent to higher centres; a form of peripheral processing.

lazy leucocyte syndrome A rare and possibly apocryphal human complaint in which neutrophils display poor locomotion towards sites of infection.

LD$_{50}$ That dose of a substance or infectious agent which causes death in 50% of the organisms to which it has been administered. Routine use of LD$_{50}$ tests in animals is now being replaced by more sensitive (and less wasteful) methods.

LDL See *low density lipoprotein*.

LE body A globular mass of nuclear material which stains with haematoxylin; associated with lesions of *systemic lupus erythematosus*.

LE cell Phagocyte which has ingested nuclear material of another cell: characteristic of *systemic lupus erythematosus*.

leader peptide See *leader sequence*.

leader sequence In the regulation of gene expression for enzymes concerned with amino acid synthesis in prokaryotes, the leader sequence codes for the leader peptide which contains several residues of the amino acid being regulated. Transcription is closely linked to translation, and if translation is retarded by limited supply of amino-acyl tRNA for the specific amino acid, the mode of transcription of the leader sequence permits full transcription of the operon genes; otherwise complete transcription of the leader sequence prematurely terminates transcription of the operon.

leading lamella Anterior region of a crawling cell, such as a fibroblast, from which most cytoplasmic granules are excluded.

lecithin Phospholipid of egg yolk (usually hen's eggs). A mixture of phosphatidyl choline and phosphatidyl ethanolamine, but usually refers to phosphatidyl choline.

lecithinase See *phospholipases*.

lectins Proteins obtained particularly from the seeds of leguminous plants, but also from many other plant and animal sources, which have binding sites for specific mono- or oligosaccharides. Named originally for the ability of some to selectively agglutinate human red blood cells of particular blood groups. Lectins such as *concanavalin A* and *wheat germ agglutinin* are widely used as analytical and preparative agents in the study of glycoproteins. See Table L1.

leghaemoglobin Form of haemoglobin found in the nitrogen-fixing root-nodules of legumes. Binds oxygen, and thus protects the nitrogen-fixing enzyme, nitrogenase, which is oxygen-sensitive.

legumin Major storage protein of the seeds of peas and other legumes.

leiotonin Smooth-muscle analogue (homologue?) of *troponin*. Two subunits, leiotonins

A and C, the latter similar in size and homologous to *calmodulin* and troponin C.

Leishman stain Romanovsky-type stain; a mixture of basic and acid dyes used to stain blood smears and that differentially stains various classes of leucocytes.

leishmaniasis Disease caused by protozoan parasites of the genus *Leishmania*. The parasite lives intracellularly in macrophages. Various forms of the disease are known, depending upon the species of parasite: in particular visceral leishmaniasis (kala-azar), and muco-cutaneous leishmaniasis.

Lentivirinae Subfamily of non-oncogenic retroviruses which cause "slow diseases", characterised by horizontal transmission, long incubation periods and chronic progressive phases. Visna virus is in this group, and there are similarities between visna, equine infectious anaemia virus, and *HIV*.

lentoid Spherical cluster of retinal cells, formed by aggregation *in vitro*, which has a core of lens-like cells inside which accumulate proteins characteristic of normal lens. The cells concerned derive from retinal glial cells.

Lepore haemoglobin Variant haemoglobin in a rare form of *thalassemia*: there is a composite δβ chain as a result of an unequal *crossing over* event. The composite chain is functional but synthesised at reduced rate.

leprosy Disease caused by *Mycobacterium leprae*, an obligate intracellular parasite which survives lysosomal enzyme attack by possessing a waxy coat. Leprosy is a chronic disease associated with depressed cellular (but not humoral) immunity; the bacterium

Table L1. Lectins

Source	Abbreviation	Sugar specificity
Bandieraea simplicifolia	BSL1	α-D-gal > α-D-GalNAc
Concanavalia ensiformis (Jack bean)	ConA	α-D-Man > α-D-Glc > α-D-GlcNAc
Dolichos biflorus	DBA	α-D-GalNAc
Lens culinaris (lentil)	LCA	α-D-Man > α-D-Glc > α-D-GlcNAc
Phaseolus vulgaris (red kidney bean)	PHA	β-D-Gal (1–4)-D-GlcNAc
Arachis hypogaea (peanut)	PNA	β-D-Gal (1–3)-D-GalNAc
Pisum sativum (garden pea)	PSA	α-D-Man > α-D-Glc
Ricinus communis (castor bean)	RCA1	β-D-Gal > α-D-Gal
Sophora japonica	SJA	β-D-GalNAc > β-D-Gal > α-D-Gal
Glycine max (soybean)	SBA	α-D-GalNAc > β-D-GalNAc
Ulex europaeus (common gorse)	UEA1	α-L-fucosyl
Triticum vulgaris (wheat germ)	WGA	β-D-GlcNAc(1–4) GlcNAc > β-D-GlcNAc(1–4)-β-D-GlcNAc

Gal = galactose; GalNAc = galactosamine; Glc = glucose; GlcNAc = glucosamine; Man = mannose.

requires a lower temperature than 37°C, and thrives particularly well in peripheral Schwann cells and macrophages. Only humans and the nine-banded armadillo are susceptible.

leptonema *Leptotene*.

Leptospira *Spirochaete* bacterium which causes a mild chronic infection in rats and many domestic animals. The bacteria are excreted continuously in the urine and contact with infected urine or water can result in infection of humans *via* cuts or breaks in the skin. Infection causes leptospirosis or Weil's disease, a type of jaundice, which is an occupational hazard for sewerage and farm workers.

leptospirosis Weil's Disease, caused by infection with **Leptospira**.

leptotene Classical term for the first stage of *prophase* I of *meiosis*, during which the chromosomes condense and become visible.

Lesch–Nyhan syndrome A sex-linked recessive inherited disease in humans which results from mutation in the gene for the purine salvage enzyme HGPRT, located on the X chromosome. Leads to severe mental retardation and distressing behavioural abnormalities, such as compulsive self-mutilation.

lethal mutation *Mutation* that eventually results in the death of an organism carrying the mutation.

LETS (large extracellular transformation/trypsin sensitive protein) Originally described as a cell-surface protein which was altered on transformation *in vitro*: now known to be *fibronectin*.

Leu enkephalin A natural peptide neurotransmitter; see *enkephalins*.

leucine (Leu; L; 2-amino-4-methylpentanoic acid; MW 131) The most abundant amino acid found in proteins. Confers hydrophobicity and has a structural rather than a chemical role. See Table A2.

leucine aminopeptidase An exopeptidase which removes neutral amino acid residues from the N-terminus of proteins.

leucocyte (USA leukocyte) Generic term for a white blood cell. See *basophil*, *eosinophil*, *lymphocyte*, *monocyte*, *neutrophil*.

leucocytosis An excess of leucocytes in the circulation.

leucopenia An abnormally low count of circulating leucocytes.

leucoplast Colourless *plastid*, which may be an *etioplast* or a storage plastid (*amyloplast*, *elaioplast* or *proteinoplast*).

leucotrienes A family of hydroxyeicosatetraenoic (HETE) acid derivatives in which the lipid moiety is conjugated to glutathione or cysteine. Members of the group are potent pharmacological mediators, eg. SRS-A, the slow reacting substance of anaphylaxis.

leukaemia (USA leukemia) Malignant neoplasia of leucocytes. Several different types are recognised according to the stem cell which has been affected, and several virus-induced leukaemias are known (eg. that caused by feline leukaemia virus). Both acute and chronic forms occur: 1. Acute lymphoblastic leukaemia — neoplastic proliferation of white cell precursors in which the blood has large numbers of primitive lymphocytes (high nuclear/cytoplasmic ratio characteristic of dividing cells and few specific surface antigens expressed); tends to be common in the young; 2. Acute myeloblastic leukaemia — more common in adults; the proliferating cells are of the *myeloid* haematopoietic series (myeloblasts) and the cells appearing in the blood are primitive *granulocytes* or *monocytes*; 3. Chronic lymphocytic leukaemia — neoplastic disease of middle or old age, characterised by excessive numbers of circulating lymphocytes of normal, mature appearance, usually B-cells; presumably a neoplastic transformation of lymphoid stem cells; 4. Chronic myelogenous leukaemia — neoplasia of myeloid stem cells, commonest in middle-aged or elderly people, characterised by excessive numbers of circulating leucocytes, most commonly neutrophils (or precursors), but occasionally eosinophils or basophils.

leuko- See *leuco-*.

Lewis blood group A pair of blood group activities associated with the A, B, H substances. Lewis Le^a is a separate gene, whereas Le^b arises from the combined activity of the enzymes specified by Le^a and H genes.

Leydig cell Interstitial cell of the mammalian testis, involved in synthesis of testosterone.

LFA-1 (CD11a/CD18; lymphocyte function-related antigen-1) Heterodimeric lymphocyte plasma-membrane protein (α_L 180kD, β 95kD) which binds *ICAM*-1, particularly involved in cytotoxic T-cell killing. One of the *integrin* superfamily of adhesion molecules Deficiency of LFA-1 in leucocyte adhesion deficiency (LAD) syndrome leads to severe impairment of normal defences and poor survival prospects. The related surface adhesion molecules (sometimes referred to as the LFA-1 class of adhesion molecules) are Mac-1 (α_M 170kD, β 95kD; CD11b/CD18) and p150,95 (α_X 150kD, β 95kD; CD11c/CD18); they are also defective in severe forms of LAD because the β subunit, which is apparently common to all three, is missing. Mac-1 (also known as Mo-1 in the earlier literature) is the complement C3bi receptor (*CR3*) and is present on mononuclear phagocytes and on neutrophils; p150,95 is less well characterised, but is particularly abundant on macrophages.

LFA-3 (lymphocyte function-related antigen-3) Ligand for the CD2 adhesion receptor which is expressed on cytolytic T-cells. LFA-3 is expressed on endothelial cells at low levels. The CD2/LFA-3 complex is an adhesion mechanism distinct from the *LFA-1*/ICAM-1 system, and binding of erythrocyte LFA-3 to T-lymphocyte CD2 is the basis of *E-rosetting*.

LH See *luteinising hormone*.

LHRF See *luteinising hormone releasing factor*.

ligand Any molecule which binds to another; in normal usage a soluble molecule such as a hormone or neurotransmitter which binds to a receptor, or a substrate or effector which binds to an enzyme. The decision as to which is the ligand and which the receptor is often a little arbitrary when the broader sense of receptor is used (where there is no implication of transduction of signal). In these cases it is probably a good rule to consider the ligand to be the smaller of the two — thus in a lectin-sugar interaction, the sugar would be the ligand (even though it is attached to a much larger molecule, recognition is of the saccharide).

ligand-gated ion channel A transmembrane *ion channel* whose permeability is increased by the binding of a specific *ligand*, typically a *neurotransmitter* at a *chemical synapse*. The permeability change is often drastic; such channels let through effectively no ions when shut, but allow passage at up to 10^7 ions/s when a ligand is bound. Recently, the receptors for both *acetylcholine* and

GABA have been found to share considerable sequence homology, implying that there may be a family of structurally related ligand-gated ion channels.

ligand-induced endocytosis The formation of coated pits and then *coated vesicles* as a consequence of the interaction of ligand with receptors, which then interact with *clathrin* and associated proteins on the cytoplasmic face of the plasma membrane and come together to form a pit. Not all coated vesicle-uptake of receptors requires receptor occupancy.

ligases (synthetases) Major class (category 6 in the *E classification*) of enzymes which catalyse the linking together of two molecules eg. DNA ligase which links two fragments of DNA by forming a *phosphodiester bond*.

ligatin Polypeptide (10kD monomer) which forms 3–4.5nm polymeric fibrils on the outside of chick neural retina cells.

light chains A non-specific term used of the smaller subunits of several multimeric proteins, for example *immunoglobulin*, *myosin*, *dynein*, *clathrin*. See also *L-chain*.

light dependent reaction The reaction taking place in the chloroplast in which the absorption of a photon leads to the formation of ATP and NADPH.

light microscopy In contrast to electron microscopy. See *bright-field, phase-contrast, interference, interference reflection, dark-field* and *fluorescence microscopy*. See also Table L2.

light scattering Particles suspended in a solution will cause scattering of light, and the extent of the scattering is related to the size and shape of the particles (in a somewhat complex relationship).

light-harvesting system Set of photosynthetic pigment molecules which absorb light and channel the energy to the photosynthetic *reaction centre*, where the light reactions of *photosynthesis* occur. In higher plants, contains *chlorophyll* and *carotenoids*, and is present in two slightly different forms in *photosystems I and II*.

lignin Complex polymer of phenylpropanoid subunits, laid down in the walls of plant cells such as *xylem* vessels and *sclerenchyma*. Imparts considerable strength to the wall, and also protects it against degradation by micro-organisms. It is also laid down as a defence reaction against pathogenic attack, as part of the *hypersensitive response* of plants.

limb bud The limbs of vertebrates start as outpushings of mesenchyme surrounded by a simple epithelium. The distal region is referred to as the progress zone. There has been extensive study of positional information within the limb-bud which determines, for example, the proximal–distal pattern of bone development and the anterior–posterior specification of digits.

limit of resolution See *resolving power*.

Limulus polyphemus Now renamed *Xiphosura*, though *Limulus* is still in common usage as a name. The king crab or horseshoe crab, found on the Atlantic coast of North America. More closely related to the arachnid arthropods than to crustacea; the only surviving representative of the subclass Xiphosura. Its compound eyes have been widely used in studies on visual systems, but it is probably better known from the "*Limulus*-amoebocyte-lysate" (LAL) test; LAL is very sensitive to small amounts of *endotoxin*, clotting rapidly to form a gel, and the test is used clinically to test for septicaemia.

lincomycin Antibiotic active against Gram positive bacteria. Acts by blocking protein synthesis by binding to the 50S subunit of the ribosome and blocking peptidyl transferase reactions. Clindamycin, a derivative of lincomycin, is used as an antimalarial drug.

linear dichroism See *circular dichroism*.

Table L2. Types of light microscopy[a]

Method	Physical parameter detected
With axial illumination: Without spatial filtration[b]	
Bright field	Absorption by specimen (May be operated in Visible, UV or IR regions of the spectrum and in quantitative microspectrophotometric modes)
Interference Transmitted	Path difference arising in specimen, qualitative or quantitative
Interference Reflected (IRM)	Path difference in films 1–10 wavelengths thick next to substrate. For cell contacts
Fluorescence	Natural fluorescence or that of probes inserted into system
Dark-field	Refractive index discontinuities revealed by scattered light
Polarisation	Birefringent and/or dichroic properties
With axial illumination: With spatial filtration[b]	
Confocal scanning microscopy	Contrast and resolution enhanced by selection of light paths modified by the object at the back focal plane of the objective. Bright-field or fluorescence modes. Usually combined with video processing of the image
Phase contrast	Path differences revealed as contrast differences non-quantitatively and non-regularly, using phase plate at back focal plane
Differential interference contrast (DIC) = Nomarski	Path difference gradients revealed as contrast or colour differences
Out-of-focus phase contrast	Path differences revealed as diffraction patterns
With anaxial illumination: With spatial filtration[b]	
Hoffman modulation contrast	Path differences
Single side-band edge enhancement (SSEE Microscopy)	Path differences from first order diffractions

[a]Nearly all systems can be run in the epi (incident) illumination mode. Video (image) processing can enhance contrast and resolution in images by the application of simple algorithms to expand the grey scale, reduce noise and subtract background. More complex processing is possible, including the extraction of further information by Fourier transforms.
[b]Spatial filtration. This is the application of methods to remove those ray paths which have not interacted with the object. This is done at the back focal plane of the objective. It can also be applied to select or remove ray paths that have interacted in some specified way with the object.

Lineweaver–Burke plot A plot of $1/v$ against $1/S$ for an enzyme-catalysed reaction, where v is the initial rate and S the substrate concentration. From the equation: $1/v = 1/V_{max}(1 + K_m/S)$ the parameters V_{max} and K_m can be determined. The equation overweights the contribution of the least accurate points and other methods of analysis are preferred; see *Eadie–Hofstee plot*.

lining epithelium An epithelium lining a duct, cavity or vessel, that is not particularly specialised for secretion or as a mechanical barrier. Not a precise classification.

linkage See *genetic linkage*.

linkage disequilibrium The occurrence of some genes together, more often than would be expected. Thus, in the HLA system of *histocompatibility antigens*, HLA-A1 is commonly associated with B8 and DR3, and A2 with B7 and DR2, presumably because the combination confers some selective advantage. See *linkage equilibrium*.

linkage equilibrium Situation which should exist in a population undisturbed by selection, migration etc., in which all possible combinations of linked genes should be present at equal frequency. The situation is no more common than such undisturbed populations.

linoleic acid (9,12,octadecadienoic acid) An essential *fatty acid*; occurs as a glyceride component in many fats and oils.

linolenic acid (9,12,15,octadecatrienoic acid) An 18-carbon *fatty acid* with three double-bonds and α- and γ-isomers. Essential dietary component for mammals.

lipaemia (USA lipemia) Presence in the blood of an abnormally large amount of lipid.

lipases Enzymes which break down mono-, di- or tri-glycerides to release fatty acids and glycerol. Calcium ions are usually required.

lipid A The lipid associated with polysaccharide in the *lipopolysaccharide* of Gram negative bacterial cell walls.

lipid bilayer See *phospholipid bilayer*.

lipidoses *Storage diseases* in which the missing enzyme is one which degrades sphingolipids (sphingomyelin, ceramides, gangliosides). In *Tay–Sachs disease* the lesion is in hexosiminidase A, an enzyme which degrades ganglioside Gm_2; in Gaucher's disease, glucocerebrosidase; in Niemann–Pick disease, sphingomyelinase.

lipids Biological molecules soluble in apolar solvents, but only very slightly soluble in water. They are a heterogenous group (being defined only on the basis of solubility) and include fats, waxes and terpenes. See Table L3.

lipoamide The functional form of lipoic acid in which the carboxyl group is attached to protein by an amide linkage to a lysine amino group.

lipoamide dehydrogenase An enzyme which regenerates lipoamide from the reduced form dihydrolipoamide.

lipocortins A family of proteins which inhibit phospholipase A either by interacting with the enzyme or by binding to the substrate surface (controversial). Lipocortin I is almost identical to *calpactin* II, lipocortin II to calpactin I. Contains both calcium- and phospholipid-binding domains, the calcium-binding domain quite unlike that of the *calmodulin* family. See *annexins*. Glucocorticoids induce enhanced synthesis and release, and this has been thought to be important in their anti-inflammatory action.

lipoic acid Coenzyme involved in oxidative decarboxylation of keto-acids such as pyruvate to acetyl-CoA.

lipomodulin The name originally given to *lipocortin* from neutrophils.

lipophorin A family of high-density lipoproteins (6–700kD) from insect haemolymph, that transport diacyl glycerols. The molecule comprises heavy (250kD) and light (85kD) subunits, the remainder of the molecular weight being accounted for by the high lipid content (40–50%, depending on insect species). Lipophorin forms large aggregates during the haemolymph clotting process.

lipopolysaccharide Major constituents of the outer membrane in the cells of Gram negative bacteria. Highly immunogenic and stimulates the production of endogenous pyrogen *interleukin-1* and *tumour necrosis factor* (TNF).

lipoproteins An important class of serum proteins in which a lipid core with a surface coat of phospholipid monolayer is packaged with specific proteins (apolipoproteins).

Table L3. Lipids

Table L3. Lipids

(i) FATTY ACIDS

These are the most important feature of the majority of biological lipids. They occur free in trace quantities and are important metabolic intermediates. They are esterified in the majority of biological lipids. Compounds are included either because they are common components of biological lipids or are used in synthetic "model" analogues of these lipids.

General formula R-COOH. Branched chain compounds are widespread, but are not found in mammalian lipids. All the examples given are straight-chain compounds.

Saturated fatty acids.

Number of carbon atoms	Name	MW (D)
2	Acetic	60
3	Propionic	74.1
4	Butyric	88.1
5	Valeric	102.1
6	Hexanoic (caproic)	116.2
7	Heptanoic	130.2
8	Octanoic (caprylic)	144.2
9	Nonanoic (pelargonic)	158.2
10	Decanoic (capric)	172.2
11	Undecanoic	186.3
12	Lauric	200.3
13	Tridecanoic	214.4
14	Myristic	228.4
15	Pentadecanoic	242.4
16	Palmitic	256.4
17	Margaric	270.7
18	Stearic	284.5
20	Eicosanoic (Arachidic)	312.5
22	Docosanoic (Behenic)	340.6

Mono-unsaturated acids.

Designation No. of C atoms : No. of double bonds (position and configuration of bonds)	Name	MW (D)
16 : 1 (cis 9)	palmitoleic	254.2
18 : 1 (cis 9)	oleic	282.5
18 : 1 (trans 9)	elaidic	282.5
18 : 1 (cis 11)	cis-vaccenic	282.5
18 : 1 (trans 11)	trans-vaccenic	282.5

Poly-unsaturated acids (all *cis* double bonds).

	Name	MW
18 : 2 (9, 12)	linoleic	280.4
18 : 3 (9, 12, 15)	α-linolenic	278.4
18 : 3 (6, 9, 12)	γ-linolenic	278.4
20 : 4 (5, 8, 11, 14)	arachidonic (eicosenoic)	304.5
22 : 6 (4, 7, 10, 13, 16, 19)	dodecosahexaenoic acid	328.6

(Continued)

Table L3. Lipids (Continued)

(ii) ACYL GLYCEROLS

Glycerol esters of fatty acids. Acyl glycerols are the parent compounds of many structural and storage lipids. Diglycerides (DG) may be considered as the parent compounds of the major family of phosphatidyl phospholipids. Triglycerides (TG) are important storage lipids.

Diglycerides

Present as trace components of membranes. They are important metabolites and second messengers in signal-response coupling.

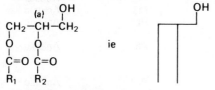

ie

^aThis carbon is asymmetric. See below under phosphatidic acid.

(iii) SPHINGOLIPIDS

Important and widespread classes of phospholipids and glycolipids.
The parent alcohol is SPHINGOSINE:

$$CH_3(CH_2)_{12}CH=CH-\underset{\underset{Y}{\overset{|}{NH}}}{\overset{\overset{OH}{|}}{CH}-CH}-CH_2-X$$

where X = OH and the primary amino-group is free.
SPHINGOSINE is normally substituted at X and Y. When Y is a long-chain unsaturated fatty acyl group the derivative is a CERAMIDE. When the CERAMIDE carries uncharged sugars as the X substituent this is a CEREBROSIDE and where the sugars include sialic acid it is a GANGLIOSIDE.

(iv) PHOSPHOLIPIDS

In animal cell membranes the major class of phospholipids are the phosphatidyl phospholipids for which phosphatidic acid can be considered as the simplest example. These are diacylglycerol (DG) derivatives and in most cases DG is the immediate metabolic precursor.
Outline structure (see diglyceride):

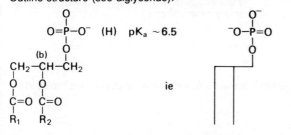

ie

^bAs in diglycerides, this carbon atom is asymmetric. The biologically important configuration is *syn*.

R_1 is usually saturated and R_2 is unsaturated in animal cell membranes.

(Continued)

Table L3. Lipids

(iv) PHOSPHOLIPIDS (continued)

Phosphatidyl phospholipids. Derived from phosphatidic acid by esterification of the phosphate group.

Base (substituent)	Phospholipid class	Abbreviation	Ionic status
None	Phosphatidic acid	PA	Anionic
Choline	Phosphatidyl choline	PC	Neutral
Ethanolamine	Phosphatidyl ethanolamine	PE	Neutral
Glycerol	Phosphatidyl glycerol	PG	Anionic
Inositol	Phosphatidyl inositol (Ptdyl. Ins.)	PI	Anionic
Inositol 4-monophosphate	Phosphatidyl inositol 4-phosphate (Ptdyl. Ins. 4-phosphate)	PIP	Anionic
Inositol 4,5 diphosphate	Phosphatidyl inositol 4,5-diphosphate (Ptdyl. Ins. 4,5 bisphosphate)	PIP_2	Anionic
Phosphatidyl glycerol	Diphosphatidyl glycerol (Cardiolipin)		Anionic
Serine	Phosphatidyl serine	PS	Anionic

Sphingomyelin. (SM) is an analogue of phosphatidyl choline in which the diacylglycerol component is replaced by a CERAMIDE.
Common variants of these structures are:-

Ether phospholipids in which the diacylglycerol structure is modified so that one or both acyl groups are replaced by ether groups.

```
CH−                    CH−
|                      |
O        becomes       O
|                      |
C=O                    R
|
R    (acyl)               (alkyl)
```

Plasmalogens in which the 1-acyl group is replaced by a 1-alkenyl group.

Plasmalogens are abundant lipid components of many membranes.

```
CH−                    CH−
|                      |
O        becomes       O
|                      |
C=O                    CH
|                      ‖
R                      CH
                       |
```

Phosphonolipids In which the ester linkage between the base (choline or ethanolamine) is replaced by a P-C (phosphono) linkage.

```
    O                         O
    ‖                         ‖
−O−P−O−CH−   becomes    −O−P−CH−
    |                         |
    O_                        O_
```

Lysophospholipids. Derivatives of phosphatidyl phospholipids in which one of the acyl groups has been removed (enzymically).

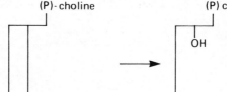

Lysophosphatidyl choline (lysolecithin) is a common, but trace component of membranes.

(V) STEROLS. Of this large class of compounds only one member, CHOLESTEROL, is an important structural lipid. It is the single most abundant lipid in the plasma membrane of many animal cell types.

Classified according to density: chylomicrons, large low density particles; very low density (VLDL); low density (LDL) and high density (HDL) species. Important in lipid transport, especially cholesterol transport.

liposomes Lipid bodies made from phospholipids which may contain other lipids (often cholesterol) in a bilayer arrangement. They may have a single bilayer (unilamellar) or many bilayers (multilamellar) and cover a very large size range. Of interest as model membranes, and used clinically to increase the therapeutic ratio of drugs by improving their targetting to specific delivery sites (usually to cells of the reticuloendothelial system).

lipoteichoic acid Compounds formed from *teichoic acid* linked to glycolipid and found in the walls of most Gram positive bacteria. The lipoteichoic acid of streptococci may function as an *adhesin*.

lipotropin (LPH; lipotropic hormone; adipokinetic hormone) Polypeptide hormone (β form: MW 9894, 91 residues; γ form has only residues 1–58 of β) from the pituitary hypophysis, that is of particular interest because it is the precursor of *endorphins*, which are released by proteolysis. Promotes lipolysis and acts through the adenylyl cyclase system. Part of the ACTH group of hormones.

lipoxygenase Enzyme which catalyzes the oxidative conversion of arachidonic acid to the hydroxyeicosatetraenoic acid (HETE) structure in the synthesis of *leucotrienes*.

Listeria monocytogenes Rod-shaped Gram positive bacterium. It is wide-spread and can grow over an unusually wide range of temperatures (0–45°C); normally *saprophytic* but is an opportunistic parasite, in that it can survive within cells (particularly leucocytes) and can be transmitted transplacentally. It has caused a number of serious outbreaks of food poisoning with a high mortality rate in recent years.

litorin A peptide which mimics *bombesin* in its mitogenic effects, and has a carboxyterminal octapeptide in common with bombesin.

liver cells Usually implies *hepatocytes*, even though other cell types are found in the liver (*Kupffer cells* for example). Hepatocytes are relatively unspecialised epithelial cells and are the biochemist's "typical animal cell".

LMM (light meromyosin) The rod-like portion of the myosin heavy chain (predominantly α-helical) which is involved in lateral interactions with other LMM to form the thick filament of striated muscle, and which is separated from heavy meromyosin (HMM) by cleavage with trypsin.

local circuit theory A generally accepted model for neuronal conduction, by which depolarisation of a small region of a neuronal plasma membrane produces transmembrane currents in the neighbouring regions, tending to depolarise them. As the *sodium channels* are *voltage gated*, the depolarisation causes further channels to open, thus propagating the action potential.

lock and key models Specific recognition in biological systems might be mediated through interactions which depend upon very precise steric matching between receptor and ligand, or between enzyme and substrate. The commonly used analogy is between lock and key, and implies a precise sterically determined interaction.

locomotion Used by some authors to distinguish movement of cells from place-to-place from movements such as flattening, shape-change, *cytokinesis* etc.

locus The site in a linkage map or on a chromosome where a particular gene is located. Any one of the alleles of that gene may occupy this site.

Loeffler's medium Coagulated-serum medium used to culture *Corynebacterium diphtheriae* in diagnostic bacteriology.

lomasome Membranous structure, often containing internal membranes, located between the plasma membrane and cell wall of plant cells. Included in the more general term, *paramural body*.

long-terminal repeats (LTR) Identical DNA sequences, several hundred nucleotides long, found at either end of *transposons* and the proviral DNA, formed by reverse transcription of retroviral RNA. They are thought to have an essential role in integrating the transposon or provirus into the host DNA. LTRs have inverted repeats, that is, sequences close to either end are identical when read in opposite directions. In proviruses the *upstream* LTR acts as a promoter and enhancer and the *downstream* LTR as a polyadenylation site.

lorica Shell or test secreted by a protozoan or unicellular alga; often vase-shaped.

low-density lipoprotein receptor (LDL-receptor) A cell-surface protein which mediates the endocytosis of LDL by cells. Genetic defects in LDL-receptors lead to abnormal serum levels of LDL and hyper-cholesterolaemia.

low-density lipoprotein (LDL) See *lipoprotein*.

LPS *Lipopolysaccharide*.

LTA4, LTB4, LTC4, etc. See *leucotrienes*.

Lubrol A non-ionic detergent.

luciferase An enzyme from firefly tails which catalyses the production of light in the reaction between luciferin and ATP. Used by the firefly for producing light to attract a mate, and used in the laboratory as a *chemiluminescent* bioassay for ATP.

lucigenin Compound used as a bystander substrate in assaying the metabolic activation of leucocytes by *chemiluminescence*. When oxidised by *superoxide* it emits light.

Lucké carcinoma A renal carcinoma, caused by a herpesvirus, in frogs; it aroused interest because its abnormal growth appears to be dependent on a restricted temperature range. Nuclei from these cells give rise to normal frogs if transplanted into enucleated eggs, giving support to the *epigenetic* theories of neoplasia.

lumen A cavity or space within a tube or sac.

lumicolchicine A derivative of *colchicine* produced by exposure to ultraviolet light and which does not inhibit tubulin polymerisation, although it has many of the non-specific effects of colchicine.

luminol Compound used as a bystander substrate in assaying the metabolic activation of leucocytes by *chemiluminescence*. When oxidised by the *myeloperoxidase/* hydrogen peroxide system, it emits light.

lumisome Sub-cellular membrane-enclosed vesicle which is the site of bioluminescence in some marine coelenterates.

lupus erythematosus Skin disease in which there are red scaly patches, especially over the nose and cheeks. May be a symptom of *systemic lupus erythematosus*.

luteinising hormone A glycoprotein hormone (26kD) and *gonadotrophin*. Made up of an α-chain (96 amino acids) identical to other gonadotrophins, and a hormone-specific β-chain. Acts with *follicle stimulating hormone* to stimulate sex hormone release.

luteinising hormone releasing factor (LHRF) A decapeptide releasing-hormone (MW 1182), synthesised and released from the hypothalamus, which stimulates release of *luteinising hormone*.

lutropin Synonym for *luteinising hormone*.

luxury protein A term sometimes used to describe those proteins which are produced specifically for the function of differentiated cells and are not required for general cell maintenance (the so-called "house-keeping" proteins).

lyases Enzymes of the EC Class 4 (see *E-classification*) which catalyse the non-hydrolytic removal of a group from a substrate with the resulting formation of a double-bond; or the reverse reaction, in which case the enzyme is acting as a synthetase. Include decarboxylases, aldolases and dehydratases.

Lyb antigen Surface antigens of mouse B-cells.

lycopene A linear, unsaturated hydrocarbon *carotenoid* (MW 536); the major red pigment in some fruit.

lymph Fluid found in the lymphatic vessels which drain tissues of the fluid that filters across the blood vessel walls from blood. Lymph carries lymphocytes which have entered the lymph nodes from the blood.

lymph node (lymph gland) Small organ made up of a loose meshwork of reticular tissue in which are enmeshed large numbers of lymphocytes, macrophages and accessory cells. Recirculating lymphocytes leave the blood through the specialised high endothelial venules of the lymph node and pass through the node before being returned to the blood through the lymphatic system. Because the lymph nodes act as drainage points for tissue fluids, they are also regions in which foreign antigens present in the tissue fluid are most likely to begin to elicit an immune response. There are distinct T- and B-cell areas within the lymph node.

lymphadenitis Inflammation of *lymph nodes*.

lymphadenopathy Pathological disorder

of lymph nodes. Lymphadenopathy associated virus (LAV) was the name originally given to *HIV* by the Pasteur Institute group.

lymphoblast Often referred to as a blast cell. Unlike other usages of the suffix -blast a lymphoblast is a further differentiation of a lymphocyte, T or B, occasioned by an antigenic stimulus. The lymphoblast usually develops by enlargement of a lymphocyte, active re-entry to the *S phase* of the cell cycle, mitogenesis and production of much mRNA and ribosomes.

lymphocyte White cell of the blood, mediator of specific immunity, derived from stem cells of the lymphoid series. Two main classes, T- and B-lymphocytes, are recognised, the latter responsible (when activated) for production of antibody, the former subdivided into sub-sets (helper, suppressor, cytotoxic T-cells), and responsible both for cell-mediated immunity and for regulation of the activation of B-cells. See also Table C1.

lymphocyte activation (lymphocyte transformation) The change in morphology and behaviour of lymphocytes exposed to a mitogen or to an antigen to which they have been primed. The result is the production of lymphoblasts, cells which are actively engaged in protein synthesis and which divide to form effector populations. Should not be confused with transformation of the type associated with oncogenic viruses, and "activation" is therefore perhaps a better term.

lymphocyte transformation See *lymphocyte activation*.

lymphocytic leukaemia See *leukaemia*.

lymphoid cell Cells derived from stem cells of the lymphoid lineage: large and small lymphocytes, plasma cells.

lymphoid tissue Tissue which is rich in lymphocytes (and accessory cells such as macrophages and reticular cells), particularly the *lymph nodes*, spleen, *thymus*, *Peyer's patches*, pharyngeal tonsils, adenoids, and (in birds) the *Bursa of Fabricius*.

lymphokine Substance produced by a leucocyte which acts upon another cell. Examples are *interleukins*, *interferon* γ, lymphotoxin (*tumour necrosis factor* β), granulocyte-monocyte colony-stimulating factor (*GM-CSF*).

lymphoma Malignant neoplastic disorder of lymphoreticular tissue which produces a distinct tumour mass, not a leukaemia (in which the cells are circulating). Includes tumours derived both from the lymphoid lineage and from mononuclear phagocytes; lymphomas arise commonly (but not invariably) in lymph nodes, spleen, or other areas rich in lymphoid tissue. Lymphomas are subclassified as Hodgkin's disease, and non-Hodgkin's lymphomas (eg. Burkitt's lymphoma, large-cell lymphoma, histiocytic lymphoma).

lymphotoxins Cytotoxic products of T-cells; term now generally restricted to lymphotoxin (*tumour necrosis factor* β).

Lyon hypothesis Hypothesis, first advanced by Lyon, concerning the random inactivation of one of the two X chromosomes of the cells of female mammals. In consequence females are chimaeric for the products of the X chromosomes, a situation which has been exploited in female Negroes (who are heterotypic for isozymes of glucose-6-phosphate dehydrogenase) as a means to confirm the monoclonal origin of papillomas and of atherosclerotic plaques.

Lyonisation See *Lyon hypothesis*.

lysine (Lys; K; MW 146) Amino acid; the only carrier of a side-chain primary amino group in proteins. Has important structural and chemical roles in proteins. See Table A2.

lysis Rupture of cell membranes and loss of cytoplasm.

lysogenic conversion See *lysogeny*.

lysogeny The ability of some phages to survive in a bacterium as a result of the integration of their DNA into the host chromosome. The integrated DNA is termed a prophage. A regulator gene produces a *repressor protein* which suppresses the lytic activity of the phage, but various environmental factors such as ultraviolet irradiation may prevent synthesis of the repressor, leading to normal phage development and lysis of the bacterium. The best example of this is bacteriophage λ.

lysophosphatides Mono-acyl derivatives of diacyl phospholipids which are present in membranes as a result of cyclic deacylation and reacylation of phospholipids. Membranolytic in high concentrations, and fusogenic at concentrations which are just sub-lytic. May have important modulatory roles.

lysosomal diseases Diseases (also known as storage diseases) in which a deficiency of a particular lysosomal enzyme leads to accumulation of the undigested substrate for that enzyme within cells. Not immediately fatal, but within a few years leads to serious neurological and skeletal disorders and eventually to death. See *Hunter* and *Sanfillipo syndrome*, *Gaucher's*, *Hurler's*, *Niemann–Pick*, *Pompe's* and *Tay–Sachs diseases*.

lysosomal enzymes A range of degradative enzymes (hydrolases), most of which operate best at acid pH. The best known marker enzymes are acid phosphatase and β-glucuronidase, but many others are known.

lysosome Membrane-bounded cytoplasmic organelle containing a variety of hydrolytic enzymes which can be released into a phagosome or to the exterior. Release of *lysosomal enzymes* in a dead cell leads to autolysis (and is the reason for hanging game, to "tenderise" the muscle), but it is misleading to refer to lysosomes as "suicide bags", since this is certainly not their normal function. Part of the *trans-Golgi network*.

lysosome–phagosome fusion A process which occurs after the internalisation of a primary phagosome. Fusion of the membranes leads to the release of lysosomal enzymes into the phagosome. Some species of intracellular parasite evade intracellular destruction by interfering with this process.

lysozyme Glycosidase which hydrolyses the bond between N-acetyl muramic acid and N-acetyl glucosamine, thus cleaving an important polymer of the cell wall of most bacteria. Present in tears, saliva and in the lysosomes of phagocytic cells, it is an important anti-bacterial defence, particularly against Gram positive bacteria.

lysyl oxidase Extracellular enzyme which deaminates lysine and hydroxylysine residues in collagen or elastin to form aldehydes, which then interact with each other or with other lysyl side chains to form crosslinks.

Lyt antigen A set of plasmalemmal surface glycoproteins on mouse T-lymphocytes. Possession of Lyt 1 partly defines a T-helper cell, and of Lyt 2,3 suppressor and cytotoxic cells. Formerly known as Ly antigens; see also *Lyb antigen*.

lytic complex (membrane attack complex) The large (2000kD) cytolytic complex formed from complement C5b6789. See *complement*.

lytic infection The normal cycle of infection of a cell by a virus or bacteriophage, in which mature virus or phage particles are produced and the cell is then usually lysed.

M

M cells Cells found amongst the other cells of the cuboidal surface epithelium of *gut-associated lymphoid tissue*; have a complex folded surface.

M phase Mitotic phase of cell cycle of eukaryotic cells, as distinct from the remainder, which is known as *interphase* (and which can be further subdivided as G1, S and G2). Beginning of M is signalled by separation of centrioles, where present, and by the condensation of chromatin into chromosomes. M phase ends with the establishment of nuclear membranes around the two daughter nuclei, normally followed immediately by cell division (*cytokinesis*).

M-band Central region of the *A-band* of the *sarcomere* in striated muscle.

M-line Central part of the *A-band* of striated muscle (and of the *M-band*): contains M-line protein (*myomesin*, 165kD), *creatine kinase* (40kD) and glycogen phosphorylase b (90kD). Involved in controlling the spacing between thick filaments.

M-protein Cell wall protein of streptococci: antibody typing of the M-protein is important in identification of different strains of Group A streptococci (at least 55 serotypes are known). The M-protein confers anti-phagocytic properties on the cell and is present as hair-like *fimbriae* on the surface. M-protein is an important virulence factor, and antibodies directed against M-protein are essential for phagocytic killing of the bacteria.

M-ring Innermost (motor) ring of the bacterial flagellar base, located in the outer leaflet of the plasma membrane. It is this ring which is linked to the hook region (and thus to the flagellum itself), and which rotates. Composed of 16 or 17 subunits (one more or less than the *S-ring*).

Machupo virus A member of the *Arenaviridae* that may cause a severe haemorrhagic fever in humans. The natural hosts are rodents and transmission from human to human is not common.

Mach–Zehnder system Interferometric system in which the original light beam is divided by a semi-transparent mirror: object and reference beams pass through separate optical systems and are recombined by a second semi-transparent mirror. Interference fringes are displaced if the optical path difference for the reference beam is greater, and this can be compensated with a wedge-shaped auxiliary object. The position of the wedge allows the phase-retardation of the object to be measured. The Mach–Zehnder system was used in a microsocope designed by Leitz.

macrocytes Abnormally large red blood cells, numerous in pernicious anaemia.

macrolides A group of antibiotics produced by various strains of *Streptomyces*, which have a complex macrocyclic structure. They inhibit protein synthesis by blocking the 50S ribosomal subunit. Include erythromycin, carbomycin. Used clinically as broad-spectrum antibiotics.

macromolecules Biological term including proteins, nucleic acids and carbohydrates, but probably not phospholipids.

macrophage Relatively long-lived phagocytic cell of mammalian tissues, derived from blood *monocyte*. Macrophages from different sites have distinctly different properties. Main types are peritoneal and alveolar macrophages, tissue macrophages (*histiocytes*), *Kupffer cells* of the liver, and osteoclasts. In response to foreign materials may become stimulated or activated. Macrophages play an important role in killing of some bacteria, protozoa and tumour cells, release substances which among other things stimulate cells of the immune system, and are involved in *antigen presentation*. May further differentiate within chronic inflammatory lesions to epithelioid cells or may fuse to form *foreign body giant cells* or *Langhans' giant cells*.

macrophage colony stimulating factor (M-CSF) Growth factor which stimulates committed stem cells of bone marrow into differentiating towards the production of monocytes (mononuclear phagocytes).

macrophage inhibition factor (MIF) A group of lymphokines (including a 14kD glycoprotein) produced by activated T-lymphocytes which reduce macrophage mobility and increase heterotypic–homotypic macrophage adhesion.

macula adhaerens Spot desmosome: see *desmosome*.

macule A spot: only commonly met in the construct "immaculate", meaning unspotted.

magainins Peptides of about 20 amino acid residues with antimicrobial activity, found in amphibian skin. Probably have membrane insertion and lytic properties. Sequence related to melittin.

magnetosome Enveloped compartment in magnetotactic bacteria containing magnetite particles. By using this organelle to detect the vertical component of the Earth's magnetic field, the bacteria swim towards the bottom of aquatic habitats.

magnocellular neuron A neuron in the magnocellular region of the brain. Perhaps the first class of neuron from the central nervous system shown to be sensitive to *nerve growth factor* (which had previously been thought only to act at the periphery).

major histocompatibility antigen See *histocompatibility antigens*, *MHC restriction*.

major histocompatibility complex (MHC) The set of gene loci specifying major *histocompatibility antigens*, eg. HLA in man, H-2 in mice, RLA in rabbits, RT-1 in rats, DLA in dogs, SLA in pigs, etc.

malabsorption syndrome A variety of conditions in which digestion by and absorption in the small intestine are impaired. The effects of failure to digest or absorb carbohydrates are attributable mainly to the osmotic pressure effects exerted by the unabsorbed sugars. Bacteria may ferment the unabsorbed sugars. Fat malabsorption leads to steatorrhoea, protein malabsorption to possible hypoproteinaemia with oedema and wasting. Vitamin and mineral malabsorptions may be detected. Multiple causes, including lymphoma, amyloid and other infiltrations, surgical resection, blind and hence stagnant loops of the bowel, *Crohn's disease*, gluten-sensitive enteropathy and the sprue syndrome in which the villi atrophy for unknown reasons.

malaria In humans, the set of diseases caused by infection by the protozoans **Plasmodium** *vivax* causing the tertian type, *P. malariae* the quartan type and *P. falciparum* the subtertian or irregular type of disease, the names referring to the frequency of fevers. The fevers occur when the merozoites are released from the erythrocytes. The organisms are transmitted by the *Anopheles* mosquito.

malate The ion from malic acid, a component of the *tricarboxylic acid cycle*.

maleate The ion from maleic acid, often used in biological buffers.

malignant As applied to tumours means that the primary tumour has the capacity to invade locally and to *metastasise*. Implies loss of both *growth control* and *positional control*.

malonate The ion from malonic acid. Malonate is a competitive inhibitor for succinate dehydrogenase in the *tricarboxylic acid cycle*. Malonyl-SCoA is an important precursor for fatty acid synthesis.

maltase Enzyme which hydrolyzes maltose (and the glucose trimer maltotriose) to glucose, during the enzymic breakdown of starch.

maltose Disaccharide, occurs particularly as an intermediate of the breakdown of starch, glucose-α(1–4)-glucose. Fermentable substrate in brewing.

maltose binding protein Protein of the bacterial (*E. coli*) surface which links with *MCP*-II and is involved in the chemotactic response to maltose; probably derived from a similar protein which links with a transmembrane transport system.

mammary gland Milk-producing gland of female mammals. An adapted sweat gland, it is made up of milk-producing alveolar cells, surrounded by contractile myoepithelial cells, together with considerable numbers of fat cells. Milk production is hormonally controlled.

mammary tumour virus (*Bittner agent*; MMTV; also known as "milk-factor") Retrovirus which induces mammary carcinoma in mice. Isolated from highly inbred strains which had very high incidence of the tumours, after the discovery that the disease was transmitted by nursing mothers in milk. Endogenous provirus present in germ-line of all inbred mice. Transcription of the provirus is regulated by a viral promoter which increases transcription in response to glucocorticoid hormones. May transform by proviral insertion activating the cellular *int-1* oncogene.

mannans Mannose-containing polysaccharides found in plants as storage material, in association with cellulose as hemicellulose, in yeasts as a wall constituent.

D-mannitol Hexitol related to D-mannose. Found in many plants, particularly fungi and seaweeds.

D-mannose Hexose identical to D-glucose except that the orientation of the –H and –OH on carbon 2 are interchanged (ie. the 2-epimer of glucose). Found as constituent of polysaccharides and glycoproteins.

mannose-6-phosphate Mannose derivative which is of particular interest when formed by phosphorylation, in the Golgi complex, of certain mannose residues on N-glycan chains of lysosomal enzymes. Believed to function as targetting signal which causes entry of these enzymes to the lysosomes.

mannosidase Enzyme catalyzing hydrolysis of the glycosidic bond between mannose residues and a variety of hydroxyl-containing groups. α-Mannosidases in rough endoplasmic reticulum and *cis*-Golgi are responsible for removing four mannose residues during the synthesis of the complex-type N-linked glycan chains of glycoproteins.

Mantoux test Test for tuberculin reactivity in which tuberculin PPD (purified protein derivative) is injected intracutaneously. The injection site is examined after 2–3 days. A positive reaction, indicating current or previous exposure to *Mycobacterium tuberculosis* (in an uninoculated individual), is a firm red nodule, a chronic inflammatory granuloma. The classical delayed-type *hypersensitivity* reaction. See also *Heaf test*.

MAPs (microtubule-associated proteins) May form part of the electron-lucent zone around a microtubule. MAP1A and 1B (approximately 350kD) from brain, form projections from microtubules; MAP2A and 2B (270kD) are also from brain microtubules and form projections. MAP3 (180kD) and MAP4 (220–240kD) have been described as co-purifying with MAPs 1 and 2. *MAP1C* is in a separate class, being a motor molecule.

MAP1C Microtubule-associated protein (2 heavy chains of 410kD associated with 6 or 7 light chains of about 50–70kD), now considered to be the two-headed cytoplasmic equivalent of ciliary dynein and to be responsible for retrograde transport (transport towards the centrosome).

Marburg virus A Filovirus that causes Marburg disease, a severe haemorrhagic fever developed in many people who work with African green monkeys.

Marek's disease Infectious cancer of the lymphoid system (lymphomatosis) in chickens, caused by a contagious herpesvirus. An effective vaccine is now available.

Marfan's syndrome Dominant disorder of connective tissue in which limbs are excessively long and loose-jointed. Probably a collagen fibril-assembly disorder since it can be mimicked in mice by aminonitriles which interfere with crosslinking.

marginal band A bundle of equatorially-located microtubules which stabilise the biconvex shape of platelets and avian erythrocytes. They are unusual in that they do not derive from the centrosomal *microtubule organising centre*.

margination Adhesion of leucocytes to the endothelial lining of blood vessels, particularly post-capillary venules; often, but not always, a prelude to leaving the circulation and entering the tissues.

masked messenger RNA Long-lived and stable mRNA found originally in the oocytes of echinoderms and constituting a store of maternal information for protein synthesis that is unmasked (derepressed) during the early stages of morphogenesis. In these early stages the rate of cell division is so rapid that transcription from the embryonic genome cannot occur. Undoubtedly not restricted to oocytes, and the term can be applied to any mRNA which is present in inactive form.

mast cell Resident cell of connective tissue which contains many granules rich in *histamine* and heparan sulphate. Release of histamine from mast cells is responsible for the immediate reddening of the skin in a weal-and-flare response. Very similar to *basophils* and possibly derived from the same stem cells. Two types of mast cells are now recognised, connective tissue mast cells and a distinct set of mucosal mast cells; the activities of the latter are T-cell dependent.

mastigonemes Lateral projections from eukaryotic flagella. May be stiff and alter the hydrodynamics of flagellar propulsion, or flexible and alter the effective diameter of the flagellum (flimmer filaments).

mastocytoma Neoplastic *mast cells*.

maternal antibody Any antibody transferred from a mammalian mother transplacentally into the foetus. See under *IgA*, *IgG* etc. for details of the classes of Ig that are transferred to the foetus.

maternal inheritance Inheritance through the maternal cell line, eg. through the oocyte and eggs. The normal mode of inheritance of mitochondrial and chloroplast (organelle) genomes. Should not be confused with maternal effect (see *masked messenger RNA* and *maternal mRNA*).

maternal mRNA Messenger RNA found in oocytes and early embryos which is derived from the maternal genome during oogenesis. See *masked messenger RNA*.

mating-type genes Genes which, in *Saccharomyces cerevisiae*, specify into which of the two mating types (a and α) a particular cell falls. Only unlike mating-type haploids will fuse. The interest derives from the way in which mating-type is controlled; the existing gene is removed from an expression site and a new gene, derived from a (silent) master copy elsewhere in the genome, is spliced in. Later this gene will in its turn be replaced by a new copy of the old gene, also derived from a silent "master". The a- and α-genes code for pheromones which affect cells of the opposite mating-type. Similar mating-type genes are known from other yeasts, and the switching mechanism ("cassette mechanism") may be used more generally.

matrix Ground substance in which things are embedded or which fills a space (as for example the space within the mitochondrion or the nucleus). A common usage is for a loose meshwork within which cells are embedded (eg. extracellular matrix), although it may also be used of filters or absorbent material.

McArdle's disease Glycogen *storage disease* in which the defective enzyme is muscle phosphorylase.

MCPs (methyl-accepting chemotaxis proteins) Proteins of the inner cytoplasmic face of the bacterial plasma membane with which the receptors of the outer face interact. Four different MCPs are known in *E. coli*, each with a separate set of receptors. Can be methylated at various sites; methylation is part of the *adaptation* to the signal. Although important intermediate integration sites, they are not directly connected to the motor.

measles virus *Paramyxovirus* that causes the childhood disease measles and is responsible for *subacute sclerosing panencephalitis*.

media Avascular middle layer of the artery wall, composed of alternating layers of elastic fibres and smooth muscle cells.

medium Shorthand for culture medium or growth medium, the nutrient solution in which cells or organs are grown.

megakaryocytes Giant polyploid cells of bone marrow which give rise to 3–4,000 platelets each.

megaspore Haploid spore produced by a plant sporophyte, which develops into a female gametophyte. The larger of the two types of spore produced by heterosporous species.

meiocyte Cell which will undergo meiosis; a little-used term.

meiosis A specialised form of nuclear division associated with sexually-reproducing organisms in which there are two successive nuclear divisions (meiosis I and II) without any chromosome replication between them. Each division can be divided into 4 phases similar to those of mitosis (pro-, meta-, ana- and telophase). Meiosis reduces the starting number of 4n chromosomes in the parent cell to n in each of the 4 daughter cells. Each cell receives one of each *homologous chromosome* pair, with the maternal and paternal chromosomes being distributed randomly between the cells. During the prophase of meiosis I (classically divided into stages: *leptotene, zygotene, pachytene, diplotene* and *diakinesis*), homologous chromosomes pair to form *bivalents*, thus allowing *crossing over*, the physical exchange of chromatid segments (*recombination*). Meiosis occurs during the formation of *gametes* in animals and the formation of the spore by the sporophyte generation in plants.

meiotic spindle The meiotic equivalent of the *mitotic spindle*.

melanin Pigments largely of animal origin, high-molecular weight polymers of indole quinone. Colours include black/brown, yellow, red and violet. Found in feathers, cuttlefish ink (sepia), human skin, hair and eyes, and in cellular immune responses and wound-healing in arthropods.

melanocyte Neural crest-derived cell found in mammalian dermis which contains melanin in melanosomes.

melanocyte stimulating hormone (MSH; melanotropin) A releasing hormone produced in the mammalian hypophysis and related structures in lower vertebrates. Made up of α-MSH (MW 1665), the same as amino acids 1–13 of *adrenocorticotrophin*, and β-MSH (18 amino acids, 22 in humans). Causes darkening of the skin by expansion of the *melanophores*, but its role in mammals is not yet clear.

melanoma Malignant neoplasm derived from *melanocytes*. Generally the cells contain melanin granules and for this reason they have been used in studies on metastasis because the secondary tumours are easily located in lung. Moles are benign neoplasias (*hamartomas*) derived from melanocytes.

melanophore Cell type found in skin of lower vertebrates (amphibian skin, fish scales), which contains granules of the black pigment melanin. The granules can be rapidly redeployed between a dispersed state (which darkens the skin) and concentration at the centre (which lightens it). One of a family of pigmented or light-diffracting, coloured cells, known collectively as *chromatophores*.

melatonin (N-acetyl 5-methoxytryptamine) A hormone secreted by the pineal gland. In lower vertebrates causes aggregation of pigment in melanophores, and thus lightens skin. In humans believed to play a role in establishment of circadian rhythms.

melittin The major component of bee venom, responsible for the pain of the sting. A 26-amino acid peptide, that has a hydrophobic and a positively-charged region. Can lyse cell membranes and activate phospholipase A_2; it has a very high affinity for calmodulin but the biological relevance of this is unclear.

membrane Generally, a sheet or skin. In cell biology the term is usually taken to mean a modified lipid bilayer with integral and peripheral proteins, as forms the plasma membrane. Because this usage is so general, it is advisable to avoid other uses where possible, particularly in histology or ultrastructure.

membrane attack complex See *complement*.

membrane capacitance The electrical capacitance of a membrane. Plasma membranes are excellent insulators and dielectrics: capacitance is the measure of the quantity of charge which must be moved across unit area of the membrane to produce unit change in membrane potential, and is measured in Farads. Most plasma membranes have a capacitance around 1μfarad/cm².

membrane depolarisation See *depolarisation*.

membrane fluidity Biological membranes are viscous two-dimensional fluids within their physiological temperature range.

membrane fracture See *freeze fracture*.

membrane potential More correctly, transmembrane potential difference: the electrical potential difference across a plasma membrane. See *resting potential*, *action potential*.

membrane protein A protein with regions permanently attached to a membrane (peripheral membrane protein), or inserted into a membrane (integral membrane protein). Insertion into a membrane implies hydrophobic domains in the protein. All *transport proteins* are integral membrane proteins. See Table E1 for erythrocyte membrane proteins.

membrane recycling The process whereby membrane is internalised, fuses with an internal membranous compartment, and is then re-incorporated into the plasma membrane. In cells which are actively secreting by an exocrine method (in which secretory granules fuse with the plasma membrane), it is obviously essential to have some way of reducing the area of the plasma membrane. The membrane can then be used to form new secretory vesicles. The converse is true for phagocytic cells.

membrane transport The transfer of a substance from one side of a membrane (eg. the plasma membrane) to the other, in a specific direction and at a rate faster than diffusion alone. See *active transport*.

membrane vesicles Closed unilamellar shells formed from membranes either in physiological transport processes or else when membranes are mechanically disrupted. They form spontaneously when membrane is broken because the free ends of a lipid bilayer are highly unstable.

membrane zippering See *zippering*.

memory cells Cells of the immune system which "remember" the first encounter with an antigen and facilitate the more rapid *secondary immune response* when the antigen is encountered on a subsequent occasion. The long-lasting immune memory is humoral and resides in B-cells, although it appears that persistence of the antigen may be essential. T-cell memory is shorter.

Mendelian inheritance In sexually-reproducing organisms, any process of heredity explicable in terms of chromosomal *segregation*, independent assortment and homologous exchange.

meningitis Inflammation of the meninges (outer covering) of the brain and spinal cord. Meningitis can arise from bacterial infection (see *meningococcus*) and can be fatal even in some treated cases. Aseptic meningitis may arise in some viral infections.

meningococcus *Neisseria meningitidis*; Gram negative non-motile pyogenic coccus which is responsible for epidemic bacterial meningitis.

mercaptoethanol A water soluble thiol, not of biological origin. Used in biochemistry to cleave disulphide bonds in proteins or to protect sulphydryl groups from oxidation.

meristem Group of actively dividing plant cells, found as apical meristems at the tips of roots and shoots and as lateral meristems in vascular tissue (vascular *cambium*) and in cork tissue (*phellogen*). Also found in young leaves, and at the bases of internodes in grasses. Consists of small undifferentiated cells with no or sparse vacuoles and possessing only proplastids.

merocrine Commonest mode of secretion in which a secretory vesicle fuses with the plasma membrane and releases its contents to the exterior.

meromyosin Fragments of myosin formed by trypsin digestion. Heavy meromyosin (HMM) has the hinge region and ATPase activity, light meromyosin (LMM) is mostly α-helical and is the portion normally laterally associated with other LMM to form the thick filament itself.

merotomy Partial cutting: used in reference to experiments in which protozoa are enucleated and the behaviour of the residual cytoplasm is studied.

merozoite Stage in the life-cycle of the malaria parasite (**Plasmodium**): formed during the asexual division of the intracellular schizont. Merozoites are released and invade other cells.

mesangial cells Cells found within the glomerular lobules of mammalian kidney, where they serve as structural supports, may regulate blood flow, are phagocytic, and may act as *accessory cells*, presenting antigen in immune responses.

mesenchyme Embryonic tissue of mesodermal origin.

mesoderm Middle of the three *germ layers*; gives rise to the musculo-skeletal, blood vascular and urinogenital systems, to connective tissue (including that of dermis) and contributes to some glands.

mesophase (smectic mesophase) Arrangement of phospholipids in water where the liquid-crystalline phospholipids form multilayered parallel-plate structures, each layer being a bilayer, the layers separated by aqueous medium.

mesophyll Tissue found in the interior of leaves, made up of photosynthetic (parenchyma) cells, also called *chlorenchyma* cells. Consists of relatively large, highly vacuolated cells, with many *chloroplasts*. Includes *palisade parenchyma* and *spongy parenchyma*.

mesosecrin Glycoprotein (46kD) secreted by mesothelial cells (including endothelium). In culture forms a fine coating on the substratum.

mesosome Invagination of the plasma membrane in some bacterial cells, sometimes with additional membranous lamellae inside. May have respiratory or photosynthetic functions.

mesothelioma Malignant tumour of the *mesothelium*, usually of lung; caused by exposure to asbestos fibres, particularly those of crocidolite, the fibres of which are thin and straight and penetrate to the deep layers of the lung. Because of their size and shape, the fibres puncture the macrophage phagosome and are released, leading to a chronic inflammatory state which is thought to contribute to development of the tumour.

mesothelium Simple squamous epithelium of mesodermal origin. It lines the peritoneal, pericardial and pleural cavities and the synovial space of joints. The cells are said to be phagocytic.

messenger RNA (mRNA) Single-stranded RNA molecule that specifies the amino acid sequence of one or more polypeptide chains. This information is *translated* during protein synthesis when ribosomes bind to the mRNA. In prokaryotes, mRNA is the primary transcript from a DNA sequence and protein synthesis starts while the mRNA is still being synthesised. Prokaryote mRNAs are usually very short-lived (average half-life 5min.). In contrast, in eukaryotes the primary transcripts (*hnRNA*) are synthesised in the nucleus and they are extensively processed to give the mRNA which is exported to the cytoplasm where protein synthesis takes place. This processing includes the addition of a 5′-5′-linked 7-methylguanylate "cap" at the 5′ end and a sequence of adenylate groups at the 3′ end, the poly-A tail, as well as the removal of any *introns* and the splicing together of *exons*; only 10% of hnRNA leaves the nucleus. Eukaryote mRNAs are comparatively long-lived with half-life ranging from 30min to 24h.

metabolic burst (respiratory burst) Response of phagocytes to particles (particularly if *opsonised*), and to agonists such as *formyl peptides* and *phorbol esters*; an enhanced uptake of oxygen leads to the production, by an NAD- or NADPH-dependent system, of hydrogen peroxide, superoxide anions and hydroxyl radicals, all of which play a part in bactericidal activity. Defects in the metabolic burst, as in *chronic granulomatous disease*, predispose to infection particularly with *catalase*-positive bacteria, and are usually fatal in childhood.

metabolic cooperation Transfer between tissue cells in contact of low-molecular weight metabolites such as nucleotides and amino acids. Transfer is *via* channels constituted by the *connexons* of gap junctions, and does not involve exchange with the extracellular medium. First observed in cultures of animal cells in which radio-labelled purines were transferred from wild-type cells to mutants unable to utilise exogenous purines.

metabolic coupling The same as *metabolic cooperation*.

metabolism Sum of the chemical changes that occur in living organisms.

metacentric Descriptive of a chromosome that has its *centromere* roughly at the centre of the strand giving two more-or-less equal arms.

metachromasia (metachromatic staining) The situation where a stain when applied to cells or tissues gives a colour different from that of the stain solution.

metachronism Type of synchrony found in the beating of cilia. A metachronal process is one which happens at a later time, and the synchronisation is such that the active stroke of an adjacent cilium is slightly delayed so as to minimise the hydrodynamic interference; coordination is by visco-mechanical coupling. Different patterns of metachronal synchronisation are recognised: in symplectic m. the wave of activity in the field passes in the same direction as the active stroke of the individual cilium, in antiplectic m. the opposite is true. In dexioplectic and laeoplectic m. the wave of activity in the field is normal to the beat axis. Symplectic and antiplectic m. are considered orthoplectic, the other forms as diaplectic.

metalloenzyme An enzyme which contains a bound metal ion as part of its structure. The metal may be required for enzymic activity, either participating directly in catalysis, or stabilising the active conformation of the protein.

metalloprotein A protein which contains a bound metal ion as part of its structure.

metamorphosis Change of body form, for example in the development of the adult frog from the tadpole or the butterfly from the caterpillar.

metaphase In *mitosis* and *meiosis*, the stage at which the condensed chromosomes are attached to the *spindle fibres* and segregation has not yet commenced.

metaphase plate The plane of the *spindle* approximately equidistant from the two poles along which the chromosomes are lined up during *mitosis* or *meiosis*. Also termed the equator.

metaplasia Change from one differentiated phenotype to another, for example, the change of simple or transitional epithelium to a stratified squamous form as a result of chronic damage.

metastasis Development of secondary tumour(s) at a site remote from the primary; a hallmark of *malignant* cells.

metastatic spread Process of development of secondary tumours. A sequence of events that involves local invasion (in most cases), passive transport, lodgement and proliferation of cells at a remote site.

met-enkephalin Amino acid sequence YGGFM; see *enkephalins*.

methaemoglobin An oxidised form of haemoglobin containing ferric iron that is produced by the action of oxidising poisons. Non-functional.

Methanobacterium A genus of strictly anaerobic bacteria that reduce CO_2 using molecular hydrogen, H_2, to give methane. They show a number of features which distinguish them from other bacteria, and are now classified as a separate group within the *Archaebacteria*. Methanobacteria are found in the anaerobic sediment at the bottom of ponds and marshes (hence marsh gas is the common name for methane) and as part of the microflora of the rumen in cattle and other herbivorous mammals.

methicillin Penicillinase-resistant penicillin antibiotic.

methionine (Met; M; MW 149) Contains the -SCH₃ group which can act as a methyl donor (see *S-adenosyl methionine*). Common in proteins but at low frequency. The met-x linkage is subject to specific cleavage by cyanogen bromide. See Table A2.

methisazone (N-methylisatin; β-thiosemicarbazone) Drug which blocks poxvirus infection at a late stage, and was used prophylactically for smallpox. Its mechanism of action is obscure.

methotrexate Analogue of dihydrofolate. Inhibits dihydrofolate reductase and kills rapidly-growing cells. Therapeutic agent for leukaemias, but has a low therapeutic ratio.

methyl- (CH₃) Specific reference to the methyl group is made when macromolecules are modified after synthesis by enzymic addition of methyl groups. The group is transferred to nucleic acids and proteins by several potent carcinogens. See *methyl transferase*.

methyl-accepting chemotaxis proteins See *MCPs*.

N-methyl-D-aspartate (NMDA) Ligand for a class of receptors found in the central nervous system: see *kainate*.

3-methylcholanthrene Carcinogenic polycyclic hydrocarbon. One of many such substances formed during incomplete combustion of organic material.

methyldopa An antihypertensive drug, preferred in pregnant patients.

methylotroph Yeasts, like *Hansenula polymorpha*, which utilise methanol as an energy source.

methyltransferase Enzyme which transfers a methyl group from S-adenosyl methionine to a substrate. Encountered in post-translational modification of proteins and nucleic acids, in the removal of methyl groups added to DNA by alkylating carcinogens, and in bacterial chemotaxis where the methyl-accepting chemotaxis proteins (*MCPs*) become methylated in the course of *adaptation*.

metorphamide Amidated opioid octapeptide from bovine brain. Derived by proteolytic cleavage from proenkephalin.

mevinolin Intermediate in terpene synthesis; an analogue of compactin, a fungal metabolite which is used to lower plasma levels of low-density lipoprotein. It acts as an inhibitor of HMG-CoA-reductase, the rate-controlling enzyme in cholesterol biosynthesis.

MHC See *major histocompatibility complex*.

MHC restriction Restriction on interaction between cells of the immune system because of the requirement to recognise foreign antigen in association with MHC antigens (major *histocompatibility antigens*). Thus, cytotoxic T-cells will only kill virally-infected cells which have the same Class I antigens as themselves, whereas helper T-cells respond to foreign antigen associated with Class II antigens.

micelle 1. One of the possible ways in which amphipathic molecules may be arranged; a spherical structure in which all the hydrophobic portions of the molecules are inwardly directed, leaving the hydrophilic portions in contact with the surrounding aqueous phase. The converse arrangement will be found if the major phase is hydrophobic. 2. *Bot.* The parallel arrangement of cellulose chains in *microfibrils* in the plant cell wall.

Michaelis constant See K_m.

Michaelis–Menten equation Equation derived from a simple kinetic model of enzyme action, which successfully accounts for the hyperbolic (adsorption–isotherm) relationship between substrate concentration S and reaction rate V. $V = V_{max} \times S/(S + K_m)$, where K_m is the Michaelis constant and V_{max} is maximum rate approached by very high substrate concentrations.

microbody See *peroxisome*.

microcentrum Obsolete name for the pericentriolar region.

microcinematography The making of films using a microscope and cine camera.

microcolliculi Broad swellings (0.5μm) on the dorsal surface of a moving epidermal cell in culture, that move rearward as the cell moves forward (as do ruffles on fibroblasts).

microcytes Abnormally small red blood cells, found in some types of anaemia.

microelectrode An electrode, with tip dimensions small enough (less than 1μm) to allow non-destructive puncturing of the plasma membrane. This allows the intracellular recording of *resting* and *action potentials*, the measurement of intracellular ion and pH levels (using *ion-selective electrodes*), or *microinjection*. Microelectrodes are generally pulled from glass capillaries, and filled with conducting solutions of potassium chloride or potassium acetate to maximise conductivity near the tip. Electrical contact, if required, is usually made with a silver chloride-coated silver wire.

microfibril Basic structural unit of the plant cell wall, made of *cellulose* in higher plants and most algae, *chitin* in some fungi, and *mannan* or *xylan* in a few algae. Higher plant microfibrils are about 10nm in diameter, and extremely long in relation to their width. The cellulose molecules are oriented parallel to the long axis of the microfibril in a paracrystalline array (micelle), which provides great tensile strength. The microfibrils are held in place by the wall matrix, and their orientation is closely controlled by the protoplast.

microfilament Cytoplasmic filament, 5–7nm thick, of F-*actin* which can be decorated with *HMM*; may be laterally associated with other proteins (tropomyosin, α-actinin) in some cases, and may be anchored to the membrane. Microfilaments are conspicuous in *adhaerens junctions*.

microglial cells Small glial cells of mesodermal origin, with scanty cytoplasm and small spiny processes. Distributed throughout grey and white matter. Derive from monocytes and invade neural tissue just before birth; capable of enlarging to become macrophages.

microglobulin Any small globular plasma protein. See $β_2$-*microglobulin*.

microinjection The insertion of a substance into a cell through a *microelectrode*. Typical applications include the injection of drugs, histochemical markers (such as *horseradish peroxidase* or lucifer yellow) and RNA or DNA in molecular biological studies. To extrude the substances through the very fine electrode tips, either hydrostatic pressure (pressure injection) or electric currents (ionophoresis) are employed.

microperoxidase Part of a cytochrome c molecule which retains its haem group and has *peroxidase* activity.

microperoxisome Small *peroxisomes* of 150–250nm diameter found in most cells.

micropinocytosis Pinocytosis by small vesicles (around 100nm in diameter). Not blocked by cytochalasins.

micropore filters Filters made of a meshwork of cellulose acetate or nitrate and with defined pore size. They can be autoclaved, and the smaller pore sizes (0.22μm, 0.45μm) are used for sterilising heat-labile materials by filtering out micro-organisms. Larger pore-size filters are used in setting up *Boyden chambers*. They are about 150μm thick and should be distinguished from *Nucleopore filters*. Millipore is a trade name for micropore filters.

microprobe See *electron microprobe*.

micropyle 1. Small hole or aperture in the protective tissue surrounding a plant ovule, through which the pollen tube enters at fertilisation. Develops into a small hole in the seed coat through which, in many cases, water enters at germination. 2. Perforation in the shell (chorion) of an insect's egg through which the sperm enters at fertilisation.

microsomal fraction See *microsomes*.

microsomes Heterogenous set of vesicles 20–200nm in diameter formed from the endoplasmic reticulum when cells are disrupted.

microspikes Projections from the leading edge of some cells, particularly, but not exclusively, nerve growth cones. They are usually about 100nm diameter, 5–10μm long, and are supported by loosely bundled microfilaments. They are referred to by some authors as filopodia. Functionally a sort of linear version of a ruffle on a leading lamella.

microspore A haploid spore produced by a plant sporophyte, which develops into a male gametophyte. In seed plants, it corresponds to the developing pollen grain at the uninucleate stage. Smaller of the spores of a heterosporous species.

microtome A device used for cutting sections from an embedded specimen, either for light- or electron-microscopy.

microtrabecular network Complex network arrangement seen, using the high voltage electron microscope, in the cytoplasm of cells prepared by very rapid freezing. The suggestion is that most cytoplasmic proteins are in fact loosely associated with one another in this fibrillar network and are separate from the aqueous phase which contains only small molecules in true solution. If it exists, then it must certainly be very labile in cells where there is cytoplasmic flow and rapid organelle movement.

microtubule Cytoplasmic tubule, 25nm outside diameter with a 5nm thick wall. Made of tubulin heterodimers packed in a three-start helix (or of 13 protofilaments, looked at another way), and associated with various other proteins (*MAPs*, *dynein*, *kinesin*). Microtubules of the ciliary *axoneme* are more permanent than cytoplasmic and spindle microtubules.

microtubule associated proteins See *MAPs*.

microtubule organising centres (MTOC) Rather amorphous region of cytoplasm fron which microtubules radiate. The pattern and number of microtubules is determined by the MTOC. The pericentriolar region is the major MTOC in animal cells; the basal body of a cilium is another example. Activity of MTOCs can be regulated, but the mechanism is unclear.

microvillus Projection from the apical surface of an epithelial cell which is supported by a central core of microfilaments associated with bundling proteins such as *villin* and *fimbrin*. In the intestinal **brush border** the microvilli presumably increase absorptive surface area, whereas the stereovilli (*stereocilia*) of the cochlea have a distinct mechanical role in sensory transduction.

midbody Dense structure formed during *cytokinesis* at the cleavage furrow. It consists of remnants of *spindle fibres* and other amorphous material and disappears before cell division is completed.

middle-lamella First part of the plant cell wall to be formed, laid down in the *phragmoplast* during cell division as the *cell plate*. Subsequently makes up the central part of the double cell wall which separates two adjacent cells, cementing together the two primary walls. Rich in *pectin*, and relatively poor in *cellulose*.

MIF See *macrophage inhibition factor*.

Millipore filter Trade name for *micropore filters*.

Mimosa pudica The "sensitive plant" whose leaflets fold inwards very rapidly when touched. A more vigorous stimulus causes the whole leaf to droop, and the stimulus can be transmitted to neighbouring leaves.

miniature end plate potential (MEPP) Small fluctuations (typically 0.5mV) in the resting potential of postsynaptic cells, caused by spontaneous release of transmitter from the presynaptic terminal. They are the same shape as, but much smaller than, the end plate potentials caused by stimulation of the presynaptic cell. MEPPs are considered as evidence for the quantal release of *neurotransmitters* at *chemical synapses*, a single MEPP resulting from the release of the contents of a single synaptic vesicle.

minimyosin Form of *myosin* isolated from *Acanthamoeba*; only 180kD, but capable of binding to actin.

minisegregant cells Human cells with small amounts of DNA and few chromosomes, obtained experimentally by perturbing cell division. Can readily be fused with whole cells.

minute mutant A class of recessive *lethal mutants* of **Drosophila**. The heterozygotes grow more slowly, are smaller and less fertile than the wild-type flies. There are about 40 loci that produce *minute* mutants.

mismatch repair A DNA repair system that detects and replaces wrongly paired, mismatched, bases in newly replicated DNA. *E. coli* has a mismatch correction enzyme coded for by three genes *mutH*, *mutL* and *mutS*, which is directed to the newly synthesised strand and removes a segment of that strand including the incorrect nucleotide. The gap is then filled by DNA polymerase.

missense mutation A mutation which alters a *codon* for a particular amino acid to one specifying a different amino acid.

mitochondrion Highly pleomorphic organelle of eukaryotic cells which varies from short rod-like structures present in high number to long branched structures. Contains DNA and ribosomes. Has a double membrane and the inner membrane may contain numerous folds (cristae). The inner fluid phase has most of the enzymes of the *tricarboxylic acid cycle* and some of the urea cycle. The inner membrane contains the components of the *electron transport chain*. Major function is to regenerate ATP by oxidative phosphorylation (see *chemiosmotic hypothesis*).

mitogen Any chemical specifically stimulating a eukaryote cell to enter S phase of the *cell cycle*, which then continues through G2 and mitosis.

mitogenesis The process of stimulating transit through the cell cycle especially as applied to lymphocytes. Concanavalin A is a mitogen for T-lymphocytes; the best mitogen for B-lymphocytes is Cowan strain *Staphylococcus aureus*.

mitomycin C Aziridine antibiotic isolated from *Streptomyces caespitosus*. Inhibits DNA synthesis by cross-linking the strands.

mitoplasts Isolated mitochondria without their outer membranes. They have finger-like processes, and retain the capacity for oxidative phosphorylation.

mitoribosomes Mitochondrial ribosomes; these more closely resemble prokaryotic ribosomes than cytoplasmic ribosomes of the cells in which they are found, though they are even smaller and have fewer proteins than bacterial ribosomes.

mitosis The usual process of nuclear division in the somatic cells of eukaryotes. Mitosis is classically divided into four stages. The chromosomes are actually replicated prior to mitosis during the S phase of the *cell cycle*. During the first stage, prophase, the chromosomes condense and become visible as double strands (each strand being termed a chromatid) and the *nuclear envelope* breaks down. At the same time the mitotic *spindle* forms by the polymerisation of microtubules and the chromosomes are attached to spindle fibres at their kinetochores. In metaphase the chromosomes align in a central plane perpendicular to the long axis of the spindle (the metaphase plate). During anaphase the paired chromatids are apparently pulled to opposite poles of the spindle by means of the spindle fibre microtubules attached to the kinetochore, though the actual mechanism for this movement is still controversial. In anaphase B the poles are further separated, possibly by a different mechanism, and the separation of chromatids is completed during telophase, when they can be regarded as chromosomes proper. The chromosomes now lengthen and become diffuse and new nuclear envelopes form round the two sets of chromosomes. This is usually followed by cell division (cytokinesis) in which the cytoplasm is also divided to give two daughter cells. Mitosis ensures that each daughter cell has a set of chromosomes which is identical in number to that of the parent cell.

mitotic apparatus See *spindle*.

mitotic death Cells fatally damaged by ionising radiation may not die until the next mitosis, at which point the radiation damage to the DNA becomes evident, particularly when there is fragmentation of chromosomes.

mitotic index The fraction of cells in a sample that are in mitosis. It is a measure of the relative length of the mitotic phase of the cell cycle.

mitotic recombination Somatic crossing over. *Crossing over* can occur between *homologous chromosomes* during mitosis, but is very rare because the chromosomes do not normally pair. When it occurs it can lead to new combinations of previously linked genes. Although infrequent, mitotic recombination has been utilised for genetic analysis in **Aspergillus** and yeast and in studies on developmental *compartments* in

Drosophila. The frequency of mitotic recombination can be increased by ultraviolet- and X-irradiation.

mitotic segregation *Mitotic recombination*.

mitotic shake-off method A method of collecting cells in mitosis, so that the chromosomes can be examined and the karyotype determined. Many cultured cells round up during mitosis and so become less firmly attached to the culture substratum. Cells in mitosis thus can be removed into suspension by gentle shaking of the culture vessel, leaving the non-mitotic cells still attached. The number of cells that are in mitosis is usually increased by using a drug, such as *colcemid* that blocks mitosis at *metaphase*.

mitotic spindle See *spindle*.

mixed lymphocyte reaction (mixed lymphocyte culture) Response of lymphocytes when co-cultured with lymphocytes from another animal. *Blast cell* transformation occurs after 3–5 days in proportion to the extent of incompatibility.

MLCK See *myosin light chain kinase*.

MN blood-group antigens A pair of blood group antigens governed by genes which segregate independently of the ABO locus. The alleles are codominant and there are three types MM, NN and MN. *Glycophorin* has M or N activity and this is associated with oligosaccharides attached to the amino-terminal portion of the molecule. M-type glycophorin differs from N-type in amino acid residues 1 and 5, although the antigenic determinants are associated with the carbohydrate side chains.

mobile genetic elements See *transposons*.

mobile ion carrier See *ionophore*.

molluscan catch muscle Muscle responsible for holding closed the two halves of the shell of bivalves. Specialised to maintain tension with low expenditure of ATP. Rich in *paramyosin*.

molluscum bodies Intracellular inclusions of a poxvirus found in human epidermis; harmless, but contagious, skin lesions (molluscum contagiosum).

Moloney murine leukaemia virus Replication-competent retrovirus (*Oncovirinae*) which causes leukaemia in mice, isolated by Moloney from cell-free extracts made from a transplantable mouse sarcoma.

Moloney murine sarcoma virus Replication-defective retrovirus, source of the *oncogene v-mos*, responsible for inducing fibrosarcomas *in vivo*, and transforming cells in culture.

Moloney test Skin test for immunity to diphtheria in which active toxin is injected into one site and toxoid into another. This is to control for pseudopositive reactions to the toxin.

monensin A sodium *ionophore* (MW 671) from *Streptomyces cinnamonensis*. Has antibiotic properties, and is used as a feed additive in chickens. Also used in *ion-selective electrodes*.

Mongolism See *Down's syndrome*.

monoamine neurotransmitters See *biogenic amines*.

monoamine oxidase (MAO; tyraminase) Enzyme catalyzing breakdown of several *biogenic amines*, such as serotonin, adrenaline, noradrenaline, dopamine.

monocentric chromosome Chromosome with a single *centromere*, ie. most chromosomes.

monocistronic RNA A *messenger RNA* that gives a single polypeptide chain when translated. All eukaryote mRNAs are monocistronic, but some bacterial mRNAs are polycistronic especially those transcribed from *operons*.

monoclonal Used of a cell line whether within the body or in culture to indicate that it has a single clonal origin. Monoclonal antibodies are produced by a single clone of *hybridoma* cells, and are therefore a single species of antibody molecule.

monocyte Mononuclear phagocyte circulating in blood which will later emigrate into tissue and differentiate into a macrophage.

monokines Soluble factors, derived from macrophages, which act on other cells (eg. *interleukin-1*, *tumour necrosis factor*).

monolayer A single layer of any molecule, but most commonly applied to polar lipids. Present on the surface of lipoproteins. Can be formed at an air/water interface in experimental systems. The term should not be used to describe one layer of a lipid bilayer, for which the term "leaflet" is generally used. See also *monolayering of cells*.

monolayering of cells Tendency of animal

tissue cells growing on solid surfaces to cover the surface with a complete layer only one cell thick, before growing on top of each other. This non-random distribution is generated by contact inhibition of loco-motion, a phenomenon in which colliding cells change direction rather than move over one another. Of the theories why some (but by no means all) types of cells stop growing when a monolayer is formed, present evidence favours limitation by supply of growth factors from the medium, rather than any inhibitory effect of contact on growth.

monopodial Adjective describing an amoeba which has only one pseudopod (as opposed to polypodial forms).

monosaccharide A simple sugar that cannot be hydrolysed to smaller sugar units. Empirical formula is $(CH_2O)_n$, and molecules range in size from trioses (n = 3) to heptoses (n = 7).

monosomy Situation in a normally *diploid* cell or organism in which one or more of the *homologous chromosome* pairs is represented by only one chromosome of the pair. For example, sex determination in grasshoppers depends on the fact that females are XX and males XO; that is, males have only one sex chromosome and are monosomic for the X chromosome.

MOPS (morpholino-propane sulphonic acid) A "biological" buffer; a synthetic zwitter-ionic compound with a pK_a of 7.2, that is non-toxic and has a low temperature coefficient. Widely used in biochemical studies, largely as a replacement for phosphate buffers.

Morbilli virus Genus of viruses (of the *Para-myxoviridae*). Type species is *measles virus*; other species include canine distemper virus and the related seal virus.

morphallaxis Regenerative process in which part of an organism is transformed directly into a new organism without replication at the cut surface.

morphine An opioid alkaloid, isolated from opium, with a complex ring structure. It is a powerful analgesic with important medical uses, but is highly addictive. Functions by occupying the receptor sites for the natural neurotransmitter peptides, endorphins and enkephalins, but is stable to the peptidases that inactivate these compounds.

morphogen Diffusible substance which carries information relating, for example, to position in the embryo, and thus determines the differentiation which cells perceiving this information will undergo. The only identi-fied morphogen in animals at present appears to be retinoic acid which is important in defining the pattern of digit differentiation in the chick limb-bud.

morphogenesis The process of "shape formation": the processes which are responsible for producing the complex shapes of adults from the simple ball of cells that derives from division of the fertilised egg.

morphogenetic movements Movements of cells or of groups of cells in the course of development. Thus the invagination of cells in gastrulation is one of the most dramatic of morphogenetic movements; another much-studied example is the migration of neural crest cells.

morphometry Method which involves measurement of shape. A variety of methods exist to enable one to examine, for example, the distribution of objects in a two-dimensional section of a cell and then to use this to predict the shapes and the distribution of these objects in three dimensions.

morula Stage of development in holoblastic embryos. The morula stage is usually likened to a spherical raspberry, a cluster of blastomeres without a cavity.

mosaic egg At one time a distinction was drawn between those organisms in which the egg seemed to have a firmly committed fate map built in and "regulating" embryos. In the former, after the first cleavage one blastomere was committed to produce one set of tissues, the other blastomere a different set, and removal of one blastomere led to the production of an incomplete embryo. This was particularly obvious in mollusc development where one blastomere had the polar lobe material. This early differentiation (or determination) of blastomeres for particular fates was in distinction to "regulating" embryos in which the removal of one blastomere did not matter, the other blastomere(s) compensating and producing a full set of tissues. The distinction is, however, only based upon the timing of differentiative events, and within a few divisions the regulating embryo also becomes a mosaic of determined cells.

motilin Peptide (22 residues) found in duodenum, pituitary and pineal that stimulates intestinal motility.

motoneuron A *neuron* which connects functionally to a *muscle fibre*.

motor end plate Synonym for *neuromuscular junction*.

motor neuron Synonym for *motoneuron*.

Mott cells Plasma cells containing large eosinophilic inclusions; found in the brain in cases of African trypanosomiasis.

mRNA See *messenger RNA*.

MTOC See *microtubule organising centre*.

mucilage Sticky mixture of carbohydrates in plants.

mucocyst Small membrane-bounded vesicular organelle in *pellicle* of ciliate protozoans which will discharge a mucus-like secretion.

mucopeptide Synonym for *peptidoglycan*.

mucopolysaccharide The polysaccharide component of proteoglycans, now more usually known as *glycosaminoglycan*.

mucopolysaccharidoses Inherited diseases in humans resulting from inability to break down glycosaminoglycans. *Hunter syndrome* and *Hurler's disease*, for example, result from defects in lysosomal enzymes needed to break down sulphated mucopolysaccharides.

mucous gland A type of *merocrine* gland which produces a thick (mucopolysaccharide-rich) secretion (as opposed to a *serous gland*).

mucus Viscous solution secreted by various membranes; rich in glycoprotein.

multienzyme complex Cluster of distinct enzymes catalyzing consecutive reactions of a metabolic pathway or associated reactions such as those involved in DNA replication, which remain physically associated through purification procedures. Multifunctional enzymes, found in eukaryotes, are a somewhat different phenomenon, since the several enzymic activities are associated with different domains of a single polypeptide.

multiple myeloma See *myeloma cell*.

multivesicular body Secondary lysosome containing many vesicles of around 50nm diameter.

muramyl dipeptide Fragment of *peptidoglycan* which is used as an adjuvant.

murein Cross-linked peptidoglycan complex from the inner cell wall of all Eubacteria. Constitutes 50% of the cell wall in Gram negative and 10% in Gram positive organisms, and comprises $\beta(1-4)$-linked N-acetyl glucosamine and N-acetyl muramic acid extensively cross-linked by peptides.

muscarinic A type of *chemical synapse* in which the neurotransmitter is *acetylcholine*, and which is distinguished from a *nicotinic* synapse by its sensitivity to the plant poison, muscarine. Nicotinic synapses are referred to as type I, muscarinic as type II.

muscle Tissue specialised for contraction: see *striated muscle*, *cardiac muscle* and *smooth muscle*.

muscle cell Cell of muscle tissue; in *striated* (skeletal) *muscle* it comprises a *syncytium* formed by the fusion of embryonic *myoblasts*, in *cardiac muscle* a cell linked to the others by specialised junctional complexes (*intercalated discs*), in smooth muscle a single cell with large amounts of actin and myosin capable of contracting to a small fraction of its resting length.

muscle fibre Component of a skeletal muscle comprising a single syncytial cell which contains *myofibrils*.

muscle spindle A specialised muscle fibre found in tetrapod vertebrates. A bundle of muscle fibres is innervated by sensory neurons and a few motoneurons. Stretching the muscle causes the sensory neurons to fire; the muscle spindle thus functions as a stretch receptor.

mutagenicity test See *Ames test*.

mutagens Agents which cause an increase in the rate of mutation; includes X-rays, ultraviolet irradiation (260nm) and various chemicals.

mutation Change. Usage usually restricted to change in the DNA sequence of an organism, which may arise in any of a variety of different ways. See *frame-shift mutation*, *nonsense codon*.

mutation rate The frequency with which a particular mutation arises in a population or the frequency with which any mutation arises in the whole genome of a population. Normally the context makes the precise use clear. See *fluctuation analysis*.

myalgia Muscle pain.

myasthenia gravis The characteristic feature of the disease is easy fatigue of certain voluntary muscle groups on repeated use. Muscles of the face or upper trunk are especially likely to be affected. In most, and perhaps all, cases due to the development of autoantibodies against the acetylcholine receptor in neuromuscular junctions. Immunisation of mice or rats with this receptor protein leads to a disease with the features of myasthenia.

mycelium Mass of *hyphae* which constitutes the vegetative part of a fungus.

mycobacteria Bacteria with unusual cell walls which are resistant to digestion, being waxy, very hydrophobic and rich in lipid, especially esterified *mycolic acids*. Staining properties differ from those of Gram negative and Gram positive organisms, being acid-fast. Many are intracellular parasites, causing serious diseases such as leprosy and tuberculosis. Cell wall has strong immuno-stimulating (*adjuvant*) properties due to muramyl dipeptide (MDP).

Mycobacterium bovis Causes tuberculosis in cattle; attenuated strain is ***Bacillus Calmette-Guerin*** (BCG), used for immunisation.

Mycobacterium leprae Causative agent of *leprosy*.

Mycobacterium microti Mycobacterium which causes tuberculosis-like disease in small rodents (*Microtus microtus* is the vole); will infect mice but not humans, and is therefore much used as a laboratory model. Releases large amounts of cyclic AMP which may inhibit *lysosome–phagosome fusion*.

Mycobacterium tuberculosis Obligate anaerobic non-motile bacterium, causative agent of tuberculosis in humans. Lives intracellularly in macrophages.

mycolic acids Saturated fatty acids found in the cell walls of *mycobacteria*, **Nocardia** and *corynebacteria*. Chain lengths can be as high as 80, and the mycolic acids are found in waxes and in glycolipids.

mycoplasmas Prokaryotic micro-organisms lacking cell walls, and therefore resistant to many antibiotics. Formerly known as pleuro-pneumonia-like organisms (PPLO). Causative agents of pneumonia in humans and some domestic animals. Troublesome contaminants of animal cell cultures, in which they may grow attached or close to cell surfaces, subtly altering properties of the cells, but escaping detection unless specifically monitored. Similar organisms, spiroplasmas, cause various diseases in plants.

mycosides Complex glycolipids found in *mycobacterial* cell wall. Non-toxic, non-immunogenic molecules which influence the form of the colony and the susceptibility of the bacteria to bacteriophages.

mycosis fungoides A human disease in which a frequent secondary feature is fungal infection of lesions in the skin. Recognised as a tumour of T-lymphocytes which accumulate in the dermis and epidermis and cause loss of the epidermis.

myelin The material making up the *myelin sheath* of nerve axons.

myelin figures Structures which form spontaneously when bilayer-forming phospholipids (eg. egg lecithin) are added to water. They are reminiscent of the concentric layer structure of myelin.

myelin sheath An insulating layer surrounding vertebrate *neurons*, which dramatically increases the speed of conduction. In the peripheral nervous system it is formed by specialised *Schwann cells*, which can wrap around neurons up to 50 times; neurons in the central nervous system are myelinated by *oligodendrocytes*. The exposed areas are called nodes of Ranvier: they contain very high densities of *sodium channels*, and *action potentials* jump from one node to the next, without involving the intermediate axon, a process known as *saltatory conduction*.

myeloid cells One of the two classes of marrow-derived blood cells; includes *megakaryocytes*, erythrocyte-precursors (though these are more commonly considered as a separate, erythroid, line), *monocytes* and all the *polymorphonuclear leucocytes*. That all these are ultimately derived from one stem cell lineage is shown by the occurrence of the Philadelphia chromosome in these, but not *lymphoid*, cells. Most authors would, however, restrict the term "myeloid" to mononuclear phagocytes and granulocyte precursors, and this is certainly the common usage.

myeloma cell Neoplastic *plasma cell*. The proliferating plasma cells often replace all the others within the marrow, leading to immune deficiency, and frequently there is destruction of the bone cortex. Because they are monoclonal in origin they secrete a monoclonal immunoglobulin. Bence–Jones proteins are monoclonal immunoglobulin light chains overproduced by myeloma cells and excreted in the urine. Myeloma cell lines are used for producing hybridomas in raising monoclonal antibodies.

myeloma proteins The immunoglobulins and Bence–Jones proteins secreted by *myeloma cells*.

myeloperoxidase Peroxidase found in the lysosomal granules of *myeloid cells*, particularly macrophages and neutrophils; responsible for generating potent bacteriocidal activity by the hydrolysis of hydrogen peroxide (produced in the *metabolic burst*) in the presence of halide ions. A metalloenzyme containing iron. Deficiency of myeloperoxidase is not fatal, and it is reportedly absent entirely in chickens.

myoblast Cell that by fusion with other myoblasts gives rise to *myotubes* which eventually develop into *skeletal muscle* fibres. The term is sometimes used for all the cells recognisable as immediate precursors of skeletal muscle fibres. Alternatively, the term is reserved for those post-mitotic cells capable of fusion, others being referred to as presumptive myoblasts.

myocarditis Inflammation of heart muscle usually due to bacterial or viral infection.

myoepithelial cell (basket cell; basal cell) Cell found between epithelium of exocrine glands (eg. salivary, sweat, mammary, mucous) and their basement membranes, which resembles a smooth muscle cell, and is thought to be contractile.

myofibril Long cylindrical organelle of striated muscle, composed of regular arrays of thick and thin filaments, and constituting the contractile apparatus.

myogenesis The developmental sequence of events leading to the formation of adult muscle which occurs in the animal and in cultured cells. In vertebrate skeletal muscle the main events are: the fusion of *myoblasts* to form *myotubes* which increase in size by further fusion to them of myoblasts, the formation of *myofibrils* within their cytoplasm, and the establishment of functional *neuromuscular junctions* with *motoneurons*.

At this stage they can be regarded as mature muscle fibres.

myoglobin Protein (17.5kD) found in red skeletal muscle. It was the first protein for which the *tertiary structure* was determined by *X-ray diffraction*, by J.C.Kendrew's group working on sperm whale myoglobin. It is a single polypeptide chain of 153 amino acids, containing a *haem* group bonded *via* its ferric iron to two histidine residues. It binds oxygen non-cooperatively and has a higher affinity for oxygen at all partial pressures than *haemoglobin*. In capillaries oxygen is effectively removed from haemoglobin and diffuses into muscle fibres where it binds to myoglobin which acts as an oxygen store.

myomesin Protein (165kD) found in the *M-line* of the *sarcomere*.

myoneme Non-actin-containing contractile organelle of ciliates and certain other protozoa (*Acantharia*); referred to as M-bands in *Stentor*, where they are composed of 8–10nm tubular fibrils. The *spasmoneme* of peritrich ciliates was originally called a myoneme.

myosin Multimeric protein (440kD) which has actin-activated ATPase activity, and which undergoes a conformational change as ATP is bound and cleaved. The conformational change is the basic movement of the actin-myosin motor. There are two heavy chains (200kD) and two pairs of light chains (17–22kD) in every hexameric unit, and the molecules pack together to form bipolar thick filaments. There is considerable diversity in the properties of myosin isolated from different tissues and organisms. See also *myosin light chains*, *meromyosin*.

myosin heavy chain See *myosin*: do not confuse with heavy meromyosin (*HMM*) which is a sub-fragment of the heavy chain.

myosin light chain kinase (MLCK) *Calmodulin*-regulated kinase of *myosin light chains*: molecular weight varies according to source, 130kD in non-muscle mammalian cells. May regulate activity of myosin in some cells.

myosin light chains Small subunit proteins (17–22kD) of *myosin*, all with sequence homology to *calmodulin*, but not all with calcium-binding activity: two pairs of different light chains are found per myosin. Several types are known: regulatory light chains (LC-2, DNTB-light chains) probably regulate the ATPase activity of the heavy chain directly (through the binding of calcium) or indirectly (activating when they themselves are phosphorylated by *myosin light chain kinase*), and essential light chains (LC-1, LC-3; alkali light chains), which have a more subtle and apparently non-essential role. In molluscan muscle the EDTA-light chains (similar to LC-2 from vertebrate muscle) confer calcium-sensitivity on the myosin itself.

myotonic dystrophy An inherited human disease classed as an autosomal dominant disease with progressive muscle weakening and wasting.

myotube Elongated multinucleate cells (three or more nuclei) which contain some peripherally located *myofibrils*. They are formed *in vivo* or *in vitro* by the fusion of *myoblasts* and eventually develop into mature muscle fibres which have peripherally-located nuclei and most of their cytoplasm filled with myofibrils. In fact, there is no very clear distinction between myotubes and muscle fibres proper.

Mytilus edulis The edible mussel, a marine bivalve mollusc. It has large ciliated gills which are used for filter-feeding; these are utilised in studies on *cilia* and *metachronism*.

myxamoebae (USA myxamebae) In the Myxomycetes, such as **Physarum**, each spore on germination produces two amoeboid cells, myxamoebae, which then transform into flagellated cells.

Myxobacteria Group of Gram negative bacteria, found mainly in soil. They are non-flagellated with flexible cell walls. They show a gliding motility, moving over solid surfaces leaving a layer of slime (myxo = slime). At some stage in their growth the cells of this group swarm together and form fruiting bodies and spores in a fashion similar to the slime moulds.

myxoedema Severe hypothyroidism usually as a result of autoimmunity to *thyroglobulin*. A variety of severe physiological problems accompany the reduction in thyroid function.

myxoma virus A poxvirus which causes myxomatosis. Originally isolated from a species of wild rabbit, *Sylvilagus* in Brazil, in which it causes a mild non-fatal disease, it was found to be 99% fatal in the European rabbit *Oryctolagus*. It causes the characteristic, sub-cutaneous gelatinous swellings, "myxomata" and usually kills in 2–5 days. It has been used to control rabbit populations in Australia and Britain, but there are signs that they have developed immunity.

N

N-lines Regions in the *sarcomere* of striated muscle. The N1 line is in the I-band near the Z-disc, the N2 line is at the end of the A-band. The N-lines may represent the location of proteins such as nebulin which contribute to the stability of the sarcomere.

N-protein 1. Anti-terminator protein of the bacteriophage λ and other phages, which plays a key role in the early stages of infection. During the early phase, only two genes N and *cro* are transcribed, by *transcription* of the DNA in opposite directions. N-protein binds to sites on the DNA (*nut* sites for N-utilisation), prevents *rho*-dependent termination and allows transcription of the genes. 2. Name which has been used for *GTP-binding proteins* (G-proteins); now obsolete and should be avoided because of confusion with N-protein of bacteriophages.

NAA See *naphthalene acetic acid*.

NAD (nicotinamide adenine dinucleotide; NAD$^+$; formerly DPN) Coenzyme in which the nicotine ring undergoes cyclic reduction to NADH and oxidation to NAD. Acts as a diffusible substrate for dehydrogenases etc. NADH is one source of reducing equivalents for the electron transport chain. NAD is of special interest as the source of ADP-ribose (see *ADP ribosylation*).

NADP (formerly TPN) Phosphorylated form of *NAD*, but NADPH is used for reductive biosynthetic processes (eg. pentose phosphate synthesis) rather than ATP generation.

Naegleria gruberi A normally amoeboid protozoan found in the soil. When it is flooded with water or a solution of low ionic strength it transforms into a swimming form with two *flagella*.

naevus (USA nevus) Tumour-like but non-neoplastic *hamartoma* of skin. A vascular naevus is a localised capillary-rich area of the skin ("strawberry birthmark"; sometimes the much more extensive "port-wine stain"). A mole is a pigmented naevus, a cluster of melanocytes containing melanin.

nagarse (nagarase) Broad-specifity protease from bacteria.

Nagler's reaction Standard method for identifying *Clostridium perfringens*. When grown on agar containing egg yolk, an opalescent halo is formed around colonies which produce α-toxin (lecithinase).

naloxone An alkaloid antagonist of morphine and of the opiate peptides.

Namalwa cells Lymphoblastoid cell line grown in suspension and used to produce *interferon* (stimulated by Sendai virus infection).

nanovid microscopy Technique of brightfield light microscopy using electronic contrast enhancement and maximum *numerical aperture*.

naphthalene acetic acid (NAA) A synthetic auxin, often used in plant physiology and in plant tissue culture media because it is more stable than *IAA*.

napthoquinones Plant pigments derived from napthoquinone, eg. juglone from walnut. Also include the K vitamins.

β-naphthylamine Potent carcinogen; used in production of aniline dyes, one of the first chemicals to be associated with a tumour (bladder cancer). The compound itself is not directly carcinogenic; a metabolite produced by hydroxylation (1-hydroxy-2-aminonaphthalene) is de-toxified in the liver by conjugation with glucuronic acid, but reactivated by a glucuronidase in the bladder.

natural killer cells (NK cells) See *killer cells*.

natural selection The hypothesis that genotype–environment interactions occurring at the phenotypic level lead to differential reproductive success of individuals and hence to modification of the gene pool of a population.

nebulin Protein found in the *N-line* of the *sarcomere*.

necrosis Death of some or all cells in a tissue as a result of injury, infection or loss of blood supply.

negative feedback This occurs where the products of a process can act at an earlier stage in the process to inhibit their own formation. The term was first used widely in conjunction with electrical amplifiers where negative feedback was applied to limit distortion of the signal by the amplification mechanism. Tends to stabilise the process. In contrast to positive feedback.

negative regulation Negative feedback in biological systems mediated by allosteric regulatory enzymes.

negative staining Microscopic technique in which the object stands out against a dark background of stain. For electron microscopy the sample is suspended in a solution of an electron-dense stain such as sodium phosphotungstate and then sprayed onto a support grid. The stain dries as a structureless solid and fills all crevices in the sample. Quite fine structural detail can be observed using negative staining and it has been used extensively to study the structure of viruses and other particulate samples.

negative-stranded RNA virus Class V *viruses* which have a single-stranded RNA genome which is complementary to the mRNA, the positive strand. They also carry the virus-specific *RNA polymerase* necessary for the synthesis of the mRNA. Includes *Rhabdoviridae*, *Paramyxoviridae* and Myoviridae (eg. the T-even phages).

Negri body Acidophilic cytoplasmic inclusion (mass of *nucleocapsids*) characteristic of rabies virus infection.

Neisseria Gram negative non-motile pyogenic coccus. Two species are serious pathogens, N. meningitidis *(see meningitis)* and *N. gonorrhoeae*. The latter associates specifically with urinogenital epithelium through surface *pili*. Both species seem to evade the normal consequences of attack by phagocytes.

nematocyst (cnidocyst) Stinging mechanism used for defence and prey capture by **Hydra** and other members of the Cnidaria (Coelenterata). Located within a specialised cell, the *nematocyte* and consists of a capsule containing a coiled tube. When the nematocyst is triggered, the wall of the capsule changes its water permeability and the inrush of water causes the tube to evert explosively, ejecting the nematocyst from the cell. The tube is commonly armed with barbs and may also contain toxin.

nematocyte (cnidoblast) Stinging cells found in *Hydra*, used for capturing prey and for defence. There are four major types, containing different sorts of *nematocysts*:

stenoteles (60%), desmonemes, holotrichous isorhizas and atrichous isorhizas. They differentiate from interstitial cells and are almost all found in the tentacles.

nematode sperm Nematode worms have unusual amoeboid spermatozoa which are actively motile yet appear to lack both actin and tubulin.

neoendorphin Opioid peptide (*endorphin*) cleaved from pro-dynorphin.

neomycin Aminoglycoside *antibiotic*.

neoplasia Literally new growth, usually refers to abnormal new growth, and thus means same as *tumour*, which may be benign or malignant. Unlike *hyperplasia*, neoplastic proliferation persists even in the absence of the original stimulus.

neoteny The persistence in the reproductively-mature adult of characters usually associated with the immature organism.

neoxanthin A *xanthophyll carotenoid* pigment, found in higher plant chloroplasts as part of the *light-harvesting system*.

nephelometry Any method for estimating the concentration of cells or particles in a suspension by measuring the intensity of scattered light, often at right-angles to the incident beam. Light scattering depends upon number, size and surface characteristics of the particles.

Nernst equation A basic equation of biophysics, which describes the relationship between the equilibrium potential difference across a *semipermeable membrane*, and the equilibrium distribution of the ionic permeant species. It is described by: $E = (RT/zF).\ln([C_1]/[C_2])$, where E is the potential on side 2 relative to side 1 (in volts), R is the gas constant (8.314 J K^{-1} mol^{-1}), T is the absolute temperature, z is the charge on the permeant ion, F is the Faraday constant (96500 C mol^{-1}) and C_1 and C_2 are the concentrations (more correctly activities) of the ion on sides 1 and 2 of the membrane. It can be seen that this equation is a solution of the more general equation of *electrochemical potential*, for the special case of equilibrium. The equation describes the voltage generated by ion-selective electrodes, like the laboratory pH electrode, and approximates the behaviour of the resting plasma membrane (see *resting potential*).

nerve cell See *neuron*.

nerve ending See *synapse*.

nerve growth cone See *growth cone*.

nerve growth factor (NGF) A peptide (MW 13259) of 118 amino acids (usually dimeric) with both chemotropic and chemotrophic properties for *sympathetic* and *sensory neurons*. Found in a variety of peripheral tissues, NGF attracts *neurites* to the tissues by chemotropism, where they form synapses. The successful neurons are then "protected" from neuronal death by continuing supplies of NGF. It is also found at exceptionally high levels in snake venom and male mouse submaxillary salivary glands, from which it is commercially extracted. NGF was the first of a family of nerve tropic factors to be discovered. Amino acids 1–81 show sequence homology with proinsulin.

nerve impulse An *action potential*.

neural cell adhesion molecule (NCAM) Cell surface glycoprotein of vertebrates initially defined by antiserum capable of inhibiting cell–cell adhesion of chicken neural retina cells. Believed to be important in calcium-independent intercellular adhesion of neural and other cells, including cells in early embryos.

neural crest A group of embryonic cells which separate from the *neural plate* during neurulation and migrate to give several different lineages of adult cells:- the spinal and *autonomic* ganglia, the *glial cells* of the peripheral nervous system, and non-neuronal cells such as those of *chromaffin tissue* and *cartilage*, and *melanocytes* and some haemopoietic cells.

neural fold A crease which forms in the *neural plate* during *neurulation*.

neural plate A region of embryonic ectodermal cells, called neuroectoderm, which lies directly above the *notochord*. During *neurulation*, the cells change shape, so as to produce an infolding of the neural plate (the neural fold) which then seals to form the neural tube.

neural retina Layer of nerve cells in the retina, embryologically part of the brain. The incoming light passes through nerve-fibres and intermediary nerve cells of the neural retina, before encountering the light-sensitive rods and cones at the interface between neural retina and the pigmented retinal epithelium.

neural tube The progenitor of the central nervous system. See *neural plate*, *neurulation*.

neuraminic acid (N-acetyl neuraminic acid) Sometimes known as *sialic acid*, but strictly one of a family of sialic acids (which includes also N-glycolyl neuraminic acid and O-substituted derivatives). It is a 9-carbon sugar formed by adding to *mannose* three carbons from pyruvate; occurs in the subset of glycolipids known as *gangliosides* and in glycoproteins. The presence of its carboxyl group in these forms at the surface of animal cells is responsible for much of their negative charge.

neuraminidase (sialidase) Enzyme catalyzing cleavage of *neuraminic acid* residues from oligosaccharide chains of glycoproteins and glycolipids. Since these residues are usually terminal, neuraminidases are generally exo-enzymes, although an endoneuraminidase is known. For use as a laboratory reagent, common sources are from bacteria such as *Vibrio* or *Clostridium*. A neuraminidase is one of the transmembrane proteins of the envelope of influenza virus.

neurite A process growing out of a neuron. As it is hard to distinguish a *dendrite* from an *axon* in culture (and sometimes *in vivo*), the term neurite is used for both.

neuroblastoma Malignant tumour derived from primitive ganglion cells. Mainly a tumour of childhood. Commonest sites are adrenal medulla and retroperitoneal tissue. The cells may partially differentiate into cells having the appearance of immature neurons.

neuroendocrine cell See *neurohormone*.

neurofibrils Filaments found in neurons; not necessarily *neurofilaments* in all cases, and in the older literature referred to fibrils composed of both microtubules and neurofilaments. Originally used by light microscopists to describe much larger fibrils seen particularly well with silver-staining methods.

neurofilament Member of the class of *intermediate filaments* found in axons of nerve cells. In vertebrates assembled from three distinct protein subunits.

neurohormone A hormone secreted by specialised *neurons* (neuroendocrine cells); eg. *releasing hormones*.

neuroleptic drugs (antischizophrenic drugs; antipsychotic drugs; tranquillisers) Literally "nerve-seizing": used of chlorpromazine-like drugs. In many cases antagonise the effects of *dopamine*.

neuromuscular junction A *chemical synapse* between a motoneuron and a *muscle fibre*. Also known as a motor endplate.

neuron (neurone; nerve cell) An *excitable cell* specialised for the transmission of electrical signals. Neurons receive input from sensory cells or other neurons, and send output to muscles or other neurons. Neurons with sensory input are called "sensory neurons", neurons with muscle outputs are called "motoneurons"; neurons which connect only with other neurons are called "interneurons". Neurons connect with each other *via synapses*. Neurons can be the longest cells known; a single *axon* can be several metres in length. Although signals are usually sent *via action potentials*, some neurons are *non-spiking*.

neuropeptides Peptides with direct synaptic effects (peptide neurotransmitters) or indirect modulatory effects on the nervous system (peptide neuromodulators).

neurophysin Carrier protein (10kD, 90–97 amino acids) which transports neurohypophysial hormones along *axons*, from the hypothalamus to the posterior lobe of the pituitary.

neurosecretory cells Cells which have properties both of electrical activity, carrying impulses, and secretory function, releasing hormones into the bloodstream. In a sense, they are behaving in the same way as any chemically-signalling neuron, except that the target is the blood (and remote tissues), not another nerve or postsynaptic region.

Neurospora A fungus of the Ascomycetes group. It is *haploid* and grows as a *mycelium*. There are two mating types, and fusion of nuclei of two opposite types leads to meiosis followed by mitosis. The resulting eight nuclei generate eight ascospores. These are arranged linearly in an ordered fashion in a pod-like ascus, so that the various products of meiotic division can be identified and isolated. Because of this, *Neurospora crassa* is one of the classical organisms for genetic research; studies on biochemical mutants led Beadle and Tatum to propose the seminal "one gene — one enzyme" hypothesis.

neurotensin Tridecapeptide hormone (sequence: ELYENKPRRPYIL) of gastrointestinal tract: has general vascular and neuroendocrine actions.

neurotoxin A substance, often exquisitely toxic, which inhibits neuronal function. Neurotoxins act typically against the *sodium channel* (eg. *tetrodotoxin*) or block or enhance *synaptic transmission* (eg. *curare*, bungarotoxin).

neurotransmitter A substance found in *chemical synapses*, which is released from the presynaptic terminal in response to *depolarisation*, diffuses across the *synaptic cleft*, and binds the receptor for a *ligand-gated ion channel* on the *postsynaptic cell*. This alters the *resting potential* of the postsynaptic cell, and thus its excitability. Examples: *acetylcholine*, *GABA*, *noradrenaline*, *serotonin*, *dopamine*. See Table N1.

neurotrophic Involved in the nutrition (or maintenance) of neural tissue. Classic example is *nerve growth factor*.

neurotropic Having an affinity for, or growing towards, neural tissue. Rabies virus, which localises in neurons, is referred to as neurotropic; can also be used to refer to chemicals.

neurotubules A term for *microtubules* in a neuron.

neurula The stage in vertebrate embryogenesis during which the neural plate closes to form the central nervous system.

neurulation The embryonic formation of the *neural tube* by closure of the *neural plate*, directed by the underlying notochord.

neutral mutation A mutation which has no selective advantage or disadvantage. Considerable controversy surrounds the question of whether such mutations can exist.

neutral protease Protease which is optimally active at neutral pH: see *protease*.

neutropenia Condition in which the number of *neutrophils* circulating in the blood is below normal.

neutrophil (neutrophil granulocyte; polymorphonuclear leucocyte; PMN or PMNL) A short-lived phagocytic cell of the *myeloid* series, which is responsible for the primary cellular response in an acute inflammatory episode, and for general tissue homeostasis by removal of damaged material. The commonest of the blood leucocytes, being present at 2500–7500/mm^3. Adheres to endothelium (*margination*) and then migrates into tissue, responding to chemotactic signals. Contains specific and azurophil granules.

neutrophilin Neutrophil-derived platelet activator, probably a serine protease.

nevus See *naevus*.

Newcastle Disease virus A paramyxovirus which causes the fatal disease, fowl-pest, in poultry.

nexin Protein (165kD) which links the adjacent microtubule doublets of the ciliary *axoneme*.

nexus A connection or link.

Nezelof syndrome Congenital T-cell deficiency associated with thymic hypoplasia.

NGF *Nerve growth factor*.

nick A point in a double-stranded DNA molecule where there is no *phosphodiester* bond between adjacent nucleotides of one strand.

nick-translation A technique used to radioactively label DNA. **E. coli** *DNA polymerase* I will add a nucleotide, copying the complementary strand, to the free 3′-OH group at a *nick*; at the same time its exonuclease activity removes the 5′-terminus. The enzyme then adds a nucleotide at the new 3′-OH and removes the new 5′-terminus. In this way one strand of the DNA is replaced starting at a nick, which effectively moves along the strand. Nick translation refers to this translation or movement and not to protein synthesis. In practice, DNA is mixed with trace amounts of *deoxyribonuclease* I to generate nicks, DNA polymerase I and labelled nucleotides. Because the nicks are generated randomly the DNA preparation can be uniformly labelled and to a high degree of specific activity.

nicotinamide adenine dinucleotide phosphate See *NADP*.

nicotinamide adenine dinucleotide See *NAD*.

nicotine A plant alkaloid from tobacco; blocks transmission at nicotinic adrenergic *chemical synapses*.

nicotinic A class of adrenergic *chemical synapse*, which is inhibited by *nicotine*: cf *muscarinic*.

nicotinic acid (pyridine 3-carboxylic acid) A precursor of NAD, that is a product of the oxidation of nicotine.

nidogen See *entactin*.

Table N1. Neurotransmitters[a]

Neurotransmitters	Peripheral nervous system	Central nervous system
Noradrenaline	Some post-ganglionic sympathetic neurons	Diverse pathways especially in arousal and blood pressure control
Dopamine	Sympathetic ganglia	Diverse; perturbed in Parkinson's disease and schizophrenia
Serotonin	Neurons in myenteric plexus Facilitator neurons in *Aplysia*	Distribution very similar to that of noradrenergic neurons. Lysergic acid (LSD) may antagonise
Acetylcholine	Neuromuscular junctions (nmj). All post-ganglionic parasympathetic and most post-ganglionic sympathetic neurons	Widely distributed, usually excitatory. Possibly antagonises dopaminergic neurons
Amino acids:		
GABA	Inhibitory at nmj of arthropods	Inhibitory in many pathways
Glutamate	Excitatory at nmj of arthropods	Widely distributed; excitatory
Glycine	—	Diverse; particularly in grey matter of spinal cord
Aspartate	Locust nmj	—
Neuropeptides	Diverse actions in both peripheral and central nervous systems; see Table H2	
Histamine	—	Minor role
Purines	Particularly neurons controlling blood vessels	Mostly inhibitory
Octopamine	Invertebrate nmj	—
Substance P	Sensory neurons of vertebrates	Sensory neurons

[a]Other substances known, or proposed to have neurotransmitter function are: adrenaline, β-alanine, taurine, proctolin, and cysteine.

Niemann–Pick disease Severe lysosomal storage disease caused by deficiency in *sphingomyelinase*; excess sphingomyelin is stored in "foam" cells (macrophages) in spleen, bone-marrow and lymphoid tissue. More common in Ashkenazi Jews than other groups.

***nif* genes** The complex of genes in nitrogen fixing bacteria, which code for the proteins required for *nitrogen fixation*, particularly the *nitrogenase*. Present as an *operon* in **Klebsiella** and carried on plasmid in **Rhizobium**.

nifedipine (BAY a 1041; Nifedin; Procardia) A calcium channel blocker (MW 346) used experimentally and as a coronary vasodilator.

Nissl granules Discrete clumps of material seen by phase contrast microscopy in the perikaryon of some neurons, particularly motoneurons. They are basophilic, containing much RNA and are regions very rich in rough endoplasmic reticulum. Their reaction following damage to neurons is characteristic; they disperse through the cytoplasm giving a general basophilia to the whole cell body.

Nitella Characean alga which has giant, multinucleate internodal cells. These show *cytoplasmic streaming* at rates of up to 100μm/sec and have been used as models for motile phenomena in cells and in studies on ionic movement.

nitroblue tetrazolium reduction Nitroblue tetrazolium (NBT), a yellow dye, is taken up by phagocytosing neutrophils, for example, and reduced to insoluble formazan, which forms deep-blue granules if the *metabolic burst* is normal. Reduction does not take place in *chronic granulomatous disease* and the NBT reduction test is useful diagnostically.

nitrocellulose paper Paper with a high non-specific absorbing power for biological macromolecules. Very important as a receptor in *blot*-transfer methods. Bands are transferred from a chromatogram or electropherogram onto nitrocellulose sheets either by blotting or by electrophoretic transfer. The replica can then be used for sensitive analytical detection methods.

nitrogen fixation The incorporation of atmospheric nitrogen into ammonia by various bacteria, catalysed by *nitrogenase*.

This is an essential stage in the nitrogen cycle and is the ultimate source of all nitrogen in living organisms. In the sea, the main nitrogen fixers are ***Cyanobacteria***. There are several free-living bacteria in soil which fix nitrogen including species of *Azotobacter*, **Clostridium** and **Klebsiella**. **Rhizobium** only fixes nitrogen when in symbiotic association, in root nodules, with leguminous plants. The oxygen-sensitive nitrogenase is protected by plant-produced *leghaemoglobin* and the plant obtains fixed nitrogen from the bacteria. See also **Frankia**.

nitrogen mustards A series of tertiary amine compounds having vesicant properties similar to those of mustard gas. They have the general formula $RN(CH_2CH_2Cl)_2$ and can alkylate compounds such as DNA. Used as the basis of cytostatic drugs for cancer chemotherapy.

nitrogenase Enzyme found in nitrogen-fixing bacteria that reduces nitrogen to ammonia (also ethylene to acetylene).

nitrosamines These molecules contain the N–N=O group (N-nitrosamines): many are carcinogens or suspected carcinogens.

NK cells See *killer cells*.

NMDA (N-methyl-D-aspartic acid) A powerful agonist for a class of receptor (NMDA-receptor) found on some vertebrate nerve cells involved in synaptic transmission.

Nocardia Genus of Gram positive bacteria that form a *mycelium* which may fragment into rod- or coccoid-shaped cells. They are very common *saprophytes* in soil but some are opportunistic pathogens of humans, causing nocardiosis. This is characterised by abscesses, particularly of the jaw, which if untreated may invade the surrounding bone.

node A point in a plant stem at which one or more leaves are attached.

node of Ranvier A region of exposed neuronal plasma membrane in a myelinated *axon*. Nodes contain very high concentrations of *voltage gated ion-channels*, and are the site of propagation of *action potentials* by *saltatory conduction*.

Nomarski differential interference contrast See *differential interference contrast*.

non-coding DNA DNA that does not code for part of a polypeptide chain or RNA. This includes *introns* and *pseudogenes*. In eukaryotes the majority of the DNA is non-coding.

non-competitive inhibitor Reversible inhibition of an enzyme by a compound which binds at a site other than the substrate-binding site.

non-cyclic photophosphorylation Process by which light energy absorbed by *photosystems I and II* in chloroplasts is used to generate ATP (and also NADPH). Involves photolysis of water by photosystem II, passage of electrons along the photosynthetic electron transport chain with concomitant phosphorylation of ADP, and reduction of NADP+ using energy derived from photosystem I.

non-disjunction The failure of homologous chromosomes or sister *chromatids* to separate at *meiosis* or *mitosis* respectively. It results in aneuploid cells. Non-disjunction of the X chromosome in *Drosophila* allowed Bridges to confirm the theory of chromosomal inheritance.

non-equivalence Term used in cell determination for cells which will give rise to the same sorts of differentiated tissues but which have different positional values (eg. cells of fore-limb and hind-limb buds).

non-histone chromosomal proteins *Chromatin* consists of DNA, *histones* and a very heterogeneous group of other proteins, which include DNA polymerases, regulator proteins and the HMG (high mobility group) proteins. They are often lumped together terminologically as non-histone proteins or, incorrectly, as acidic proteins, to distinguish them from the basic histones.

non-ionic detergent Detergent in which the hydrophilic head-group is uncharged. In practice hydrophilicity is usually conferred by -OH groups. Examples are the polyoxyethylene p-t-octyl phenols known as Tritons, and octyl glucoside. Non-ionic detergents can be used to solubilise intrinsic membrane proteins with less tendency to denature them than charged detergents. They do not usually cause disassembly of structures such as microfilaments and microtubules which depend on protein-protein interactions.

non-Mendelian inheritance In eukaryotes, patterns of gene transmission not explicable in terms of segregation, independent assortment and linkage. May be due to *cytoplasmic inheritance*, *gene conversion*, meiotic drive etc.

nonpermissive cell Originally a cell of a tissue type or species which does not permit replication of a particular virus. Early stages of the virus cycle may be possible in such a cell, which in the case of tumour viruses may become transformed. Now used in a more general sense, of agents and treatments other than viruses.

nonpolar group (hydrophobic group) Group in which the electronic charge density is essentially uniform, and which cannot therefore interact with other groups by forming hydrogen bonds, or by strong dipole–dipole interactions. In an aqueous environment, non-polar groups tend to cluster together, providing a major force for the folding of macromolecules and formation of membranes. Clusters are formed chiefly because they cause a smaller increase in water structure (decrease in entropy) than dispersed groups. (Non-polar groups interact with each other only by the relatively weak London–van der Waals forces).

non-reciprocal contact inhibition Collision behaviour between different cell types in which one cell shows *contact inhibition of locomotion*, and the other does not. An example is the interaction between sarcoma cells and fibroblasts (the former not being inhibited).

nonsense codon (nonsense triplet) The three *codons*, UAA (known as ochre), UAG (amber) and UGA (opal), that do not code for an amino acid but act as signals for the *termination* of protein synthesis. Any mutation which causes a base change which produces a nonsense codon results in premature termination of protein synthesis and a probably non-functional or nonsense protein.

non-spiking The ability of a neuron to convey information without generating *action potentials*. As passive electrical potentials are attenuated over distances greater than the space constant for a neuron (typically 1mm), this implies that most non-spiking neurons are involved in signalling over relatively short distances. Typical examples are invertebrate stretch receptors and *interneurons* in the central nervous system.

noradrenaline (norepinephrine; arterenol)

Catecholamine neurohormone, the neurotransmitter of most of the sympathetic nervous system (of so-called adrenergic neurons): binds more strongly to α-adrenergic receptors. Stored and released from chromaffin cells of the adrenal medulla.

norepinephrine See *noradrenaline*.

normoblast Nucleated cell of the myeloid series found in bone marrow, that gives rise to red blood cells. See *erythroblast*.

normocyte Erythrocyte of normal size and shape.

Northern blot An electroblotting method in which RNA is transferred to a filter and detected by hybridisation to ^{32}P-labelled RNA or DNA. See *blots*.

Norwalk virus Unclassified single-stranded RNA virus causing common acute infectious gastroenteritis.

nosocomial infections Hospital-acquired infections: commonest are due to *Staphylococcus aureus*, *Pseudomonas aeruginosa*, *E. coli*, *Klebsiella pneumoniae*, *Serratia marcescens* and *Proteus mirabilis*.

notochord An axial mesodermal tissue found in embryonic stages of all chordates and protochordates, often regressing as maturity is approached. Typically a rod-shaped mass of vacuolated cells. It lies immediately below the nerve cord and may provide mechanical strength to the embryo.

nuclear envelope Membrane system which surrounds the nucleus of eukaryotic cells. Consists of inner and outer membranes separated by perinuclear space and perforated by nuclear pores. The term should be used in preference to the term "nuclear membrane" which is potentially very confusing.

nuclear lamina A fibrous protein network lining the inner surface of the nuclear envelope. The extent to which this system also provides a scaffold within the nucleus is controversial. Proteins of the lamina are *lamins* A, B and C, which have sequence homology to proteins of *intermediate filaments*.

nuclear membrane A term to avoid; see *nuclear envelope*.

nuclear pore Opening in the nuclear envelope, diameter about 10nm, through which molecules such as nuclear proteins (synthesised in the cytoplasm) and mRNA must pass. Pores are generated by a large protein assembly.

nuclear RNA The nucleus contains RNA which has just been synthesised, but in addition there is some which seems not to be released, or is only released after further processing, the heterogenous nuclear RNA (*hnRNA*) and small RNA molecules associated with protein to form small nuclear ribonucleoproteins (snRNPs).

nuclear transplantation Experimental approach in study of nucleo–cytoplasmic interactions, in which a nucleus is transferred from one cell to the cytoplasm (which may be anucleate) of a second.

nuclear transport Passage of molecules in and out of the nucleus, presumably *via* nuclear pores. Passage of proteins into the nucleus may depend on possession of a nuclear location sequence containing five consecutive positively-charged residues (PKKKRKV).

nuclease An enzyme capable of cleaving the phosphodiester bonds between nucleotide subunits of nucleic acids.

nucleation A general term used in polymerisation or assembly reactions where the first steps are energetically less favoured than the continuation of growth. Polymerisation is much faster if a pre-formed seed is used to nucleate growth; eg. microtubule growth is nucleated from the *microtubule organising centre*, although the nature of this nucleation is not known.

nucleic acids Linear polymers of nucleotides, linked by 3′5′ phosphodiester linkages. In DNA, deoxyribonucleic acid, the sugar group is deoxyribose, and the bases of the nucleotides adenine, guanine, thymine and cytosine. RNA, ribonucleic acid, has ribose as the sugar, and uracil replaces thymine. DNA functions as a stable repository of genetic information in the form of base sequence. RNA has a similar function in some viruses but more usually serves as an informational intermediate (mRNA), a transporter of amino acids (tRNA), in a structural capacity or, in some newly discovered instances, as an enzyme.

nucleocapsid The coat (*capsid*) of a virus plus the enclosed nucleic acid genome.

Table N2. Nucleotides

Table N2. Nucleotides

Phosphate esters of nucleosides, which are themselves conjugates between the biological bases and sugars, either ribose or 2-deoxyribose.

Nucleosides are derived from the bases by the addition of a sugar in the position indicated (H).

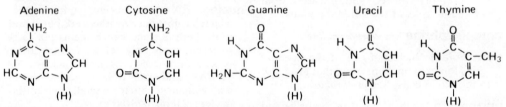

Adenine   Cytosine   Guanine   Uracil   Thymine

The structure of nucleotides, exemplified by adenine derivatives is:

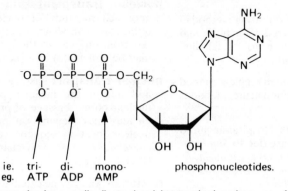

ie. tri- di- mono-   phosphonucleotides.
eg. ATP ADP AMP

The phosphate may also be a cyclic diester involving two hydroxyl groups of the sugar.

eg. 3',5' cyclic AMP

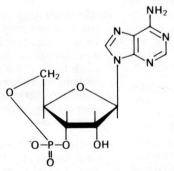

Ribonucleotides are precursors of RNA and also common metabolic intermediates and regulators;

Examples of the shorthand nomenclature are given.

	Adenine	Cytosine	Guanine	Uracil
Mononucleotide	AMP	CMP	GMP	UMP
Dinucleotide	ADP	CDP	GDP	UDP
Trinucleotide	ATP	CTP	GTP	UTP
Cyclic nucleotide	3',5' cyclic AMP		3',5' cyclic GMP	

Deoxyribonucleotides, required for the synthesis of DNA, are made by the biological reduction of the corresponding ribose dinucleotides and the deoxyribonucleotides are phosphorylated to give the triphospho-nucleotides.

(Continued)

Table N2. Nucleotides (Continued)

dTMP is made by methylation of dUMP, which is then phosphorylated to give dTTP.

	Adenine	Cytosine	Guanine	Thymidine
Dinucleotide	dADP	dCDP	dGDP	—
Trinucleotide	dATP	dCTP	dGTP	dTTP

Nucleotides occur as part of other biological molecules, eg NAD is the ADP-ribose derivative of nicotinamide. Nucleotide adducts are important intermediates in anabolic processes. CDP derivatives occur in the biosynthesis of lipids. UDP and TDP derivatives are important in sugar metabolism.

nucleoid Region of cell in a bacterium which contains the DNA.

nucleolar organiser Loop of DNA which has multiple copies of ribosomal RNA genes. See *nucleolus*.

nucleolus A small dense body (sub-organelle) within the nucleus of eukaryotic cells, visible by phase contrast and interference microscopy in live cells throughout interphase. Contains RNA and protein, and is the site of synthesis of ribosomal RNA. The nucleolus surrounds a region of one or more chromosomes (the nucleolar organiser) in which are repeated copies of the DNA coding for ribosomal RNA.

nucleoplasm By analogy with cytoplasm, that part of the nuclear contents other than the nucleolus.

Nucleopore filter Filter of defined pore size made by etching a polycarbonate filter which has been bombarded by neutrons, the extent of etching determining the pore size. Very thin, with neat circular holes going right through the membrane, not a complex meshwork like *micropore filters*.

nucleoproteins Structures containing both nucleic acid and protein. Examples are chromatin, ribosomes, certain virus particles.

nucleoside Purine or pyrimidine base linked glycosidically to ribose or deoxyribose, but lacking the phosphate residues which would make it a nucleotide. Ribonucleosides are adenosine, guanosine, cytidine and uridine. Deoxyribosides are deoxyadenosine, deoxyguanosine, deoxycytidine and deoxythymidine (the latter is almost universally referred to as thymidine).

nucleoskeletal DNA DNA which is proposed to exist mostly to maintain nuclear volume and not for coding protein.

nucleosome Repeating units of organisation of chromatin fibres in chromosomes, consisting of around 200 base pairs, and two molecules each of the *histones* H2A, H2B, H3 and H4. Most of the DNA (around 140 base pairs) is believed to be wound around a core formed by the histones, the remainder joins adjacent nucleosomes, thus forming a structure reminiscent of a string of beads.

nucleotide Phosphate esters of *nucleosides*. The metabolic precursors of nucleic acids are monoesters with phosphate on carbon 5 of the pentose (known as 5′ to distinguish sugar from base numbering). However many other structures, such as adenosine 3′5′-cyclic monophosphate (cAMP), and molecules with 2′ or 3′ phosphates or multiple phosphates are also known as nucleotides. See Table N2.

5′-nucleotidase Enzyme which cleaves the 5′ monoester linkage of nucleotides, and thus converts them to the corresponding nucleoside.

nucleus The major organelle of eukaryotic cells, in which the chromosomes are separated from the cytoplasm by the *nuclear envelope*.

nude mice Strains of athymic mice, bearing the recessive allele *nu/nu*, which are largely hairless and lack all or most of the T-cell population. Show no rejection of either allografts or xenografts. *nu/nu* alleles on some backgrounds have near normal numbers of T-cells.

numerical aperture (N.A.) For a lens the resolving power depends upon the wavelength of light being used and inversely upon the numerical aperture. The N.A. is the product of the refractive index of the medium (1 for air, 1.5 for immersion oil) and the sine of the angle, i, the semi-angle of the cone formed by joining objects to the perimeter of the lens. The larger the value of N.A., the better resolving power of the lens;

most objectives have their N.A. value engraved on the barrel and this should be quoted when describing an optical system.

nurse cells Cells accessory to egg and/or sperm formation in a wide variety of organisms. Usually thought to synthesise special substances and to export these to the developing gamete.

O

O-antigens Tetra- and penta-saccharide repeat units of the cell walls of Gram negative bacteria. They are a component of *lipopolysaccharide*.

obelin Calcium-activated photoprotein in the photocyte of the colonial hydroid coelenterate, *Obelia geniculata*.

occludens junction (tight junction) See *zonula occludens*.

ochre codon The *codon* UAA, one of the three that causes termination of protein synthesis. The most frequent *termination codon* in *E. coli* genes.

ochre mutation Mutation that changes any *codon* to the *termination codon* UAA.

ochre suppressor A gene which codes for an altered transfer RNA so that its *anticodon* can recognise the *ochre codon* and thus allows the continuation of protein synthesis. A suppressor of an *ochre mutation* is a tRNA which is charged with the amino acid corresponding to the original codon or a neutral substitute. Ochre suppressors will also suppress amber mutants.

ochronosis Deposition of dark brown pigment in cartilage, joint capsules and other tissues, usually as a result of *alkaptonuria*.

octopamine A biogenic amine found in both vertebrates and invertebrates (identified first in the salivary gland of *Octopus*). Octopamine can have properties both of a hormone and a neurotransmitter, and acts as an adrenergic *agonist*.

odontoblasts Columnar cells derived from the dental papilla after *ameloblasts* have differentiated, and which give rise to the dentine matrix which underlies the enamel of a tooth.

odontogenic epithelial cells Epithelial layer which will give rise to teeth.

oedema (USA edema) Swelling of tissue: a result of increased permeability of vascular endothelium.

oestradiol (USA estradiol; follicular hormone) A hormone (MW 272) synthesised mainly in the ovary, but also in the placenta, testis, and possibly adrenal cortex. A potent *oestrogen*.

oestrogen (USA estrogen) A type of hormone which induces oestrus ("heat") in female animals. It controls changes in the uterus that precede ovulation, and is responsible for development of secondary sexual characteristics in pubescent girls. Some tumours are sensitive to oestrogens. See *oestradiol*.

Okazaki fragments Short fragments of newly synthesised DNA strands produced during DNA replication. All the known *DNA polymerases* can only synthesise DNA in one direction, the 5' to 3' direction. However as the strands separate, replication forks will be moving along one parental strand in the 3' to 5' direction and 5' to 3' on the other parental strand. On the former, the leading strand, DNA can be synthesised continuously in the 5' to 3' direction. On the other, the lagging strand, DNA synthesis can only occur when a stretch of single-stranded DNA has been exposed and proceeds in the direction opposite to the movement of the replication fork (still 5' to 3'). It is thus discontinuous and the series of fragments are then covalently linked by *ligases* to give a continuous strand. Such fragments were first observed by Okazaki using *pulse-chase* labelling with radioactive *thymidine*. In eukaryotes, Okazaki fragments are typically a few hundred nucleotides long, whereas in prokaryotes they may contain several thousands of nucleotides.

oleic acid See *fatty acids* and Table L3.

oleosome Plant spherosome rich in lipid that serves as a storage granule in seeds and fruits. There are none of the enzymes characteristic of lysosomes.

olfactory epithelium The *epithelium* lining the nose.

olfactory neurons *Sensory neurons* from the lining of the nose. They are the only neurons that continue to divide and differentiate throughout an organism's life.

oligodendrocyte A class of *glial cell* that myelinates axons in the central nervous system.

oligomycin A bacterial toxin inhibitor of oxidative phosphorylation which acts on a small subunit of the F_1-ATPase.

oligonucleotide Linear sequence of up to about 20 nucleotides joined by phosphodiester bonds. Above this length the term polynucleotide is used, but the boundary is not strictly defined.

oligopeptide A peptide of a small number of component amino acids as opposed to a polypeptide. Exact size range is a matter of opinion but peptides from 3 to about 40 member amino acids might be so described.

oligosaccharide A saccharide of a small number of component sugars, either O- or N-linked to the next sugar. Number of component sugars not rigorously defined.

oligosaccharin An oligosaccharide derived from the plant cell wall which in small quantities induces a physiological response in a nearby cell of the same or a different plant, and thus acts as a molecular signal. Sometimes considered to be a plant hormone (*plant growth substance*). The best authenticated examples are involved in host-pathogen interactions and in the control of plant cell expansion.

ommatidium A single facet of an invertebrate compound eye.

oncogen Synonym for *carcinogen*, an agent causing cancer.

oncogene Mutated and/or overexpressed version of a normal gene of animal cells (the *proto-oncogene*) which in a dominant fashion can release the cell from normal restraints on growth, and thus alone, or in concert with other changes, convert a cell into a tumour cell. See Table O1.

oncogenic virus (*tumour virus*) A virus capable of causing cancer in animals or in humans. These include DNA viruses, ranging in size from *Papovaviridae* to *Herpetoviridae*, and the RNA-containing *[Retroviridae]*. See *Oncovirinae*.

oncostatin M Polypeptide (18kD) which inhibits the replication of A375 melanoma cells and of other human tumour cells, but not of normal fibroblasts. Produced by the monocyte-like cell line U-937.

Oncovirinae The family of retroviruses (Retroviridae) which can cause tumours. They are enveloped by membrane derived from the plasma membrane of the host cell, from which they are released by budding without lysing the cell. Within each virion is a pair of single-stranded RNA molecules.

Replication involves a DNA intermediate made on an RNA template by the enzyme *reverse transcriptase*.

ontogeny The total of the stages of an organism's life history.

oocyte The developing female gamete before completion and release.

oogenesis The process of egg formation.

oogonium In certain algae and fungi, the structure containing the ovum or ova. After fertilisation the oogonium contains the oospore.

oomycetes Group of fungi in which the mycelium is non-septate, ie. lacks cross-walls, and the nuclei are diploid. Sexual reproduction is oogamous.

opal codon The *codon* UGA, one of the three that causes termination of protein synthesis.

opal mutation *Mutation* that changes any *codon* to the *termination codon* UGA.

opal suppressor A gene which codes for an altered transfer RNA so that its *anticodon* can recognise the *opal codon* and thus allows the continuation of protein synthesis. A suppressor of an *opal mutation* is a tRNA which is charged with the amino acid corresponding to the original codon or a neutral substitute. Some eukaryote cells normally synthesise opal suppressor tRNAs. The function of these is not clear and they usually do not prevent normal termination of protein synthesis at an opal codon.

Opalina A genus of parasitic protozoans found in the guts of amphibia and fishes. They look superficially like ciliates, but are classified in a separate group as they have a number of similar nuclei.

open reading frame A *reading frame* in a sequence of nucleotides in DNA that contains no *termination codons* and so can potentially translate as a polypeptide chain.

operator The site on DNA to which a specific repressor binds and prevents the initiation of transcription at the adjacent *promoter*.

operon Group of bacterial genes with a common promotor, that are controlled as a unit and produce mRNA as a single piece, polycistronic messenger. An operon consists of two or more structural genes, which usually code for proteins with related metabolic

Table O1. Oncogenes and tumour viruses

Acronym	Virus	Species	Tumour origin	Comments
abl	Abelson leukaemia	mouse	Chronic myelogenous leukaemia	TyrPK(src)
erbA	Erythroblastosis	chicken		Homology to human glucocorticoid receptor
erbB	Erythroblastosis	chicken		TyrPK EGF/TGFα receptor
ets	E26 myeloblastosis	chicken		nuclear
fes(fps)[a]	Snyder–Theilen sarcoma Gardner–Arnstein sarcoma	cat		TyrPK(src)
fgr	Gardner–Rasheed sarcoma	cat		TyrPK(src)
fms	McDonough sarcoma	cat		TyrPK CSF-1 receptor
fps(fes)[a]	Fujinami sarcoma	chicken		TyrPK(src)
fos	FBJ osteosarcoma	mouse		Nuclear, TR
hst	NVT	human	stomach tumour	FGF homologue
int1	NVT	mouse	MMTV-induced carcinoma	Nuclear, TR
int2	NVT	mouse	MMTV-induced carcinoma	FGF homologue
jun	ASV17 sarcoma	chicken		Nuclear, TR
kit	Hardy–Zuckerman 4 sarcoma	cat		TyrPK GFR L?
B-lym	NVT	chicken	Bursal lymphoma	
mas	NVT	human	Epidermoid carcinoma	Angiotensin II receptor
met	NVT	mouse	Osteosarcoma	TyrPK GFR L?
mil (raf)[b]	Mill Hill 2 acute leukaemia	chicken		Ser/ThrPK
mos	Moloney sarcoma	mouse		Ser/ThrPK
myb	Myeloblastosis	chicken	Leukaemia	Nuclear, TR
myc	MC29 Myelocytomatosis	chicken	Lymphomas	Nuclear TR?
N-myc	NVT	human	Neuroblastomas	Nuclear
neu (ErbB2)	NVT	rat	Neuroblastoma	TyrPK GFR L?
raf (mil)[b]	3611 sarcoma	mouse		Ser/ThrPK
Ha-ras	Harvey murine sarcoma	rat	Bladder, mammary and skin, carcinomas	GTP-binding
Ki-ras	Kirsten murine sarcoma	rat	Lung, colon carcinomas	GTP-binding
N-ras	NVT	human	Neuroblastomas Leukaemias	GTP-binding
rel	Reticuloendotheliosis	turkey		
ros	UR2	chicken		TyrPK GFR L?
sis	Simian sarcoma	monkey		One chain of PDGF
src	Rous sarcoma	chicken		TyrPK
ski	SKV770	chicken		Nuclear
trk	NVT	human	Colon carcinoma	TyrPK GFR L?
yes	Y73, Esh sarcoma	chicken		TyrPK(src)

[a]fps/fes are species equivalents.
[b]mil/raf are species equivalents.
GFR L? = From sequence, a growth-factor receptor for unknown ligand;
MMTV = Mouse mammary tumour virus;
NVT = Isolated from non-retroviral tumour. In most cases detected by transfection of 3T3 cells; Ser/ThrPK = Serine, Threonine protein kinase; TyrPK = Tyrosine protein kinase; TR = Transcriptional regulator.

functions, and associated control elements that regulate the transcription of the structural genes. The first described example was the **lac** *operon*.

opiates Naturally occurring basic (alkaloid) molecules with a complex fused ring structure. Have high pharmacological activity. See *morphine*.

opines Compounds produced by plant cells containing T-DNA: induce transfer genes on *Ti plasmids*.

opioid receptors Membrane proteins, widely distributed in animal cells, but especially in the brain (enkephalin receptors) and gut. The natural ligands are the opiate peptide neurotransmitters, but the name is given because opiates are potent agonists which occupy the receptors and mimic the action of the natural transmitters.

opportunistic pathogen (secondary pathogen) Pathogenic organism which is often normally commensal, but which gives rise to infection in immunocompromised hosts.

opsonin Substance which binds to the surface of a particle and enhances the uptake of the particle by a phagocyte. Probably the most important in mammals are immunoglobulins (which are bound through the *Fc receptor*) and those derived from complement (C3b or C3bi).

opsonisation Process of coating with an *opsonin*. Often done simply by incubating particles (eg. *zymosan*) with fresh serum.

optic tectum A region of the midbrain in which input from the optic nerve is processed. Because the retinally derived neurons of the optic nerve "map" onto the optic tectum in a defined way, the question of how this specificity is determined has been a longstanding problem in cell biology. Although there is some evidence for adhesion gradients and for some adhesion specificity, the problem is unresolved.

optical diffraction A technique used to obtain information about repeating patterns. Diffraction of visible light can be used to calculate spacings in the object.

optical isomers Isomers differing only in the spatial arrangement of groups around a central atom. Optical isomers rotate the plane of polarised light in different directions. In most instances in which the possibility of optical isomerism exists, only one of the isomers is biologically functional.

orbivirus Genus of *Reoviridae* that infects a wide range of vertebrates and insects.

organelle A structurally discrete component of a cell.

organising centre See *microtubule organising centre*.

organogenesis The process of formation of specific organs in a plant or animal involving morphogenesis and differentiation.

Oriental sore Skin disease caused by the flagellate protozoan, *Leishmania tropica*, which is parasitic in macrophages of the dermis.

orientation chamber Chamber designed by Zigmond in which to test the ability of cells (*neutrophils*) to orient in a gradient of chemoattractant. The chamber is similar to a haemocytometer, but with a depth of only ca 20μm. The gradient is set up by diffusion from one well to the other, and the orientation of cells towards the well containing chemoattractant is scored on the basis of their morphology.

ornithine decarboxylase The enzyme which converts ornithine to putrescine (dibasic amine) by decarboxylation. Rate-limiting in the synthesis of the polyamines spermidine and spermine.

orosomucoid (α_1-seromucoid; α_1-acid glycoprotein) Plasma protein of mammals and birds, 38% carbohydrate. In humans a single-chain glycoprotein of 39kD. Increased levels are associated with inflammation, pregnancy, and various diseases.

orotic acid (orotate) Intermediate in the *de novo* synthesis of pyrimidines. Linked glycosidically to ribose 5′-phosphate, orotate forms the pyrimidine nucleotide OMP, which on decarboxylation at position 5 of the pyrimidine ring yields the major nucleotide uridylate (uridine 5′-phosphate).

orthogonal arrays Arrays which are at (approximately) right angles to one another. Confluent fibroblasts often become organised into such arrays; other examples are the packing of collagen fibres in the cornea, and cellulose fibrils in the plant cell wall.

orthograde transport Axonal transport from the cell body of the neuron towards the synaptic terminal. Opposite of retrograde transport and probably dependent on a different mechanochemical protein (almost definitely *kinesin*) interacting with microtubules.

orthokinesis *Kinesis* in which the speed or frequency of movement is increased or decreased.

Orthomyxoviridae Class V *viruses*. The genome consists of a single negative strand of RNA which is present as several separate segments each of which acts as a template for mRNA. The *nucleocapsid* is helical and has a viral-specific RNA polymerase for the synthesis of the mRNAs. They leave cells by budding out of the plasma membrane and are thus enveloped. They usually have two classes of spike protein in the envelope; one has *haemagglutinin* activity and the other acts as a *neuraminidase* and both are important in the invasion of cells by the virus. The major viruses of this group are the influenza viruses.

Orthopoxviruses Genus of double-stranded DNA viruses (250–390 × 200–260nm) which preferentially infect epithelial cells. Includes variola (smallpox) and vaccinia.

oscillator Something which changes regularly or cyclically. Examples: oscillator neurons, which generate regular breathing or locomotory rhythms; slime moulds which secrete cyclic AMP in regular pulses.

Oscillatoria princeps Large cyanobacterium which exhibits gliding movements, possibly involving the activity of helically-arranged cytoplasmic fibrils of 6–9nm diameter.

osmiophilic Having an affinity for *osmium tetroxide*.

osmium tetroxide (OsO_4) Used as a cytological fixative and especially these days as a post-fixative/stain in electron microscopy. Membranes in particular are osmiophilic, ie. bind osmium tetroxide.

osmosis The movement of solvent through a membrane impermeable to solute, in order to balance the chemical potential due to the concentration differences on each side of the membrane. Term frequently mis-used in the popular press.

osmotic pressure See *osmosis*. The pressure required to prevent osmotic flow across a semi-permeable membrane separating two solutions of different solute concentration. Equals the pressure that can be set up by osmotic flow in this system.

osmotic shock Passage of solvent into a membrane-bound structure due to osmosis, causing rupture of the membrane. A method of lysing cells or organelles.

osteoblast Mesodermal cell which gives rise to bone.

osteocalcin (bone γ-carboxyglutamic acid protein; BGP) Polypeptide of 50 residues formed from a 76–77 amino acid precursor, and found in the extracellular matrix of bone. Binds hydroxyapatite. Has limited homology of its leader sequence with that of other vitamin K-dependent proteins such as *prothrombin* and *Factors IX and X*.

osteoclast Large multinucleate cell formed from differentiated *macrophage*, responsible for breakdown of bone.

osteocyte *Osteoblast* which is embedded in bony tissue and which is relatively inactive.

osteogenin Bone-inducing protein (less than 50kD) associated with extracellular matrix. Binds heparin.

osteoid Uncalcified bone matrix, the product of osteoblasts. Consists mainly of collagen, but has *osteonectin* present.

osteoma Benign tumour of bone.

osteomalacia Softening of bone caused by vitamin D deficiency: adult equivalent of rickets.

osteomyelitis Inflammation of bone tissue caused by infection.

osteonectin Bone-specific protein which binds both to collagen and to hydroxyapatite (the form of calcium phosphate which mineralises the bone matrix).

osteopontin Bone-specific sialoprotein (57kD: probably two similar peptides) which links cells and the hydroxyapatite of mineralised matrix; has RGD sequence and therefore is probably bound by an *integrin*. Found only in calcified bone, probably produced by osteoblasts.

osteoporosis Loss of bony tissue; associated with low levels of *oestrogen* in older women.

osteosarcoma Malignant tumour of bone (probably neoplasia of *osteocytes*).

ouabain (Strophanthin G) A plant alkaloid from *Strophanthus gratus* that specifically binds to and inhibits the *sodium-potassium ATPase*.

Ouchterlony assay Immunological test for antigen–antibody reactions in which diffusion of soluble antigen and antibody in a gel leads to precipitation of an antigen–antibody complex, visible usually as a whitish band. The system has the advantage that, because of radial diffusion of the reagents, a very wide range of ratios of antigen to antibody concentration develop; thus it is likely that precipitation will occur somewhere in the gel even when no care is taken with quantitation of the system. It also has the advantage that it reveals the purity of the antibody and the antigen; in complex mixtures separate precipitate lines form for each antibody–antigen complex.

outside-out patch A variant of *patch-clamping* technique, in which a disc of plasma membrane covers the tip of the electrode, with the outer face of the plasma membrane facing outward, to the bath.

ovalbumin A major protein constituent of egg white. A phosphoprotein of 386 amino acids (44kD) with one N-linked oligosaccharide chain. Synthesis in the oviduct is stimulated by oestrogen. The gene, of which there is only one in the chicken genome, has eight exons and is of 7.8 kilobase pairs; it was one of the first genes to be studied in this sort of detail.

ovarian follicle In mammals the group of cells around the primary oocyte proliferate and form a surrounding non-cellular layer. A space opens up in the follicle cells and the whole structure is then the ovarian (Graafian) follicle.

ovarian granulosa cells During oogenesis in mammals the ovarian (Graafian) follicle, in which the developing ovum lies, is lined with follicle cells; the peripheral follicle cells form the stratum granulosum or ovarian granulosa.

overlap index A measure of the extent to which a population of cells in culture forms multilayers. The predicted amount of overlapping is calculated knowing the cell density, the projected area of the nucleus (usually), and assuming a Poisson distribution. The actual overlap is measured on fixed and stained preparations and the ratio of actual/predicted is derived. A value of 1 implies a random distribution with no constraint on overlapping; normal fibroblasts may have values as low as 0.05. Although a useful measure it does not unambiguously indicate the reason for the effect which may be *contact inhibition of locomotion* or differential adhesion of cells between substratum and other cells.

overlapping Situation in which the *leading lamella* of one cell moves actively over the dorsal surface of another cell — should be distinguished from *underlapping*.

overlapping genes Different genes whose nucleotide coding sequences overlap to some extent. The common nucleotide sequence is read in two or three different reading frames thus specifying different polypeptides.

oviduct The tubular tract in female animals through which eggs are discharged either to the exterior or, in mammals, to the uterus.

ovomucoid Egg-white protein produced in tubular gland cells in the epithelium of the chicken oviduct in response to progesterone or oestrogen.

ovum An egg cell.

owl-eye cells Enlarged cells infected with *cytomegalovirus* which contain large inclusion bodies surrounded by a halo, hence the name.

oxalic acid Occurs in plants, and is toxic to higher animals by virtue of its calcium-binding properties; it causes the precipitation of calcium oxalate in the kidneys, prevents calcium uptake in the gut, and is not metabolised.

oxaloacetate Metabolic intermediate. Couples with acetyl CoA to form citrate, ie. the entry point of the *tricarboxylic acid cycle*. Formed from aspartic acid by transamination.

oxidation Occurs when a compound donates electrons to an oxidising agent. Also combination with oxygen or removal of hydrogen in reactions where there is no overt passage of electrons from one species to another.

β-oxidation The process whereby fatty acids are degraded in steps, losing two carbons as (acetyl)-CoA. Involves CoA ester formation, desaturation, hydroxylation and oxidation before each cleavage.

oxidation-reduction potential See *redox potential*.

oxidative phosphorylation The phosphorylation of ADP to form ATP, coupled to the respiratory chain.

oxidoreductase An oxidase which uses molecular oxygen as the electron acceptor.

oxygen-dependent killing One of the most important microbicidal mechanisms of mammalian phagocytes involves the production of various toxic oxygen species (hydrogen peroxide, superoxide, singlet oxygen, hydroxyl radicals) through the *metabolic burst*. Although oxygen-independent killing is possible, the oxygen-dependent mechanism is crucial for normal resistance to infection, and a defect in this system is usually fatal within the first decade of life (*chronic granulomatous disease*). See *myeloperoxidase*, *chemiluminescence*.

oxygen electrode A sensitive method to detect oxygen consumption; involves a PTFE (Teflon) membrane.

oxygenase Enzyme catalyzing the incorporation of the oxygen of molecular oxygen into organic substrates. Dioxygenases (oxygen transferases) catalyze introduction of both atoms of molecular oxygen; monoxygenases (mixed function oxygenases) introduce one atom, the other becomes reduced to water, so that these enzymes require a second substrate, acting as oxygen donor. Both types are used by bacteria in degradation of aromatic compounds. Dioxygenases generally contain iron, eg. tryp-2,3 dioxygenase. Examples of mono-oxygenases are the enzymes which hydroxylate proline and lysine of collagen, using α-ketoglutarate.

oxyntic cell (parietal cell) Cell of the gastric epithelium which secretes hydrochloric acid.

oxytocin A peptide hormone (MW 1007) from hypothalamus: transported to the posterior lobe of the pituitary (see *neurophysin*). Induces smooth muscle contraction in uterus and mammary glands. Related to *vasopressin*.

P

P antigen Antigenic determinant on the surface of human red blood cells to which the Donath–Landsteiner antibody reacts. This antibody binds in the cold (a "cold IgG"), but elutes from red cells at 37°C, is particularly associated with tertiary syphylis, and its binding causes paroxysmal nocturnal haemoglobinuria.

P-face See *freeze fracture*.

P-light chain (DNTB light chain) Myosin light chain which can be phosphorylated by *myosin light chain kinase*; as a result of phosphorylation, the myosin is activated.

P-protein Protein found in large amounts in phloem sieve tubes. Appears as thin strands when seen in the electron microscope.

P-ring One of the bushes at the base of the flagellum of Gram negative bacteria, anchoring it in the peptidoglycan layer of the cells wall. Lies below the *L-ring*.

P-site The peptidyl-tRNA binding site on the ribosome, the site to which the growing chain is attached; the incoming aminoacyl-tRNA attaches to the A-site.

P680 Form of chlorophyll that has its absorption maximum at 680nm. See *photosystem II*.

P700 Form of chlorophyll that has its absorption maximum at 700nm. See *photosystem I*.

pachynema Synonym for *pachytene*.

pachytene Classical term for the third stage of prophase I of meiosis, during which the homologous chromosomes are closely paired and *crossing over* takes place.

PAGE See *polyacrylamide gel electrophoresis*.

pagoda cells Ganglion cells, from the central nervous system of a leech, with a spontaneous firing pattern which can look a little like a pagoda on an oscilloscope.

PAL See *phenylalanine ammonia lyase*.

palindromic sequence Nucleic acid sequence which is identical to its complementary strand when each is read in the correct direction (eg. TGGCCA). Palindromic sequences are often the recognition sites for *restriction enzymes*. Degenerate palindromes with internal mismatching can lead to loops or hairpins being formed (as in tRNA).

palisade parenchyma Tissue found in the upper layers of the leaf *mesophyll*, consisting of regularly-shaped, elongated parenchymal cells, orientated perpendicular to the leaf surface, which are active in *photosynthesis*.

palmitic acid (n-hexadecanoic acid) One of the most widely distributed of *fatty acids*.

pancreatic acinar cells Cells of the pancreas which secrete digestive enzymes; the archetypal secretory cells upon which much of the early work on the sequence of events in the secretory process was done.

pancytopenia Simultaneous decrease in the numbers of all blood cells: can be caused by aplastic anaemia, hypersplenism, or tumours of the marrow.

Paneth cells Coarsely granular secretory cells found in the basal regions of crypts in the small intestine.

pannus 1. Vascularised granulation tissue rich in fibroblasts, lymphocytes, and macrophages, derived from synovial tissue; overgrows the bearing surface of the joint in rheumatoid arthritis and is associated with the breakdown of the articular surface. 2. Granulation tissue which invades the cornea from the conjunctiva in response to inflammation.

pantonematic flagella See *hispid flagella*.

pantothenic acid Vitamin of the B₂ group. See Table V1.

papain (EC 3.4.22.2) Thiol protease from *Carica papaya* (pawpaw). Thermostable and will act in the presence of denaturing agents. Although it will cleave a variety of peptide bonds there is greatest activity one residue towards the C-terminus from a phenylalanine.

paper chromatography Separation method in which filter paper is used as the support. Not a very sensitive method, but historically important as one of the first methods available for separating natural compounds.

papilla 1. A projection occurring in various

animal tissues and organs. 2. A small blunt hair on plants.

papilloma Benign tumour of epithelium. Warts (caused by Papillomavirus) are the most familiar example, and each is a clone derived from a single infected cell.

Papillomavirus Genus of *Papovaviridae*. See *papilloma*, *Shope papilloma virus*.

Papovaviridae Family of oncogenic DNA viruses including papilloma, polyoma, and simian vacuolating virus (SV40). Non-enveloped, small viruses which mainly infect mammals.

papule Small raised spot on skin (as in the rash of chickenpox).

parabiosis Surgical linkage of two organisms so that their circulatory systems interconnect.

paracortex Mid-cortical region of lymph node; area which is particularly depleted of T-lymphocytes in thymectomised animals, and is referred to as the thymus-dependent area.

paracrine Form of signalling in which the target cell is close to the signal-releasing cell. Neurotransmitters and some neurohormones are usually considered to fall into this category.

parainfluenza virus Species of Paramyxoviridae; there are four types: Type 1 is also known as Sendai virus or Haemagglutinating virus of Japan (HVJ), and the inactivated form is used to bring about cell fusion. Types 2–4 cause mild respiratory infections in humans.

Paramecium Genus of ciliate protozoans. The "slipper animalcule" is cigar-shaped, covered in rows of cilia and about 250μm long. Free-swimming, common in freshwater ponds: feeds on bacteria and other particles. Reproduces asexually by binary fission, and sexually by conjugation involving the exchange of "gamete" nuclei. See also *kappa particle*.

paramural body Membranous structure located between the plasma membrane and *cell wall* of plant cells. If it contains internal membranes, it may be called a *lomasome*; if not, it may be termed a plasmalemmasome.

paramylon Storage polysaccharide of *Euglena* and related algae, present as a discrete granule in the cytoplasm and consisting of β(1–3)-glucan.

paramyosin Protein (200–220kD) which forms a core in the thick filaments of invertebrate muscles. The molecule is rather like the rod part of myosin and has a two-chain coiled-coil α-helical structure, 130nm × 2nm. Paramyosin is present in particularly high concentration in the "catch" muscle of bivalve molluscs, where it forms the almost crystalline core of the thick filaments.

Paramyxoviridae Class V viruses of vertebrates. The genome consists of a single negative strand of RNA as one piece, The helical nucleocapsid has a virus-specific RNA polymerase (transcriptase) associated with it. They are enveloped viruses: main members of the Family are *Newcastle Disease virus*, *measles virus* and the *parainfluenza viruses*.

parasympathetic nervous system One of the two divisions of the vertebrate *autonomic nervous sytem*. Parasympathetic nerves emerge cranially as pre-ganglionic fibres from oculomotor, facial, glossopharyngeal and vagus, and from the sacral region of the spinal cord. Most neurons are *cholinergic* and responses are mediated by *muscarinic* receptors. The parasympathetic system innervates, for example, salivary glands, thoracic and abdominal viscera, bladder and genitalia. cf *sympathetic nervous system*.

parathormone (parathyrin; parathyroid hormone) A peptide hormone of 84 amino acids (MW 9402). Stimulates osteoclasts to increase blood calcium levels, the opposite effect to *calcitonin*.

parathyroid hormone See *parathormone*.

parenchyma Type of unspecialised cell making up the ground tissue of plants. The cells are large and usually highly vacuolated, with thin, unlignified walls. They are often photosynthetic, in which case they may be termed *chlorenchyma*.

Parkinsonism (paralysis agitans) Disease (Parkinson's disease) characterised by tremor and associated with the under-production of L-DOPA (dihydroxyphenylalanine) by dopaminergic neurons and their death, particularly in the substantia nigra of the brain. Can be treated quite successfully in many cases by administration of L-DOPA.

parthenocarpy Fruit formation without fertilisation; results in seedless fruits. Occurs spontaneously in some plants, eg. banana, and in other plants can be induced by application of *auxin* or *gibberellins*, eg. seedless grapes.

parthenogenesis Development of an ovum without fusion of its nucleus with a male pronucleus to form a zygote.

partition coefficient Equilibrium constant for the partitioning of a molecule between hydrophobic (oil) and hydrophilic (water) phases. A measure of the affinity of the molecule for hydrophobic environments, and thus, for example, a rough guide to the ease with which a molecule will cross the plasma membrane.

parvalbumins Calcium binding proteins (12kD), found in teleost and amphibian muscle; sequence homology with *calmodulin* but only two calcium binding sites.

Parvoviridae Class II viruses, responsible for a number of important diseases of humans and domestic animals. Simple viruses with single-stranded DNA and an icosahedral nucleocapsid. The autonomous parvoviruses have a negative strand DNA and include viruses of vertebrates and arthropods. The defective Adeno-associated viruses cannot replicate in the absence of helper adenoviruses and have both positive and negative stranded genomes, but packaged in separate virions.

PAS See *periodic acid Schiff's* staining.

passage Term which derives originally from maintenance of, for example, a parasite by serially infecting host animals, passaging the parasite each time. Subsequently also used to describe the sub-culture of cells in culture, and therefore not equivalent to cell division number.

passive transport The movement of a substance, usually across a plasma membrane, by a mechanism which does not require metabolic energy. See *active transport*, *facilitated diffusion*, *ion channel*.

Pasteur effect Decrease in the rate of carbohydrate breakdown that occurs in yeast and other cells when switched from anaerobic to aerobic conditions. Results from a relatively slow flux of material through the biochemical pathways of respiration compared with those of fermentation.

patch-clamping A specialised and powerful type of *voltage clamp*, in which a patch electrode of relatively large tip diameter (5μm) is pressed tightly against the plasma membrane of a cell, forming an electrically tight, "gigohm" seal. The current flowing through individual *ion channels* can then be measured. Different variants on this technique allow different surfaces of the plasma membrane to be exposed to the bathing medium: the contact just described is a "cell-attached patch". If the electrode is pulled away, leaving just a small disc of plasma membrane occluding the tip of the electrode, it is called an "inside-out patch". If suction is applied to a cell-attached patch, bursting the plasma membrane under the electrode, a "whole cell patch" (similar to an intracellular recording) is formed. If the electrode is withdrawn from the whole-cell patch, the membrane fragments adhering to the electrode reform a seal across the tip, forming an "outside-out patch".

patching Passive process in which integral membrane components become clustered following cross-linking by an external or internal polyvalent ligand.

pathogenic Capable of causing disease.

pattern formation One of the classic problems in developmental biology is the way in which complex patterns are formed from an apparently uniform field of cells. Various hypotheses have been put forward, and there is now evidence in some cases for the existence of gradients of diffusible substances (*morphogens*) specifying the differentiative pathway that should be followed according to the concentration of the morphogen around the cell.

pavementing Term used to describe the *margination* of leucocytes on the endothelium near a site of damage.

"pawn" Mutant of *Paramecium* which, like the chess-piece, can only move forward and is unable to reverse to escape noxious stimuli. Defect is apparently in the voltage-sensitive *calcium channel* of the ciliary membrane.

PC12 A rat *phaeochromocytoma* cell line from adrenal medulla. Widely used in the study of *stimulus-secretion coupling*, and because it differentiates to resemble sympathetic neurons on application of *nerve growth factor*.

Pecten (scallop) A bivalve mollusc. The

catch muscle has been a favourite with muscle physiologists and biochemists as well as with gourmets.

pectin Class of plant cell wall polysaccharide, soluble in hot aqueous solutions of chelating agents or in hot dilute acid. Includes polysaccharides rich in galacturonic acid, *rhamnose*, *arabinose* and *galactose*, eg. the *polygalacturonans*, *rhamnogalacturonans*, and some arabinans, galactans and *arabinogalactans*. Prominent in the *middle-lamella* and *primary cell wall*.

pedicels See *podocytes*.

PEG See *polyethylene glycol*.

pellicle The outer covering of a protozoan: the plasma membrane plus underlying reinforcing structures, eg. membrane-bounded spaces (alveoli) in ciliates.

penetrance The proportion of individuals with a specific genotype who express that character in the phenotype.

penicillamine (dimethyl cysteine) Product of acid hydrolysis of *penicillin* that chelates heavy metals (lead, copper, mercury) and assists in their excretion in cases of poisoning. Also used in treatment of rheumatoid arthritis although its mode of action as an anti-rheumatic drug is not clear.

penicillin Probably the best known of the antibiotics, derived from the mould *Penicillium notatum*. It blocks the cross-linking reaction in *peptidoglycan* synthesis, and therefore destroys the bacterial cell wall making the bacterium very susceptible to damage.

pentobarbital An anticonvulsant and anaesthetic, usually used as the sodium or calcium salt.

pentoses Sugars (monosaccharides) with five carbon atoms. Include *ribose* and *deoxyribose* of nucleic acids, and many others such as the aldoses *arabinose* and *xylose*, and the ketoses ribulose and *xylulose*.

pentose phosphate pathway (pentose shunt; hexose monophosphate pathway; phosphogluconate oxidative pathway) An important metabolic pathway in the cornea, lens, liver and lactating mammary gland; a common pathway in invertebrates, plants and bacteria. Alternative metabolic route to *Embden-Meyerhof pathway* for breakdown of glucose. Diverges from this when *glucose-6-phosphate* is oxidised to ribose-5-phosphate

by the enzyme glucose-6-phosphate dehydrogenase. This step reduces *NADP* to NADPH, generating a source of reducing power in cells for use in reductive biosyntheses. In plants, part of the pathway functions in the formation of hexoses from CO_2 in *photosynthesis*. The main metabolic pathway in activated *neutrophils*, rendering them relatively insensitive to inhibitors of oxidative phosphorylation; congenital deficiency of the first enzyme in the pathway, glucose-6-phosphate dehydrogenase, produces a sensitivity to infection similar to that seen in *chronic granulomatous disease*. The pathway is also important as source of pentoses, eg. for nucleic acid biosynthesis.

PEP *Phosphoenolpyruvate*.

PEP carboxylase Enzyme responsible for the primary fixation of CO_2 in *C4 plants*. Carboxylates PEP (*phosphoenolpyruvate*) to give oxaloacetate. Also important in *crassulacean acid metabolism*, since it is responsible for CO_2 fixation in the dark.

pepsin (EC 3.4.23.1) Acid protease from stomach of vertebrates. Cleaves preferentially between two hydrophobic amino acids (eg. F-L, F-Y,), and will attack most proteins except protamines, keratin and highly glycosylated proteins. A single-chain phosphoprotein (327 amino acids; 34.5kD) released from the enzymatically inactive zymogen, pepsinogen, by autocatalysis at acid pH in the presence of hydrochloric acid. One of the peptides cleaved off in this process is a pepsin inhibitor and has to be further degraded to allow the pepsin to have full activity.

pepsinogen The inactive precursor (42.5kD) of *pepsin*.

peptidase Alternative name for a *protease*.

peptide bond The amide linkage between the α-carboxyl group of one amino acid and the α-amino group of another. The linkage does not allow free rotation and can occur in *cis* or *trans* configuration, the latter the most common in natural peptides, except for links to the amino group of proline, which are always *cis*.

peptide map Proteases will produce fragments of a characteristic size from a particular protein, and this can be used as a test for the identity or otherwise of two similar-sized proteins. It is possible to produce a peptide fragment map from a single gel band.

peptidoglycan (murein) Cross-linked polysaccharide-peptide complex of indefinite size found in the inner cell wall of all bacteria (50% of the wall in Gram negative, 10% in Gram positive). Consists of chains of approximately 20 residues of $\beta(1-4)$-linked N-acetyl glucosamine and N-acetyl muramic acid cross-linked by small peptides (4–10 residues).

peptidyl transferase (EC 2.3.2.12) Integral enzymic activity of the large subunit of a ribosome, catalysing the formation of a peptide bond between the carboxy-terminus of the nascent chain and the amino group of an arriving tRNA-associated amino acid.

Percoll Trademark for colloidal silica coated with polyvinylpyrrolidone which is used for density gradients. Inert and will form a good gradient rapidly when centrifuged. Useful for the separation of cells, viruses, and subcellular organelles.

perforins Molecules, released by cytotoxic lymphocytes, which form tubular transmembrane complexes (16nm diameter) at the sites of target cell lysis.

peribacteroid membrane Membrane derived from the plasma membrane of a plant cell and which surrounds the nitrogen-fixing bacteroids in legume root nodules. Has a high lipid content and may regulate the passage of material from the plant cell cytoplasm to the symbiotic bacterial cell. The idea that it restricts *leghaemoglobin* to the peribacteroid space seems untenable since leghaemoglobin is found in the cytoplasm of some cells.

pericanicular dense bodies Electron-dense membrane-bounded cytoplasmic organelles found near the canaliculi in liver cells: lysosomes.

pericentric inversion *Chromosomal inversion* in which the region that is inverted includes the centromere.

pericentriolar region Rather amorphous region of electron-dense material surrounding the centriole in animal cells: the major *microtubule organising centre* of the cell.

perichondrial cell Cell of the perichondrium, the fibrous connective tissue surrounding cartilage.

pericyte Cell associated with the walls of small blood vessels: not a smooth muscle cell, nor an endothelial cell.

periderm The outer cork layer of a plant which replaces the epidermis of primary tissues. Cells have their walls impregnated with *cutin* and *suberin*.

perinuclear space Gap 10–40nm wide separating the two membranes of the *nuclear envelope*.

periodic acid–Schiff reaction (PAS) A method for staining carbohydrates: adjacent hydroxyl groups are oxidised to form aldehydes by periodic acid (HIO_4) and these aldehyde groups react with Schiff's reagent (basic fuchsin decolourised by sulphurous acid) to give a purple colour. Used in histochemistry and in staining gels on which glycoproteins have been run.

peripheral lymphoid tissue Lymphoid tissue such as lymph nodes, spleen etc. containing functional mature lymphocytes which respond to antigen (unlike *central lymphoid tissue*). Also known as secondary lymphoid tissue.

peripheral membrane protein Membrane proteins which are bound to the surface of the membrane and not integrated into the hydrophobic region. Usually soluble and were originally thought to bind to integral proteins by ionic and other weak forces (and could therefore be removed by high ionic strength, for example). Now clear that some peripheral membrane proteins are covalently linked to molecules which are part of the membrane bilayer (see *glypiation*), and that there are others which fit the original definition but are perhaps more appropriately considered proteins of the cytoskeleton (eg. *spectrin*) or extracellular matrix (eg. *fibronectin*).

periplasmic binding proteins Transport proteins located within the *periplasmic space*. Some act as receptors for bacterial chemotaxis, interacting with *MCPs*. Their mode of action is unclear.

periplasmic space Structureless region between the plasma membrane and the cell wall of Gram negative bacteria.

periseptal annulus Organelle associated with cell division in Gram negative bacteria. There are two circumferential zones of cell envelope in which membranous elements of the envelope are closely associated with *murein*. The annuli appear early in division and in the region between them, the periseptal compartment, the division septum is formed.

peritoneal exudate A term most commonly used to describe the fluid drained from the peritoneal cavity some time after the injection of an irritant solution. For example, a standard method for obtaining neutrophil leucocytes is to inject intraperitoneally saline with glycogen (to activate complement) and drain off the leucocyte-rich peritoneal exudate some hours later.

permease General term for a membrane protein which increases the permeability of the plasma membrane to a particular molecule, by a process not requiring metabolic energy. See *facilitated diffusion*.

permissive cells Cells of a type or species in which a particular virus can complete its replication cycle.

peroxidase A *haem* enzyme which catalyzes reduction of hydrogen peroxide by a substrate which loses two hydrogen atoms. Within cells, may be localised in peroxisomes. Coloured reaction-products allow detection of the enzyme with high sensitivity, so peroxidase-coupled antibodies are widely used in microscopy and *ELISA*. *Lactoperoxidase* is used in the catalytic surface-labelling of cells with radioactive iodine.

peroxisome Organelle bounded by a single membrane and containing peroxidase and often catalase, sometimes as a large crystal. A site of oxygen utilisation, but not of ATP synthesis. In plants, associated with *chloroplasts* in *photorespiration* and considered to be part of a larger group of organelles, the *microbodies*.

persistence Term used in cell behaviour to describe the tendency of a cell to continue moving in one direction: an internal bias on the random walk behaviour which cells exhibit in isotropic environments.

pertussis toxin Protein complex (ca 117kD). An *AB toxin*, the active subunit is a single polypeptide (28kD), the binding subunit a pentamer (two heterodimers, 23 + 11.7kD, 11.7 + 22kD, and a monomer (9.3kD) which binds the heterodimers). The active subunit *ADP-ribosylates* the α subunit of the inhibitory *GTP-binding protein* (Gi). Crucial to the pathogenicity of *Bordetella pertussis*, the causative agent of whooping cough.

petite mutants A class of yeast mutants, most studied in *Saccharomyces cerevisiae*. They occur spontaneously and can be induced with agents such as acridine orange and *ethidium bromide*. Mutants grow slowly and rely on anaerobic respiration: mitochondria, although present, have reduced cristae and are functionally defective (termed promitochondria). There are three types of petite mutant: (i) Segregational mutants, which show Mendelian behaviour and result from mutations in chromosomal genes, (ii) Neutral petites, which are recessive genotypes and result from the complete absence of mitochondrial DNA, (iii) Suppressive petites, in which most of the mitochondrial DNA is lost (60–99%), though what remains is often amplified. When crossed with wild-type yeast a certain proportion of the progeny (depending upon the strain) have only petite mitochondria, and the wild-type is lost because there is preferential replication of the petite mitochondrial DNA.

Petromyzon (lamprey) Primitive marine vertebrate (Class Agnatha) with eel-like body and lacking true jaws. The relatively simple nervous system has been studied in some detail.

Peyer's patches Lymphoid organs located in the sub-mucosal tissue of the mammalian gut containing very high proportions of IgA-secreting precursor cells. The patches have B- and T-dependent regions and germinal centres. A specialised epithelium lies between the patch and the intestine. Involved in gut-associated immunity.

Pfr The form of *phytochrome* that absorbs light in the far red region, 730nm, and is thus converted to *Pr*. It slowly and spontaneously converts to Pr in the dark.

pH ($-\log [H^+]$) A logarithmic scale for the measurement of the acidity or alkalinity of an aqueous solution. Neutrality corresponds to pH 7, whereas a 1 molar solution of a strong acid would approach pH 0, and a 1 molar solution of a strong alkali would approach pH 14.

PHA See *phytohaemagglutinin*.

phaeochromocytoma (USA pheochromocytoma) A normally benign cancer (*neuroblastoma*) of the *chromaffin tissue* of the adrenal medulla. In culture, the cells secrete enormous quantities of *catecholamines*, and can be induced to form neuron-like cells on addition of (for example) *cyclic AMP* or *nerve growth factor*. Excessive production of adrenaline and noradrenaline leads to secondary hypertension.

Phaeophyta (USA Pheophyta; brown algae) Division of algae, generally brown in colour, with multicellular, branched thalluses. Includes large seaweeds such as *Laminaria* and *Fucus*. The brown colour is due to the **xanthophylls**, fucoxanthin and lutein. Many have **laminarin** as a food reserve and alginic acid as a wall component.

phage See *bacteriophage*.

phagocyte A cell which is capable of phagocytosis. The main mammalian phagocytes are **neutrophils** and **macrophages**.

phagocytic vesicle Membrane-bounded vesicle enclosing a particle internalised by a phagocyte. The primary phagocytic vesicle (phagosome) subsequently fuses with lysosomes to form a secondary phagosome in which digestion will occur.

phagocytosis Uptake of particulate material by a cell (endocytosis). See *opsonisation*, *phagocyte*.

phalangeal cells Cells of the organ of Corti (in the inner ear).

phalloidin Cyclic peptide (789D) from the Death Cap fungus (*Amanita phalloides*) that binds to, and stabilises, F-actin. Fluorescent derivatives are used to stain actin in fixed and permeabilised cells, although there is some uptake by live cells.

pharmacodynamics The study of how drugs affect the body: contrast with *pharmacokinetics*.

pharmacokinetics The study of what the body does to drugs, in contrast to *pharmacodynamics*.

phase separation The separation of fluid phases which contain different concentrations of common components. Occurs with partially miscible solvents used in many biochemical separation methods. Also temperature-dependent phase separation occurs with some detergent solutions. With reference to membranes means the segregation of lipid components into "domains" which have different chemical composition.

phase variation Alteration in the expression of surface antigens by bacteria. For example, **Salmonella** can express either of two forms of *flagellin*, H1 and H2, which are coded by different genes. Control of which

form is expressed is brought about by inversion of the promoter for the H2 gene, which if functional (non-inverted) is associated with the expression of H2 and the production of a repressor of the H1 gene. Inversion occurs about every 1000 bacterial divisions, and is under the control of another gene, *hin*, which is within the invertable sequence.

phase-contrast microscopy A simple non-quantitative form of *interference microscopy* of great utility in visualising live cells. Small differences in optical path length due to differences in refractive index and thickness of structures are visualised as differences in light intensity.

phaseollin A *phytoalexin* produced by *Phaseolus* (bean) plants in response to pathogenic attack or other stress.

phasic See *adaptation*.

phasing of nucleosomes A non-random arrangement of *nucleosomes* on DNA, in which, at certain segments of the genome, nucleosomes are positioned in the same way relative to the nucleotide sequence in all cells. Most nucleosomes are arranged randomly, but phasing has been detected in some genes.

phelloderm (secondary cortex) Tissue containing parenchyma-like cells, in the bark of tree roots and shoots. Produced by cell division in the *phellogen*.

phellogen *Meristematic* tissue in plants, giving rise to cork phellem and *phelloderm* cells. Also termed "cork cambium".

phencyclidine (1-(1-phenylcyclohexyl)-piperidine; Angel dust; PCP) An anaesthetic which can produce marked behavioural effects. Interacts with the *NMDA* receptor.

phenocopy An environmentally produced phenotype simulating the effect of a particular genotype.

phenothiazines A group of antipsychotic drugs, thought to act by blocking dopaminergic transmission in the brain. Examples are *chlorpromazine* and *trifluoperazine*. Trifluoperazine binds to and inhibits *calmodulin* and has been used experimentally to block calcium/calmodulin-controlled reactions.

phenotype The characteristics displayed by an organism under a particular set of environmental factors, regardless of the actual genotype of the organism.

phenyalanine (Phe; F; MW 165) An essential dietary amino acid with an aromatic side chain. See Table A2.

phenylalanine ammonia lyase (PAL) Enzyme involved in the deamination of phenylalanine to produce the phenyl-propane moiety which acts as a precursor for plant phenolic acids and their derivatives, from which *lignin* is derived.

phenylephrine hydrochloride An α_1-adrenergic agonist (MW 204).

phenylketonuria (PKU) Congenital absence of phenylalanine hydroxylase (an enzyme that converts phenylalanine into tyrosine). Phenylalanine accumulates in blood and seriously impairs early neuronal development. The defect can be controlled by diet and is not serious if treated in this way. Incidence highest in Caucasians.

pheo- See *phaeo-*.

Philadelphia chromosome Characteristic chromosomal abnormality of chronic myelogenous *leukaemia* in which a portion of chromosome 22 is translocated to chromosome 9.

phiX-174 (φX-174) Bacteriophage of *E. coli* with a single-stranded DNA genome and an icosahedral shell. This was the first DNA phage to be fully sequenced: the genome consists of 10 genes, some of which are *overlapping genes*.

phloem Tissue forming part of the plant vascular system, responsible for the transport of organic materials, especially sucrose, from the leaves to the rest of the plant. Consists of *sieve tubes*, *companion cells*, *fibre cells* and *parenchyma*.

phorbol esters Polycyclic componds isolated from croton oil in which two hydroxyl groups on neighbouring carbon atoms are esterified to fatty acids. The commonest of these derivatives is phorbol myristoyl acetate (PMA). Potent cocarcinogens or tumour promotors, they are diacyl glycerol analogues and activate protein kinase C irreversibly.

phosphatases Enzymes which hydrolyse phosphomonoesters. Acid phosphatases are specific for the single-charged phosphate group and alkaline phosphatases for the double-charged group. These specificities do not overlap. The phosphatases comprise a very wide range of enzymes including broad and narrow specificity members. Phosphoprotein phosphatases specifically dephosphorylate a particular protein and are essential if phosphorylation is to be used as a reversible control system.

phosphatides The family of phospholipids based on 1,2 diacyl 3-phosphoglyceric acid. See *phospholipids*.

phosphatidic acid (diacyl glycerol 3-phosphate; 1,2 diacyl 3-phosphoglyceric acid; PA) The "parent" structure for phosphatidyl phospholipids, present in low concentrations in membranes. The acyl groups are derived from long-chain fatty acids. An intermediate in the synthesis of diacyl glycerol, the immediate precursor of most of the phosphatidyl phospholipids (except phosphatidyl inositol) and of triacyl glycerols.

phosphatidyl choline (PC) The major phospholipid of most mammalian cell membranes where the 1-acyl residue is normally saturated and the 2-acyl residue unsaturated. Choline is attached to phosphatidic acid by a phosphodiester linkage. Major synthetic route is from diacyl glycerol and CDP-choline. Forms monolayers at an air water interface, and forms bilayer structures (liposomes) if dispersed in aqueous medium. A zwitterion over a wide pH range. Readily hydrolysed in dilute alkali.

phosphatidyl ethanolamine (PE) A major structural phospholipid in mammalian systems. Tends to be more abundant than phosphatidyl choline in the internal membranes of the cell, and is an abundant component of prokaryotic membranes. Ethanolamine is attached to phosphatidic acid by a phosphodiester linkage. Synthesis from diacyl glycerol and CDP-ethanolamine.

phosphatidyl inositol (PI) Very important minor phospholipid in eukaryotes, involved in signal transduction processes. Contains myo-inositol linked through the 1-hydroxyl group to phosphatidic acid. The 4-phosphate (PIP) and 4,5 bisphosphate (PIP$_2$) derivatives are formed and broken down in membranes by the action of cyclic kinase and phosphatase (futile cycles). Signal-sensitive phospholipase C enzymes remove the inositol moiety, in particular as the 1,4,5 trisphosphate (InsP$_3$)). Both the diacyl glycerol and InsP$_3$ products act as *second messengers*.

phosphatidyl serine (PS) An important

minor species of phospholipid in membranes. Serine is attached to phosphatidic acid by a phosphodiester linkage. Synthesis is from phosphatidyl ethanolamine by exchange of ethanolamine for serine. Distribution is asymmetric, as the molecule is only present on the cytoplasmic side of cellular membranes. It is negatively charged at physiological pH and interacts with divalent cations; involved in calcium-dependent interactions of proteins with membranes (eg. protein kinase C).

phosphocreatine Present in high concentration (about 20mM) in striated muscle, and is synthesised and broken down by creatine phosphokinase to buffer ATP concentration. It acts as an immediate energy reserve for muscle.

phosphodiester bond Not a precise term. Refers to any molecule in which two parts are joined through a phosphate group. Examples are found in RNA, DNA, phospholipids, cyclic nucleotides, nucleoside diphosphates and triphosphates.

phosphodiesterase An enzyme which cleaves phosphodiesters to give a phosphomonoester and a free hydroxyl group. Examples include RNAase, DNAase, phospholipases C and D and the enzymes which convert cyclic nucleotides to the monoester forms.

phosphoenolpyruvate (PEP) An important metabolic intermediate. The enol (less stable) form of pyruvic acid is trapped as its phosphate ester, giving the molecule a high phosphate transfer potential. Formed from 2-phosphoglycerate by the action of enolase.

phosphofructokinase The pacemaker enzyme of glycolysis. Coverts fructose-6-phosphate to fructose 1,6-bisphosphate. A tetrameric allosteric enzyme which is sensitive to the ATP/ADP ratio.

phosphoglycerate The molecules 2-phosphoglycerate and 3-phosphoglycerate are intermediates in glycolysis. 3-phosphoglycerate is the precursor for synthesis of phosphatidic acid and diacyl glycerol, hence of phosphatidyl phospholipids.

phospholipases Enzymes which hydrolyse ester bonds in phospholipids. They comprise two types:- aliphatic esterases (phospholipase A_1, A_2 and B) which release fatty acids, and phosphodiesterases (phospholipase C and D) which release diacyl glycerol or phosphatidic acid respectively. Type A_2 is widely distributed in venoms and digestive secretions. Types A_1, A_2 and C (specific for phosphatidyl inositol) are present in all mammalian tissues. Type C is also found as a highly toxic secretion product of pathogenic bacteria. Type B attacks monoacyl phospholipids and is poorly characterised. Type D is largely of plant origin.

phospholipid The major structural lipid of most cellular membranes (except the chloroplast which has galactolipids). Contain phosphate, usually as a diester. Examples include phosphatidyl phospholipids, plasmalogens and sphingomyelins. See Table L3.

phospholipid bilayer A lamellar organisation of phospholipids which are packed as a bilayer with hydrophobic acyl tails inwardly directed and polar head groups on the outside surfaces. It is this bilayer that forms the basis of membranes in cells, though in most cellular membranes a very substantial proportion of the area may be occupied by integral proteins. The triple-layered appearance of membranes seen in electron microscopy is thought to arise because the *osmium tetroxide* binds to the polar regions leaving a central, unstained, hydrophobic region.

phospholipid transfer protein Cytoplasmic proteins which bind phospholipids and facilitate their transfer between cellular membranes. May also cause net transfer from the site of synthesis.

phosphomannose See *mannose-6-phosphate*.

phosphoprotein Proteins which contain phosphate groups esterified to serine, threonine or tyrosine. The phosphate group usually regulates protein function.

phosphorylase (glycogen phosphorylase) Enzyme which catalyses the sequential removal of glycosyl residues from glycogen to yield one glucose-1-phosphate per reaction. Its activity is controlled by phosphorylation (by *phosphorylase kinase*).

phosphorylase kinase The enzyme which regulates the activity of phosphorylase and glycogen synthetase by addition of phosphate groups. A large and complex enzyme, itself regulated by phosphorylation. Integrates the hormonal and calcium signals in muscle.

phosphorylation of proteins Addition of phosphate groups to hydroxyl groups on proteins (side chains S, T or Y) catalysed by a protein kinase (often specific) with ATP as phosphate donor. Activity of proteins is often regulated by phosphorylation.

phosphotransferase An enzyme which transfers a phosphate group from a donor to an acceptor. Very important in metabolism.

phosphotyrosine Tyrosine phosphate, but normally refers to the phosphate ester of a protein tyrosine residue. Present in very small amounts in tissues, but believed to be important in systems which regulate growth control, and is therefore of interest in studies of malignancy. The *src* gene product (pp60^{src}) was one of the first kinases shown to phosphorylate at a tyrosine residue.

photoaffinity labelling A technique for covalently attaching a label or marker molecule onto another molecule such as a protein. The label, which is often fluorescent or radioactive, contains a group which becomes chemically reactive when illuminated (usually with ultraviolet light) and will form a covalent linkage with an appropriate group on the molecule to be labelled: proximity is essential. The most important class of photoreactive groups used are the aryl azides, which form short-lived but highly reactive nitrenes when illuminated.

photobleaching Light-induced change in a *chromophore*, resulting in the loss of its absorption of light of a particular wavelength. Photobleaching is a problem in fluorescence microscopy where prolonged illumination leads to progressive fading of the emitted light because less of the exciting wavelength is being absorbed.

photodynesis Initiation of cytoplasmic streaming by light. Uncommon usage.

photolysis Light-induced cleavage of a chemical bond, as in the process of photosynthesis.

photophosphorylation The synthesis of ATP that takes place during photosynthesis. In *non-cyclic photophosphorylation* the photolysis of water produces electrons which generate a proton motive force which is used to produce ATP, the electrons finally being used to reduce NADP$^+$ to NADPH. When the cellular ratio of reduced to non-reduced NADP is high, *cyclic photophosphorylation* occurs and the electrons pass down an electron transport system and generate additional ATP, but no NADPH.

photopigment Pigment involved in *photosynthesis* in plants. Includes *chlorophyll*, *carotenoids* and *phycobilins*.

photoreceptor A specialised cell type in a multicellular organism which is sensitive to light. This definition excludes single-celled organisms, but includes non-eye receptors, such as snake infra-red (heat) detectors or photosensitive pineal gland cells. See *retinal rod*, *retinal cone*.

photorespiration Increased respiration that occurs in photosynthetic cells in the light, due to the ability of *RuDP carboxylase* to react with oxygen as well as carbon dioxide. Reduces the photosynthetic efficiency of *C3 plants*.

photosynthesis Process by which green plants, algae, and some bacteria absorb light energy and use it to synthesise organic compounds (initially carbohydrates). In green plants, occurs in *chloroplasts*, which contain the photosynthetic pigments. Occurs by slightly different processes in *C3* and *C4 plants*.

photosynthetic bacteria Bacteria which are able to carry out *photosynthesis*. Light is absorbed by *bacteriochlorophyll* and *carotenoids*. Two principal classes are the green bacteria and the purple bacteria.

photosynthetic unit Group of photosynthetic pigment molecules (*chlorophylls* and *carotenoids*) which supply light to one *reaction centre* in *photosystem I or II*.

photosystem I Photosynthetic system in *chloroplasts* in which light of up to 700nm is absorbed and its energy used to bring about charge separation in the *thylakoid* membrane. The electrons are passed to ferredoxin and then used to reduce NADP$^+$ to NADPH (non-cyclic photophosphorylation) or to provide energy for the phosphorylation of ADP to ATP (cyclic photophosphorylation).

photosystem II Photosynthetic system in *chloroplasts* in which light of up to 680nm is absorbed and its energy used to split water molecules, giving rise to a high energy reductant, Q$^-$, and oxygen. The reductant is the starting point for an electron transport chain which leads to *photosystem I* and which is coupled to the phosphorylation of ADP to ATP.

phototaxis Movement of a cell or organism towards (positive phototaxis) or away from (negative phototaxis) a source of light.

phototropism Movement or growth of part of an organism (eg. a plant shoot) towards (positive phototropism) a source of light, without overall movement of the whole organism.

phragmoplast Central region of *mitotic spindle* of a plant cell at *telophase*, in which vesicles gather and fuse to form the *cell plate*, apparently guided by spindle microtubules.

phragmosome In plant cells, the region of the *cytoplasm* in which the nucleus is located during nuclear division. Can also refer to *microbodies* associated with the developing *cell plate* after nuclear division.

phycobilins Photosynthetic pigments found in certain algae, especially red algae (Rhodophyta) and *cyanobacteria*.

phycocyanin Blue *phycobilin* found in some algae, especially *cyanobacteria*.

phycoerythrin Red *phycobilins* found in some algae, especially red algae (Rhodophyta).

Phycomycetes A group of fungi possessing hyphae which are usually non-septate (without cross walls).

physaliphorous cells Cells of chordoma (tumour derived from notochordal remnants) which appear vacuolated because they contain large intracytoplasmic droplets of mucoid material.

Physarum Genus of Myxomycetes or acellular slime moulds. Normally exists as a multinucleate plasmodium which may be many centimetres across, but if starved and stimulated by light will produce spores that later germinate to produce amoeboid cells, myxamoebae, which may transform into flagellated swarm cells. Either of these cell types may fuse to produce a zygote which forms the plasmodium by synchronous nuclear division. Easily grown in the laboratory and much used for studies on cytoplasmic streaming and on the cell cycle (because there is synchronous DNA synthesis and nuclear division).

phytic acid Inositol hexaphosphate, found in plant cells, especially in seeds, where it acts as a storage compound for phosphate groups.

phytoalexins Toxic compounds produced by higher plants in response to attack by pathogens and to other stresses. Sometimes referred to as plant antibiotics, but rather non-specific, having a general fungicidal and bacteriocidal action. Production is triggered by *elicitors*. Examples: *pisatin*, *phaseollin*.

phytochrome Plant pigment protein, which absorbs red light and then initiates physiological responses governing light-sensitive processes such as germination, growth and flowering. Exists in two forms, *Pr* and *Pfr*, which are interconverted by light.

phytohaemagglutinin (PHA) Sometimes used as a general term for *lectins* produced by leguminous plants (eg. concanavalins), but more usually refers to lectin from seeds of the red kidney bean *Phaseolus vulgaris* that binds to oligosaccharides containing N-acetyl galactosyl residues. Binds to both B- and T-lymphocytes, but acts as a *mitogen* only for T-cells.

phytohormones See *plant growth substances*.

phytol Long-chain fatty alcohol (C20) forming part of *chlorophyll*, attached to the protoporphyrin ring by an ester linkage.

Picornaviridae Class IV viruses, with a single positive strand of RNA and an icosahedral capsule. Two main classes: enteroviruses, which infect the gut and include poliovirus, and rhinoviruses which infect the upper respiratory tract (common cold virus, Coxsackie A and B, Foot-and-Mouth disease virus and hepatitis A).

pigment cells Cells which contain pigment: see *melanocytes*, *chromatophores*.

pigmented retinal epithelium (PRE; retinal pigmented epithelium, RPE) Layer of unusual phagocytic epithelial cells lying below the photoreceptors of the vertebrate eye. The dorsal surface of the PRE cell is closely apposed to the ends of the rods, and as discs are shed from the rod outer segment they are internalised and digested. Do not have *desmosomes* or *cytokeratins*.

PIIF (proteinase-inhibitor inducing factor) Factor produced by a plant in response to attack by insects. Induces the formation of a substance that inhibits the proteinase which the insect secretes to digest plant tissues.

May be mobile within the plant, thus inducing inhibitor formation away from the site of original attack.

pilin 1. General term for the protein subunit of *pilus*. 2. Protein subunit (7.2kD) of F-pili, *sex-pili* coded for by the F-plasmid.

pilus (*pl.* pili) (fimbria) Hair-like projection from surface of some bacteria. Involved in adhesion to surfaces (may be important in virulence), and specialised *sex-pili* are involved in conjugation with other bacteria. Major constituent is a protein, *pilin*.

pinacocyte A cell type forming the surface layers of a sponge. Capable of synthesising collagen.

pinocytosis Uptake of fluid-filled vesicles into cells (endocytosis). Macropinocytosis and micropinocytosis are distinct processes, the latter (also called *receptor mediated endocytosis*) being energy independent and involving the formation of receptor-ligand clusters on the outside of the plasma membrane, and *clathrin* on the cytoplasmic face.

pinocytotic vesicle Fluid-filled endocytotic vesicle.

PIP$_2$ (phosphatidyl inositol 4,5,bisphosphate) Formed by linked "futile cycles" from *phosphatidyl inositol via* phosphatidyl inositol phosphate (PIP). Chiefly important because a ligand-activated PIP$_2$-specific phosphodiesterase (a phospholipase C) breaks down PIP$_2$ to form diacyl glycerol, which activates protein kinase C, and inositol 1,4,5,trisphosphate (InsP$_3$), which releases calcium from the endoplasmic store.

pi protein (π protein) Polypeptide (35kD) which is required for the initiation of DNA replication in the R6K antibiotic-resistance plasmid, of which there are 12–18 copy equivalents in the *E. coli* chromosome.

piroplasm Member of the Kingdom Protista, Phylum Apicomplexa (Sporozoa or Telosporidea), which includes *Babesia*.

pisatin *Phytoalexin* produced by peas.

pit Region of the plant cell wall in which the *secondary wall* is interrupted, exposing the underlying *primary cell wall*. One or more *plasmodesmata* are usually present in the primary wall, communicating with the other half of a pit pair. May be simple or bordered; in the latter case, the secondary wall overarches the pit field.

pituicytes Dominant intrinsic cells of the neural lobe of the hypophysis. Have long branching processes and resemble *glial cells*: secrete *antidiuretic hormone*.

placode Area of thickened ectoderm in the embryo from which a nerve ganglion, or a sense organ will develop.

plakalbumin Fragment of ovalbumin produced by *subtilisin* cleavage: more soluble than ovalbumin itself.

plakoglobin Polypeptide (83kD) present at cell–cell but not cell–substratum contacts. Associated with desmosomes and with adherens junctions: soluble 7S form present in cytoplasm.

plant growth substances Substances which, at low concentration, influence plant growth and differentiation. Formerly referred to as plant hormones or phytohormones, these terms are now suspect because some aspects of the "hormone concept", notably action at a distance from the site of synthesis, do not necessarily apply in plants. Also known as "plant growth regulators". The major classes are *abscisic acid, auxin, cytokinin, ethylene* and *gibberellin*; others include steroid and phenol derivatives.

plaque assay 1. Assay for virus in which serial dilutions of the virus are applied to a culture dish containing a layer of the host cells and spread is restricted. After incubation the "plaques", areas in which cells have been killed (or transformed), can be recognised, and the number of infective virus particles in the original suspension estimated. 2. Assay for cells producing antibody against erythrocytes or against antigen which has been bound to the erythrocytes. The cell is surrounded by a clear plaque of haemolysis, provided that complement is present. Basic principle behind the assay is the same as for the virus plaque assay.

plaque forming unit (PFU) Number of infectious virus particles/unit volume: see *plaque assay*.

plasma The fluid component of blood. Serum obtained by defibrinating plasma (plasma-derived serum) lacks platelet-released factors and is less suitable to support the growth of cells in culture.

plasma cell A terminally differentiated cell of the B-cell lineage. Rich in rough endoplasmic reticulum and secretes large amounts of antibody. Has a characteristic eccentric "cartwheel" or "clockface" nucleus and deeply basophilic cytoplasm.

plasma membrane The external, limiting lipid bilayer membrane of cells.

plasmal reaction Long-chain aliphatic aldehydes occurring in *plasmalogens* react with Schiff's reagent in the so-called plasmal reaction, to form eg. palmitaldehyde, stearaldehyde.

plasmacytoma Malignant tumour of *plasma cells*, very similar to a *myeloma* (plasmacytomas usually develop into multiple myeloma). Can easily be induced in rodents by the injection of complete Freund's adjuvant. Plasmacytoma cells are fused with primed lymphocytes in the production of monoclonal antibodies.

plasmalemma Archaic name for the plasma membrane of a cell (the term often included the cortical cytoplasmic region). Adjectival derivative (plasmalemmal) still current.

plasmalogens A group of glycerol-based phospholipids in which the aliphatic side chains are not attached by ester linkages. Widespread distribution. Less easily studied than the acyl phospholipids.

plasmid (episome) A small, independently-replicating, circular piece of cytoplasmic DNA which can be transferred from one organism to another. Plasmids can become incorporated into the genome of the host, or can remain independent. An example is the *F-factor* of *E. coli*. May transfer genes, and plasmids carrying antibiotic-resistant genes can spread this trait rapidly through the population. Described largely from bacteria and protozoa. Widely used in genetic engineering as vectors of genes (cloning vectors).

plasmin (fibrinolysin) Trypsin-like *serine protease* which is responsible for digesting *fibrin* in blood clots. Generated from *plasminogen* by the action of another protease, *plasminogen activator*. Also acts on activated *Hagemann factor* and on complement.

plasminogen Inactive precursor of *plasmin*; occurs at 200mg/l in blood plasma.

plasminogen activator *Serine protease* which acts on *plasminogen* to generate *plasmin*. Has also been implicated in invasiveness, and is produced by many normal and invasive cells. The vascular form (tPA; 55kD) is very similar to tissue plasminogen activator (uPA; 70kD) and to *streptokinase* and *urokinase*.

plasmodesma (*pl.* plasmodesmata) Narrow tube of cytoplasm penetrating the plant cell wall, linking the protoplasts of two adjacent cells. A desmotubule runs down the centre of the tube, which is lined by plasma membrane.

Plasmodium Genus of parasitic protozoa that cause *malaria*, which is responsible for more deaths than any other infectious disease. The life-cycle is complex, involving several changes in cellular morphology and behaviour. Intermediate host is mosquito (*Anopheles*) which infects vertebrate host when taking a blood meal. Predominant form of the organism in humans is the intracellular parasite in the erythrocyte, where it undergoes a form of multiple cell division termed schizogony. As a result the erythrocyte bursts and the progeny infect other erythrocytes. Eventually some cells develop into gametes which, when ingested by a female mosquito, will fuse in her gut to form a zygote (ookinete). Multiple cell division within the resultant oocyte, attached to the gut wall, gives rise to infective sporozoites; these migrate to the salivary glands and are ejected with the saliva the next time the mosquito takes a blood-meal. Sporozoite invades liver parenchyma cells for a round of exo-erythrocytic schizogony before the parasite invades the erythrocytes.

plasmodium Multinucleate mass of protoplasm bounded only by plasma membrane; the main vegetative form of *acellular slime moulds*. See **Physarum**.

plasmolysis Process by which the plant cell protoplast shrinks, so that the plasma membrane becomes partly detached from the wall. Occurs in solutions of high osmotic potential, due to water moving out of the protoplast by osmosis.

plastid Type of plant cell organelle, usually surrounded by a double membrane and often containing elaborate internal membrane systems. Partially autonomous, containing some DNA, RNA and ribosomes, and reproducing itself by binary fission. Includes *amyloplasts*, *chloroplasts*, *chromoplasts*, *elaioplasts*, *etioplasts*, *leucoplasts*, and *proteinoplasts*. Develop from *proplastids*.

plastocyanin An electron-carrying protein present in chloroplasts, forming part of the electron transport chain. Contains two copper atoms per molecule. Associated with *photosystem I*.

plastoglobuli Globules found in plastids, containing principally lipid, including *plastoquinone*.

plastoquinone A *quinone* present in chloroplasts, forming part of the photosynthetic electron transport chain. Closely associated with *photosystem II*. May be stored in *plastoglobuli*.

platelet Anucleate discoid cell (3μm diameter) found in large numbers in blood; important for blood coagulation and for haemostasis. Platelet α granules contain lysosomal enzymes; dense granules contain ADP (a potent platelet aggregating factor), and *serotonin* (a vasoactive amine). They also release *platelet-derived growth factor* which presumably contributes to later repair processes by stimulating fibroblast proliferation.

platelet activating factor (PAF; PAFacether; 1–0-hexadecyl-2-acetyl-sn-glycero-3-phosphorylcholine) Potent activator of many leucocyte functions, not just platelet activation.

platelet factor 3 Phospholipid associated with the platelet plasma membrane which contributes to the blood clotting cascade by forming a complex (thromboplastin) with other plasma proteins and activating *prothrombin*.

platelet factor 4 Platelet-released protein which promotes blood clotting by neutralising *heparin*.

platelet-derived growth factor (PDGF) The major mitogen in serum for growth in culture of cells of connective tissue origin. It consists of 2 different but homologous polypeptides A and B (approximately 30kD) linked by disulphide bonds. Believed to play a role in wound healing. The B chain is almost identical in sequence to p28^{sis}, the transforming protein of *Simian virus 40*, which can transform only those cells that express receptors for PDGF, suggesting that transformation is caused by *autocrine* stimulation.

pleiotropic Having multiple effects. For example, changes in the cyclic AMP concentration in a cell will have a variety of effects because the cAMP acts to control a protein

kinase which in turn affects a variety of proteins.

pleomorphism Having different forms at different stages of the life cycle.

Pleurobrachia Small free-swimming marine organism, member of the Phylum Ctenophora. Roughly spherical and transparent with most of the body made up from transparent jelly-like material. The animal has two long tentacles for catching prey, and is interesting for cell biologists because it swims by means of 8 rows of *comb plates* (made of fused cilia) which run along the body.

pleuropneumonia-like organism See *mycoplasmas*.

pluripotent stem cell Cells in a stem cell line capable of differentiating into several final differentiated types.

PMA Phorbol myristate acetate, one of the most commonly used *phorbol esters*.

PMF See *proton motive force*.

PMN (PMNL) Polymorphonuclear leucocyte: could be an eosinophil, basophil or *neutrophil* granulocyte, but usually intended to mean the latter (an idle habit).

pneumococci Gram positive pyogenic organisms (about 1μm diameter), usually encapsulated, closely related to streptococci; associated with diseases of the lung.

podocalyxin Major sialoprotein (140kD) of renal glomerular epithelial cells (podocytes).

podocytes Cells of the visceral epithelium which closely invest the network of glomerular capillaries in the kidney. Most of the cell body is not in contact with the *basement membrane*, but is separated from it by trabeculae which branch to give rise to club-shaped protrusions, known as pedicels, interdigitating with similar processes on adjacent cells. The complex interdigitation of these cells produces thin filtration slits which seem to be bridged by a layer of material (of unknown composition), that acts as a filter for large macromolecules.

podophyllotoxin Plant-derived toxin (414D) which binds to tubulin and prevents microtubule assembly.

poikilocytosis Irregularity of red cell shape.

point mutation *Mutation* which causes the

replacement of a single base pair with another pair.

polar body In animals each meiotic division of the oocyte leads to the formation of one large cell (the egg) and a small polar body as the other cell. Polar body formation is a consequence of the very eccentric position of the nucleus and the spindle.

polar granules Granules containing a basic protein found in insect eggs which induce the formation of and become incorporated into germ cells.

polar group Any chemical grouping in which the distribution of electrons is uneven enabling it to take part in electrostatic interactions.

polar lobe In some molluscs a polar lobe appears as a clear protrusion close to the vegetal pole of the cell prior to the first cleavage, and becomes associated with only one of the daughter cells. Removal of the first polar lobe, or of any polar lobe that forms at a subsequent mitosis, leads to defects in the embryo; it seems that the polar lobe contains special morphogenetic factors.

polar plasm Differentiated cytoplasm associated with the animal or vegetal pole of an oocyte, egg or early embryo.

polarity Literally "having poles" (like a magnet), but used to describe cells which have one or more axes of symmetry. In epithelial cells, the polarity meant is between apical and baso-lateral regions; in moving cells, having a distinct front and rear. Some cells seem to show multiple axes of polarity (which will hinder forward movement).

polarisation microscopy Any form of microscopy capable of detecting birefringent objects. Usually performed with a polarising element (such as a Nicol quartz prism) below the stage to produce plane-polarised light, and an analyser which is set to give total extinction of the background, and thus to detect any birefringence.

pole cell A cell at or near the animal or vegetal pole of an embryo.

pole fibres Microtubules inserted into the pole regions of the mitotic spindle and interacting with microtubules from the opposite pole (each pole is the product of the division of the centrioles and constitutes a *microtubule organising centre*).

poliovirus A member of the enterovirus group of the ***Picornaviridae*** which causes poliomyelitis.

pollen mother cell A diploid plant cell that forms four *microspores* by meiosis; the microspores give rise to pollen grains in seed plants.

polyacrylamide gel electrophoresis (PAGE) Analytical and separative technique in which molecules, particularly proteins, are separated by their different electrophoretic mobilities in a hydrated gel. The gel suppresses convective mixing of the fluid phase through which the electrophoresis takes place, and contributes molecular sieving. Commonly carried out in the presence of the anionic detergent sodium dodecylsulphate (***SDS***). SDS denatures proteins so that non-covalently associating subunit polypeptides migrate independently, and by binding to the proteins confers a net negative charge roughly proportional to the chain weight. The technique is also used to separate oligonucleotides eg. in sequencing gels.

poly-A See *polyadenylic acid*.

poly-A tail Polyadenylic acid sequence of varying length found at the 3′ end of most eukaryotic *messenger RNAs*. Histone mRNAs do not have poly-A tail. The poly-A tail is added post-transcriptionally to the primary transcript as part of the nuclear processing of RNA, yielding *hnRNAs* with 60–200 adenylate residues in the tail. In the cytoplasm the poly-A tail on mRNAs is gradually reduced in length. The function of the poly-A tail is not clear but it is the basis of a useful technique for the isolation of eukaryotic mRNAs. The technique uses *affinity chromatography* with oligo(U) or oligo(dT) immobilised on a solid support. If cytoplasmic RNA is applied to such a column, poly-A-rich RNA (mRNA) will be retained.

polyadenylic acid Polynucleotide chain consisting entirely of residues of adenylic acid (ie. the base sequence is AAAA....AAAA). Polyadenylic chains of various lengths are found at the 3′ end of most eukaryotic mRNAs, the *poly-A tail*.

polyanion Macromolecule carrying many negative charges. The commonest in cell-biological systems is nucleic acid.

polycation Macromolecule with many positively-charged groups. At physiological pH the most commonly used in cell biology is *polylysine*; this is often used to coat surfaces thereby increasing the adhesion of cells (which have net negative surface charge). See also *cationised ferritin*.

polycistronic mRNA A single *messenger RNA* molecule that is the product of the *transcription* of several tandemly arranged genes; typically the mRNA transcribed from an *operon*.

polyclonal antibody An antibody produced by several clones of B-lymphocytes as would be the case in a whole animal. Usually refers to antibodies raised in immunised animals, whereas a *monoclonal* antibody is the product of a single clone of B-lymphocytes, usually maintained *in vitro*.

polyclonal compartment When the progeny of several cells occupy an area or volume with a defined boundary, it is referred to as a polyclonal compartment, eg. clones lying close to the mid-line of the wing of *Drosophila*.

polycythemia Increase in the haemoglobin content of the blood, either because of a reduction in plasma volume or an increase in red cell numbers. The latter may be a result of abnormal proliferation of red cell precursors (polycythemia vera, Vaquez–Osler disease).

polyelectrolyte An ion with multiple charged groups.

polyendocrine syndrome Autoimmune disorder (the antigen to which the response is mounted is in the *B cells* of the pancreas) in which there is involvement of several organ systems.

polyethylene glycol (PEG) A hydrophilic polymer which interacts with cell membranes and promotes fusion of cells to produce viable hybrids. Often used in producing *hybridomas*.

polygalacturonan Plant cell wall polysaccharide consisting predominantly of galacturonic acid. May also contain some rhamnose, arabinose and galactose. Those with significant amounts of rhamnose are termed *rhamnogalacturonans*. Found in the *pectin* fraction of the wall.

polygalacturonase Enzyme which degrades *polygalacturonan* by hydrolysis of the glycosidic bonds which link galacturonic acid residues. Important in fruit ripening and in fungal and bacterial attack on plants.

polylysine A polymer of *lysine*, it carries multiple positive charges and is used to mediate adhesion of living cells to synthetic culture substrates, or of fixed cells to glass slides (for observation by fluorescence microscopy, for example).

polymer A macromolecule made of repeating (monomer) units or *protomers*.

polymerisation The process of polymer formation; the joining together of *protomers* to form a multimeric complex. In many cases this requires *nucleation* and will only occur above a certain critical concentration.

polymorphism 1. The existence, in a population, of two or more alleles of a gene, where the frequency of the rarer alleles is greater than can be explained by recurrent mutation alone (typically greater then 1%). HLA alleles of the *major histocompatibility complex* are very polymorphic. 2. The differentiation of various individual units of colonial animals into different types of unit specialised for different purposes, eg. as in the feeding and reproductive *polyps* of the colonial hydroid *Obelia*.

polymorphonuclear leucocyte (PMNL; PMN) Mammalian blood leucocyte (granulocyte) of myeloid series in distinction to mononuclear leucocytes: see *neutrophil*, *eosinophil*, *basophil*.

polynucleotide Linear sequence of *nucleotides*, in which the 5'-linked phosphate on one sugar group is linked to the 3' position on the adjacent sugars. In the polynucleotide *DNA* the sugar is *deoxyribose* and in *RNA*, *ribose*. The molecules may be double-stranded or single-stranded with varying amounts of internal folding.

polyomavirus A genus of DNA tumour viruses, of the *Papovaviridae*, with a small genome. Polyoma was first isolated from mice, in which it causes no obvious disease, but when injected at high titre into baby rodents, including mice, causes tumours of a wide variety of histological types (hence poly-oma). *In vitro*, infected mouse cells are permissive for virus replication, and thus are killed, whilst hamster cells undergo *abortive infection*, and at a low frequency become transformed.

polyp 1. Growth, usually benign, protruding from a mucous membrane. 2. The sessile stage of the Cnidarian (coelenterate) life-cycle; the cylindrical body is attached to the substratum or the stalk of the colony at its lower end, and may have a mouth surrounded by tentacles bearing *nematocysts* at the upper end; *Hydra* and the feeding-polyps of the colonial *Obelia* are examples.

polypeptide Chains of α-*amino acids* joined by peptide bonds. Distinction between peptides, oligopeptides and polypeptides is arbitrarily by length; a polypeptide is perhaps more than 10 residues.

polyploid Of a nucleus, cell or organism that has more than two *haploid* sets of *chromosomes*. A cell with three haploid sets (3n) is termed triploid, four sets (4n) tetraploid and so on.

polypodial Adjective describing an amoeba with several pseudopods.

polyposis coli Hereditary disorder (Mendelian dominant) characterised by the development of hundreds of adenomatous *polyps* in the large intestine, which show a tendency to progress to malignancy.

polyprotein Protein which, after synthesis, is cleaved to produce several functionally distinct polypeptides. Some viruses produce such proteins, and some polypeptide hormones seem to be cleaved from a single precursor polyprotein (see *pro-opimelanocortin*, for example).

polyribosome Functional unit of protein synthesis consisting of several *ribosomes* attached along the length of a single molecule of *messenger RNA*.

polysaccharide Polymer of (arbitrarily) more than about ten monosaccharide residues linked glycosidically in branched or unbranched chains.

polysome See *polyribosome*.

polysomy Situation in which all chromosomes are present, and some are present in greater than the diploid number, for example, trisomy 21.

polyspermy Penetration of more than one *spermatozoon* into an *ovum* at time of *fertilisation*. Occurs as normal event in very yolky eggs (eg. bird), but then only one sperm fuses with egg nucleus. Many other eggs have mechanisms to block polyspermy.

Polysphondylium A genus of *Acrasidae*, the cellular slime moulds.

polytene chromosomes Giant chromosomes produced by the successive replication of pairs of *homologous chromosomes*, joined together (synapsed) without chromosome separation or nuclear division. They thus consist of many (up to 1000) identical chromosomes (strictly chromatids) running parallel and in strict register. The chromosomes remain visible during *interphase* and are found in some ciliates, ovule cells in angiosperms, and in larval Dipteran tissue. The best known polytene chromosomes are those of the salivary gland of the larvae of **Drosophila** which appear as a series of dense bands interspersed by light interbands, in a pattern characteristic for each chromosome. The bands, of which there are about 5,000 in *D. melanogaster*, contain most of the DNA (ca 95%) of the chromosomes, and each band roughly represents one *gene*. The *banding* pattern of polytene chromosomes provides a visible map to compare with the linkage map determined by genetic studies. Some segments of polytene chromosome show chromosome *puffs*, areas of high transcription.

polyuridylic acid Homopolymer of uridylic acid. Historically, was used as an artificial mRNA in cell-free *translation* systems, where it coded for polyphenylalanine; thus began the deciphering of the genetic code.

polyvinylpyrrolidone Polymer used to bind phenols in plant homogenates, and hence to protect other molecules, especially enzymes, from inactivation by phenols. Also occasionally used to produce viscous media for gradient centrifugation.

Pompe's disease Severe glycogen *storage disease* caused by deficiency in α(1–4)-glucosidase, the lysosomal enzyme responsible for glycogen hydrolysis. Even though the non-lysosomal glycogenolytic system is normal, glycogen still accumulates in the lysosomes.

population diffusion coefficient Coefficient which describes the tendency of a population of motile cells to diffuse through the environment. Its use presupposes that the cells move in a random-walk.

porins Transmembrane matrix proteins (37kD) found in the outer membranes of Gram negative bacteria. Associate as trimers to form channels (1nm diameter, ca 10^5 per bacterium) through which hydrophilic molecules of up to 600D can pass.

porphyrins Pigments derived from porphin: all are chelates with metals (Fe, Mg, Co, Zn, Cu, Ni). Constituents of haemoglobin, chlorophyll, cytochromes.

positional control See *positional information*.

positional information The instructions that are interpreted by cells to determine their differentiation in respect of their position relative to other parts of the organism, eg. digit formation in the limb bud of vertebrates. The loss of growth control leads to proliferation; the loss of positional control allows invasion and often metastasis.

positive control Mechanism for gene regulation that requires that a regulatory protein must interact with some region of DNA before transcription can be activated.

positive feedback See *feedback*.

positive strand RNA viruses Class IV and VI *viruses* which have a single-stranded RNA genome which can act as mRNA (plus strand) and in which the virus RNA is itself infectious. Includes *Picornaviridae*, *Togaviridae* and *Retroviridae*.

postcapillary venule That portion of the blood circulation immediately downstream of the capillary network; the region having the lowest wall-shear stress, and the most common site of leucocytic *margination* and *diapedesis*.

postsynaptic cell In a *chemical synapse*, the cell which receives a signal (binds neurotransmitter) from the presynaptic cell and responds with depolarisation or hyperpolarisation. In an *electrical synapse*, the postsynaptic cell would be downstream, but since many electrical synapses are rectifying (see *synapse*), one of the two cells involved will always be postsynaptic.

postsynaptic potential In a *synapse*, a change in the *resting potential* of a postsynaptic cell following stimulation of the presynaptic cell. For example, in a cholinergic synapse such as the neuromuscular junction, the release of acetylcholine from the presynaptic cell causes channels to open in the postsynaptic (muscle) cell. Each channel opening causes a small depolarisation, known as a *miniature end-plate potential*; these sum to produce an excitatory postsynaptic potential.

post-translational modification Change made to a protein after peptide bonds have been formed. Examples include glycosylation, acylation, limited proteolysis, phosphorylation.

potato lectin *Lectin* from the potato, *Solanum tuberosum*. Binds to *N-acetyl glucosamine* residues.

Poxviridae Class I *viruses* with double-stranded DNA genome which codes for more than 30 polypeptides. They are the largest viruses and their shell is complex, consisting of many layers, and includes lipids and enzymes, amongst which is a DNA-dependent RNA polymerase. Uniquely among the DNA viruses they multiply in the cytoplasm of the cell, establishing what is virtually a second nucleus. The most important poxviruses are *vaccinia*, *variola* (smallpox) and *myxoma virus*.

PPLO See *pleuropneumonia-like organisms*.

Pr The form of *phytochrome* which absorbs light in the red region (660nm), and is thus converted to *Pfr*. In the dark the equilibrium between Pr and Pfr favours Pr, which is therefore more abundant.

prazosin Antagonist of α-adrenergic receptors.

pre-pro-protein A pre-protein is a form which contains a signal sequence which specifies its insertion into or through membranes. A pro-protein is one which is inactive; the full function is only present when an inhibitory sequence has been removed by proteolysis. A pre-pro-protein has both sequences still present. Pre-pro-proteins usually only accumulate as products of *in vitro* protein synthesis.

precipitin Any antibody which forms a precipitating complex (a precipitin line) with an appropriate multivalent soluble antigen.

prednisone Synthetic corticosteroid with powerful anti-inflammatory and anti-allergic activity.

presecretory granules Vesicles near the maturation face of the *Golgi apparatus*. Also known as Golgi condensing vacuoles.

prespore cells Cells in the rear portion of the migrating slug (grex) of a cellular slime mould, which will later differentiate into spore cells. Can be recognised as having different proteins by immunocytochemical methods.

prestalk cells Cells at the front of the migrating grex of cellular slime moulds which will form the stalk upon which the *sorocarp* containing the spores is borne. See *prespore cells*.

presynaptic cell In a *chemical synapse*, the cell which releases neurotransmitter that will stimulate the *postsynaptic cell*. In an electrically-synapsed system, the cell which has the first action potential, but since *synapses* are rectifying, one of the two cells involved is always presynaptic.

Pribnow box See *promoter*.

prickle cell Large flattened polygonal cells of the stratum germinosum of the epidermis (just above the basal stem cells), which appear in the light microscope to have fine spines projecting from their surfaces; these terminate in *desmosomes* which link the cells together, and have many *tonofilaments* of *cytokeratin* within them.

primaquine An 8-aminoquinoline drug used to treat malaria. Affects the mitochondria of the exo-erythrocytic stages (see **Plasmodium**), but the mechanism is not understood. The most effective drug at preventing spread of all four species of human malaria.

primary cell culture Of animal cells, the cells taken from a tissue source and their progeny grown in culture before subdivision and transfer to a subculture.

primary cell wall A plant cell wall which is still able to expand, permitting cell growth. Growth is normally prevented when a *secondary wall* has formed. Primary cell walls contain more *pectin* than secondary walls, and no lignin is present until a secondary wall has formed on top of them.

primary immune response The immune response to the first challenge by a particular antigen. Usually less extensive than the *secondary immune response*, being slower and shorter-lived with smaller amounts of antibody of lower affinity being produced.

primary lysosome A lysosome before it has fused with a vesicle or vacuole.

primary meristem Synonym for an *apical meristem*.

primary oocyte The enlarging ovum before maturity is reached, as opposed to the secondary oocyte or polar body.

primary spermatocyte A stage in the differentiation of the male germ cells. Spermatogonia differentiate into primary spermatocytes, showing a considerable increase in size in doing so; primary spermatocytes divide into secondary spermatocytes.

primitive erythroblast Large cell with euchromatic nucleus found in mammalian embryos. In the mouse, the cells are located in the yolk sac and are responsible for early production of erythrocytes with foetal haemoglobin.

prions Suggested as the causative agents of several infectious diseases such as scrapie (in sheep), kuru and Creutzfeld-Jacob disease in man. Prions (proteinaceous infective particles) apparently contain no nucleic acid. The 27kD protein of scrapie is related to a normal cell protein and may possibly cause its over-production.

pro-enzyme Enzyme which does not have full (or any) function until an inhibitory sequence has been removed by limited proteolysis. See also *zymogen*.

procaine Organic base (MW 234). Procaine butyrate, borate and hydrochloride are used as local anaesthetics.

procambium Region in the subapical differentiating zone of plant axes which gives rise to primary vascular tissues.

procardia See *nifedipine*.

procentriole The forming centriole composed of microtubules. Multiple procentrioles are present in some cells as a structure called the blepharoplast.

procollagen Triple-helical trimer of collagen molecules in which the terminal extension peptides are linked by disulphide bridges; the terminal peptides are later removed by specific proteases to produce a *tropocollagen* molecule.

procollagen peptidases The proteases which remove the terminal extension peptides of *procollagen*; deficiency of these enzymes leads to *dermatosparaxis* or Ehlers–Danlos syndrome.

profilin Actin-binding protein (15kD) which forms a complex with G-actin rendering it incompetent to nucleate F-actin formation. The profilin-actin complex seems to interact with inositol phospholipids which may regulate the availability of nucleation-competent G-actin.

progesterone (luteohormone) Hormone (MW 314) produced in the corpus luteum, as an antagonist of *oestrogen*. Promotes proliferation of uterine mucosa and the implantation of the blastocyst; prevents further follicular development.

programmed cell death The concept that certain cells are determined to die at specific stages and specific sites during development, eg. cells dying in the spaces between the developing digits of vertebrates, thus dividing them.

progress zone An undifferentiated population of mesenchyme cells beneath the apical ectodermal ridge of the chick limb bud from which the successive parts of the limb are laid down in a proximo-distal sequence.

prohormone A protein hormone before processing to remove parts of its sequence and thus make it active.

prokaryotes Organisms, namely bacteria and cyanobacteria (formerly known as blue-green algae), characterised by the possession of one (or occasionally two) simple naked DNA chromosome(s), usually of circular structure, without a nuclear membrane and possessing a very small range of organelles, generally only a plasmalemma and ribosomes.

prolamellar body The disorganised membrane aggregations in chloroplasts that have been deprived of light (*etioplasts*).

proliferative unit (epidermis) The basal layer of the mammalian epidermis contains cells that undergo repeated divisions. The cells outwards from a particular basal cell are often derived from this cell or a nearby one, so that columns of cells exist running outwards from the stem cell from which they were derived. Such columns of cells are referred to as proliferative units.

proliferin A hormone, related to prolactin, associated with the induction of cell division which is triggered by serum.

proline (Pro; P; MW 115) One of the 20 amino acids directly coded for in proteins. Structure differs from all the others, in that its side chain is bonded to the nitrogen of the α-amino group, as well as the α-carbon. This makes the amino group a secondary amine, and so proline is described as an imino acid.

Has strong influence on secondary structure of proteins and is much more abundant in collagens than in other proteins, occurring especially in the sequence glycine-proline-hydroxyproline. See Table A2.

promoter A region of DNA to which *RNA polymerase* binds before initiating the *transcription* of DNA into RNA. The nucleotide at which transcription starts is designated +1 and nucleotides are numbered from this with negative numbers indicating upstream, positive indicating downstream, nucleotides. Most bacterial promoters contain two *consensus sequences* which seem to be essential for the binding of the polymerase. The first, the Pribnow box, is at about −10 and has the consensus sequence 5′-TATAAT-3′. The second, the −35 sequence, is centred about −35 and has the consensus sequence 5′-TTGACA-3′. Most factors which regulate gene transcription do so by binding at or near the promoter and affecting the initiation of transcription. Much less is known about eukaryote promoters; each of the three RNA polymerases has a different promoter. RNA polymerase I recognises a single promoter for the precursor of *ribosomal RNA*. RNA polymerase II, which transcribes all genes coding for polypeptides, recognises many thousands of promoters. Most have the Goldberg-Hogness or TATA box which is centred around position −25 and has the consensus sequence 5′-TATAAAA-3′. Several promoters have a CAAT box around −90 with the consensus sequence 5′-GGCCAATCT-3′. There is increasing evidence that all promoters for "housekeeping" genes contain multiple copies of a GC-rich element which includes the sequence 5′-GGGCGG-3′. Transcription by polymerase II is also affected by more distant elements known as enhancers. RNA polymerase III synthesises 5s ribosomal RNA, all *transfer RNAs*, and a number of small RNAs. The promoter for RNA polymerase III is located within the gene either as a single sequence, as in the 5s RNA gene, or as two blocks, as in all tRNA genes.

promyelocytes Cells of the bone marrow which derive from myeloblasts and will give rise to myelocytes; precursors of the *myeloid cells*, *neutrophil* granulocytes.

pronase Mixture of proteolytic enzymes from *Streptomyces griseus*. At least four enzymes are present, including trypsin and chymotrypsin-like proteases.

pronucleus *Haploid* nucleus resulting from *meiosis*. In animals the female pronucleus is the nucleus of the *ovum* before fusion with the male pronucleus. The male pronucleus is the sperm nucleus after it has entered the ovum at *fertilisation* but before fusion with the female pronucleus. In plants the pronuclei are the two male nuclei found in the pollen tube.

pro-opimelanocortin *Polyprotein* produced by the anterior pituitary which is cleaved to yield *adrenocorticotrophin*, α, β and γ *melanocyte stimulating hormones*, lipotropic hormones, β-*endorphin*, and other fragments.

properdin (Factor P) Component of the alternative pathway for *complement* activation: complexes with C3b and stabilises the alternative pathway C3 convertase (C3b-BbP) which cleaves C3.

prophage The genome of a *lysogenic* bacteriophage when it is integrated into the chromosome of the host bacterium. The prophage is replicated as part of the host chromosome.

prophase Classical term for the first phase of *mitosis* or of either of the divisions of *meiosis*. During this phase the chromosomes condense and become visible.

prophylaxis Preventative action which will, for example, prevent infection; thus, vaccination is a prophylactic treatment.

proplastid Small, colourless *plastid* precursor, capable of division. It can develop into a chloroplast or other form of plastid, and has little internal structure. Found in cambial and other young cells.

propranolol Potent adrenergic antagonist acting at β_1- and β_2-adrenergic receptors.

prorenin Inactive precursor of *renin*.

prosome Raspberry-shaped ribonucleoprotein particle (19S) composed of small cytoplasmic RNA (15%) and heat-shock proteins, thought to be involved in post-transcriptional repression of mRNA translation: found in both nucleus and cytoplasm.

prospherosome Proposed stage in the development of *spherosomes* in plant cells. There is an accumulation of lipid in the prospherosome which is mobilised at a later stage.

prostacyclin (PGI$_2$) Unstable *prostaglandin* released by mast cells and endothelium, a potent inhibitor of platelet aggregation; also causes vasodilation and increased vascular permeability. Release enhanced by *bradykinin*.

prostaglandins (PGs) Group of compounds derived from arachidonic acid by the action of *cyclo-oxygenase* which produces cyclic endoperoxides (PGG$_2$ and PGH$_2$) that can give rise to *prostacyclin* or *thromboxanes* as well as prostaglandins. Were originally purified from prostate (hence the name), but are now known to be ubiquitous in tissues. PGs have a variety of important roles in regulating cellular activities, especially in the inflammatory response where they may act as vasodilators in the vascular system, cause vasoconstriction or vasodilation together with bronchodilation in the lung, and act as hyperalgesics. Prostaglandins are rapidly degraded in the lungs, and will not therefore persist in the circulation. Prostaglandin E$_2$ (PGE$_2$) acts on *adenylate cyclase* to enhance the production of *cyclic AMP*.

prosthetic group A tightly bound non-polypeptide structure required for the activity of an enzyme or other protein, eg. the *haem* of *haemoglobin*.

protamine Highly basic (arginine-rich) protein which replaces *histone* in sperm heads, enabling DNA to pack in an extremely compacted form, eg. clupein, iridin (4K).

protease See also *peptidase*. The term is normally reserved for endopeptidases which have very broad specificity and will cleave most proteins into small fragments. These are usually the digestive enzymes, eg. trypsin, pepsin etc., or enzymes of plant origin (eg. ficin, papain) or bacterial origin (eg. pronase, proteinase K). Proteases are involved in activating and degrading proteins and are widely used experimentally for peptide mapping and for structural studies. See Table P1.

protein A linear polymer of amino acids joined in a specific sequence by peptide bonds.

protein A Protein obtained from *Staphylococcus aureus* (Cowan strain) which binds *immunoglobulin* molecules without interfering with their binding to antigen. Widely used in purification of immunoglobulins, and in antigen detection, eg. by *immunoprecipitation*. A very effective B-cell mitogen.

protein kinase Enzyme catalyzing transfer of phosphate from *ATP* to hydroxyl side chains on proteins, causing changes in function. Most phosphate on proteins of animal cells is on *serine* residues, less on *threonine*, with a very small amount on *tyrosine* residues.

protein S Vitamin K-dependent co-factor.

proteinase-inhibitor inducing factor See *PIIF*.

proteinoid droplets Membrane-bounded droplets supposed to have been formed in "primaeval soup" as an early stage in the evolution of cells.

proteinoplast (proteoplast) Form of *plastid* adapted as a protein storage organelle; the protein may be crystalline.

proteoglycan A high-molecular weight complex of protein and polysaccharide, characteristic of structural tissues of vertebrates, such as bone and *cartilage*, but also present on cell surfaces. Important in determining viscoelastic properties of joints and other structures subject to mechanical deformation. *Glycosaminoglycans* (GAGs), the polysaccharide units in proteoglycans, are polymers of acidic disaccharides containing derivatives of the amino sugars glucosamine or galactosamine.

proteolipid Obsolete term for hydrophobic integral membrane protein.

proteolysis Cleavage of proteins by proteases. Limited proteolysis occurs where proteins are functionally modified (activated in the case of zymogens) by highly specific proteases.

proteolytic enzyme See *protease* or *peptidase*.

Proteus 1. Genus of highly motile Gram negative bacteria, found largely in soil but also in the intestine of humans. They are opportunistic pathogens; *P. mirabilis* is a major cause of urinary tract infections. 2. A urodele amphibian; a cave dweller that is blind, has external gills and lacks any pigment.

prothrombin Inactive precursor of *thrombin*, found in blood plasma.

protirelin See *thyrotrophic releasing hormone*.

Protista The kingdom of eukaryotic unicellular organisms; includes the *Protozoa*, unicellular eukaryotic *algae* and some fungi (myxomycetes, acrasiales and oomycetes).

proto-oncogene See *oncogene*.

protofilaments One way of viewing microtubule structure is to consider it to be built of (usually) 13 protofilaments arranged in parallel.

protolignin An immature form of lignin, which can be extracted from the plant cell wall with ethanol or dioxane.

protolysosome Primary lysosome, which has not been involved in fusion with another vesicle or in digestive activity.

protomers Subunits from which a larger structure is built. Thus the tubulin *heterodimer* is the protomer for microtubule assembly, G-actin the protomer for F-actin. Because it avoids the difficulty which arises with, for example, dimers which serve as

Table P1. Proteases[a]. Proteolytic enzymes (proteases; proteinases) can be divided into "mechanistic" sets according to their mode of action. Most inhibitors tend to be specific for one set alone (the important exception being the plasma inhibitor α_2-macroglobulin). Alternatively, proteases can be classified simply according to whether they act on terminal amino acids (exopeptidases; aminopeptidases act at the N-terminal, carboxypeptidases at the C-terminus) or on peptide bonds within the chain (endopeptidases).

Set	Feature	Inhibitors	Examples
Serine	serine at active site	organic phosphate esters (DFP, PMSF)	trypsin, chymotrypsin, thrombin, plasmin, elastase, subtilisin
Metallo-exopeptidase	Metal ion, often zinc	o-phenanthroline, EDTA	carboxypeptidase
Metallo-endopeptidase	Metal ion, often zinc	o-phenanthroline, EDTA	Collagenase, thermolysin
Sulphydryl (Thiol)	CysSH at active site	iodoacetate	papain, cathepsin B bromelain
Acid	acid pH optimum	diazoketones	pepsin

[a]Based on Walsh, 1975 (Cold Spring Harbor Conferences on Cell Proliferation, Vol. 2).

subunits for assembly, it is a useful term which deserves wider currency.

proton motive force (PMF) The proton gradient across a prokaryote membrane which provides the coupling between oxidation and ATP synthesis, and is used to drive the flagellar motor.

proton pump See F_1F_0 *ATPase*.

protonophore *Ionophore* which carries protons. Many *uncouplers* are protonophores.

protoplast A bacterial cell deprived of its cell wall, for example by growth in an isotonic medium in the presence of antibiotics which block synthesis of the wall *peptidoglycan*. Alternatively, a plant cell similarly deprived by enzymic treatment.

protoporphyrin Porphyrin ring structure lacking metal ions. The most abundant is protoporphyrin IX, the immediate precursor of *haem*.

Protozoa A very diverse group comprising some 50,000 eukaryotic organisms which consist of one cell. Because most of them are motile and heterotrophic, the Protozoa were originally regarded as a phylum of the animal kingdom. However it is now clear that they have only one common characteristic, they are not multi-cellular (or they are studied by protozoologists) and Protozoa are now usually classed as a Sub-Kingdom of the Kingdom *Protista*. On this classification the Protozoa are grouped into several phyla, the main ones being the Sarcomastigophora (flagellates, heliozoans and amoeboid-like protozoa), the Ciliophora (ciliates) and the Apicomplexa (sporozoan parasites such as **Plasmodium**).

provacuoles In plant cells provacuoles are budded directly from the *rough endoplasmic reticulum* and fuse with other provacuoles to form vacuoles. Since vacuoles may contain hydrolytic enzymes, it is therefore possible to consider them as analogues of primary lysosomes in animal cells.

provirus Viral DNA which has become integrated into the host chromosomal DNA. In the case of the *Retroviridae*, their RNA genome has first to be transcribed to DNA by *reverse transcriptase*. The genes of the provirus may be transcribed and expressed, or the provirus may be maintained in a latent condition. The integration of the *oncogenic* viruses, such as *Papovaviridae* and retroviruses, can lead to cell transformation.

prozymogen granule (condensing vacuole) Stage in the development of a mature *secretory vesicle* (zymogen granule).

pseudogene Non-functional DNA sequences that are very similar to the sequences of known genes. Examples are those found in the β-like globin gene cluster. Some probably result from gene duplications which become non-functional because of the loss of *promoters*, accumulation of *stop codons*, mutations which prevent correct processing etc. Some pseudogenes contain a poly-A stretch (see *poly-A tail*) suggesting that a mRNA, at some point, was copied into DNA which was then integrated into the genome.

Pseudomonas Genus of Gram negative bacteria; rod-shaped, motile with one or more polar *flagella*. Several species produce characteristic water-soluble fluorescent pigments. They are found in soil and water. *P. syringae* is a plant pathogen causing leaf spot and wilt. *P. aeruginosa*, normally a soil bacterium, is an opportunistic pathogen of humans who are immunocompromised. It can infect the wounds of victims with severe burns, causing the formation of blue pus.

pseudopod Blunt-ended projection from a cell — usually found in cells which have an amoeboid pattern of movement.

pseudopterosins Class of natural compounds (diterpene-pentose glycosides) isolated from the soft coral *Pseudopterogorgonia elisabethae*, and which interfere with arachidonic acid metabolism. Have anti-inflammatory and analgesic properties.

pseudospatial gradient sensing Mechanism for sensing a gradient of a diffusible chemical in which the cell sends protrusions out at random; up-gradient protrusions are stabilised by positive feedback (because receptor occupancy is rising with time) and others are transitory because of adaptation. Possibly the mechanism by which *neutrophils* sense chemotactic gradients.

pseudouridine (5-β-D-ribofuranosyluracil) Unusual nucleotide found in some tRNA: glycosidic bond is associated with position 5 of *uracil*, not position 1.

psoralens Drugs capable of forming photoadducts with nucleic acids if ultraviolet-irradiated.

Pteridophyta Division of the plant kingdom

which includes ferns, horsetails and clubmosses.

PtK2 cells Cell line from *Potorous tridactylis* (potoroo or kangaroo rat) kidney. Often used in studies on mitosis because there are only a few large chromosomes and the cells remain flattened during mitosis.

puffs Expanded areas of a *polytene chromosome*. At these areas the *chromatin* becomes less condensed and the fibres unwind, though they remain continuous with the fibres in the chromosome axis. A puff usually involves unwinding at a single band, though they can include many bands as in *Balbiani rings*. Puffs represent sites of active RNA *transcription*. The pattern of puffing observed in the larvae of **Drosophila**, in different cells, and at different times in development provides possibly the best evidence that *differentiation* is controlled at the level of transcription.

pulse-chase An experimental protocol used to determine cellular pathways, such as precursor–product relationships. A sample (organism, cell or cellular organelle) is exposed for a relatively brief time to a radioactively-labelled molecule, the pulse. The radioactive precursor is then replaced with an excess of the unlabelled molecule, the chase (cold chase). The sample is then examined at various later times to determine the fate of radioactivity incorporated during the pulse.

pulse-field electrophoresis A method used for high-resolution electrophoretic separation of very large (megabase) fragments of DNA. Electric fields 100° apart (the angle may vary) are applied to the separation gel alternately. The continuous change of direction prevents the molecules aligning in the electric field and greatly improves resolution on the axis between the two fields.

purine A heterocyclic compound with a fused pyrimidine/imidazole ring. Planar and aromatic in character. The parent compound for the purine bases of nucleic acids.

Purkinje cell A class of *neuron* in the cerebellum; the only neurons which convey signals away from the cerebellum.

puromycin An antibiotic which acts as an aminoacyl-tRNA analogue. Binds to the A site on the ribosome, forms a peptide linkage with the growing chain and then causes premature termination.

purple membrane Plasma membrane of *Halobacterium* and *Halococcus*, which contains a protein-bound carotenoid pigment that absorbs light and uses the energy to translocate protons from the cytoplasm to the exterior. The proton gradient then provides energy for ATP synthesis. The binding protein is called *bacteriorhodopsin*, or purple membrane protein.

purpurin Heparin-binding protein (20kD) released by chick neural retina cells in culture.

putrescine An amine associated with putrifying tissue. Associates strongly with DNA. Has been suggested as a growth factor for mammalian cells in culture.

pyaemia (USA pyemia) Invasion of bloodstream by pyogenic organisms.

pyknosis Contraction of nuclear contents to a deep-staining irregular mass; sign of cell death.

pyocins *Bacteriocins* produced by bacteria of the genus *Pseudomonas*.

pyocyanin Blue-green phenazine pigment produced by *Pseudomonas aeruginosa*; has antibiotic properties.

pyogenic Causing the formation of pus, a thick yellow or greenish liquid at a site of infection, which contains dead leucocytes, bacteria and tissue debris.

pyramidal cell Most prevalent type of neuron found in the cerebral cortex.

pyranose Sugar structure in which the carbonyl carbon of aldo-sugars is condensed with a hydroxyl group (ie. in a hemi-acetal link), forming a ring of five carbons and one oxygen. Most hexoses exist in this form, although in sucrose, *fructose* is found with the smaller (four carbon) furanose ring.

pyrenoid Small body found within some chloroplasts, that may contain protein. In green *algae* may be involved in starch synthesis.

pyridoxal phosphate The coenzyme derivative of vitamin B_6. Forms Schiff's bases of substrate amino acids during catalysis of transamination, decarboxylation and racemisation reactions.

pyrimidine (1,3 diazine) A heterocyclic 6-membered ring, planar and aromatic in character; the parent compound of the pyrimidine bases of nucleic acid.

pyrogen Substance or agent which produces fever. The major endogenous pyrogens in mammals are probably *interleukin-1* and *tumour necrosis factor*.

pyrophosphate Two phosphate groups linked by esterification. Released in many of the synthetic steps involving nucleotide triphosphates (eg. protein and nucleic acid elongation). Rapid cleavage by enzymes which have high substrate affinity ensures that the synthetic reactions are essentially irreversible.

pyrrole rings A heterocyclic ring structure, found in many important biological pigments and structures which involve an activated metal ion, eg. chlorophyll, haem.

pyruvate carboxylase An enzyme which catalyses the formation of oxalacetate from pyruvate, CO_2 and ATP in gluconeogenesis.

pyruvate dehydrogenase A complex multienzyme system which catalyses the conversion of (pyruvate + CoA + NAD^+) to (acetyl CoA + CO_2 + NAD).

Q

Q banding See *quinacrine*.

Q enzyme Enzyme involved in the formation of (1–6)-linkages at amylopectin branch points.

QH$_2$-cytochrome c reductase Membrane-bound complex in the mitochondrial inner membrane, responsible for electron transfer from reduced coenzyme Q to cytochrome c. Contains cytochromes b and c_1, and iron-sulphur proteins.

quail Small galliform bird. Quail embryos are often used in developmental studies because quail cells can be distinguished from chicken cells, yet the two are sufficiently closely related that it is possible to graft embryonic tissue from one to the other to obtain chimeras.

quantal mitosis A controversial concept in cellular *differentiation* proposed by H. Holtzer and defined by him as a mitosis "that yields daughter cells with metabolic options very different from those of the mother cell as opposed to proliferative mitoses in which the daughter cells are identical to the mother cell". Implicit in this is the idea that the changes in cell *determination* that occur during development take place at these special quantal mitoses.

quantum yield (quantum requirement) The number of photons required for the formation of one oxygen molecule in photosynthesis. Varies from 8–14 depending on the system used to measure it.

Quellung reaction Swelling of the capsule surrounding a bacterium as a result of inter-action with anticapsular antibody; consequently the capsule becomes more refractile and conspicuous.

quercetin Mutagenic flavonol pigment found in many plants. Inhibits *F_1F_0-ATPases*.

quiescent stem cell A stem cell which is not at that time undergoing repeated cell cycles but which might be stimulated so to do later. For example, the satellite cells in the skeletal muscles of mammals are quiescent myoblasts that will proliferate after wounding and give rise to more muscle cells by fusion.

quin2 A fluorescent calcium indicator. Resembles the chelator *EGTA* in its ability to bind calcium much more tightly than magnesium. Binding of calcium ions causes large changes in ultraviolet absorption and fluorescence.

quinacrine A fluorescent dye which intercalates into DNA helices. Chromosomes stained with quinacrine show typical *banding patterns* of fluorescence, Q bands, at specific locations used to recognise chromosomes and their abnormalities. See also *C banding*, *G banding*.

quinate: NAD oxidoreductase A plant enzyme converting hydroquinic acid (a derivative of the shikimate pathway) to quinic acid. The enzyme is activated by a calcium- and calmodulin-dependent phosphorylation.

quinone Aromatic dicarbonyl compound derived from a dihydroxy aromatic compound. Ubiquinone (coenzyme Q) is a dimethoxy-dicarbonyl derivative of benzene involved in electron transport. Other quinones may act as tanning agents.

quinone reductase Enzymes that reduce quinones to phenols usually using NADH or NADPH as a source of reductant.

quisqualate An agonist of the Q-type *excitatory amino acid* receptor. See *kainate*.

R

R17 bacteriophage Bacteriophage with RNA genome which codes for the enzyme RNA synthetase and for the coat protein, a protein to which the RNA is attached and which is involved in attachment to the bacterium.

R banding A method for identification of chromosomes in which they are treated with a hot alkaline solution before staining with Giemsa; the staining pattern is the reverse of that seen in *G banding*.

R point of cell cycle See *restriction point*.

rabies virus Species of the *Rhabdoviridae* that causes rabies in humans. The virus produces widespread infection, particularly in the brain, causing a fatal encephalomyelitis. It is found all over the world, but strict quarantine regulations have excluded it from Britain and Australia. The virus infects a number of domestic and wild mammals, whose saliva is infective; dogs and cats are the main source of infection in towns and in rural areas foxes, wolves, weasels, skunks and bats are the main source of danger. Some bats and small mammals can carry the virus without showing any symptoms of disease.

radial glial cell A type of glial cell, organised as parallel fibres joining the inner and outer surfaces of the developing cortex. Thought to play a role in neuronal guidance in development. See *contact guidance*.

radial spoke The structure that links the outer microtubule doublet of the ciliary axoneme with the sheath that surrounds the central pair of microtubules. The spokes are arranged periodically along the axoneme every 29nm, have a stalk about 32nm long and a bulbous region adjacent to the sheath. At least 17 different polypeptides are associated with the spokes. Spokes are thought to restrict the sliding of doublets relative to one another; digestion of the radial spokes will allow sliding apart of the doublets.

radioautography See *autoradiography*.

radioimmunoassay An assay for testing antigen–antibody reactions in which specific binding of radioactively-labelled antigen or antibody is measured.

radioisotope An unstable isotope (form) of a chemical element which undergoes spontaneous disintegration, usually associated with the loss of an α particle (helium nucleus), β particle (electron) or γ-radiation. Used in biology to trace the fate of atoms or molecules which follow the same metabolic pathway or enzymic fate as the normal stable isotope, but which can be detected with high sensitivity by their emission of radiation. Also used to locate the position of the radioactive metabolite, as in *autoradiography*, and to measure relative rates of synthesis of compounds (eg. DNA) from radioactive precursors.

raffinose (mellitose) A non-reducing trisaccharide found in sugar beet and many seeds, consisting of the disaccharide *sucrose* bearing a D-galactosyl residue linked $\alpha(1\text{–}6)$ to its glucose group.

Raji cell binding test A test for the detection of soluble IgG-antigen complexes. Raji cells are a line of *Epstein Barr virus*-transformed lymphocytes with surface Fc receptors. Complexes are detected by their ability to compete with a radiolabelled aggregated IgG for binding to the cells.

random walk A description of the path followed by a cell or particle when there is no bias in movement. The direction of movement at any instant is not influenced by the direction of travel in the preceding period. If changes of direction are very frequent, then the displacement will be small, unless the speed is very great, and the object will appear to vibrate on the spot. Although the behaviour of moving cells in a uniform environment can be described as a random walk in the long term, this is not true in the short term because of *persistence*.

***ras* gene** One of a family of oncogenes, first identified as transforming genes of *Harvey* and *Kirsten sarcoma viruses*. (Their name derives from "rat sarcoma" because Harvey virus, though a mouse virus, obtained its transforming gene during passage in a rat). Transforming protein coded is p21^{ras}, a *GTP-binding protein* with GTPase activity.

ratio-imaging fluorescence microscopy A method of measurement of intracellular pH or intracellular calcium levels, using a fluorescent probe molecule (see *fura-2*), in which the two different excitation wavelengths are used, and the emitted light levels compared. If emission at one wavelength is sensitive to the intracellular ion level, and emission at the other wavelength is not, then standardisation for intracellular probe concentration, efficiency of light collection, inactivation of probe and thickness of cytoplasm can all be performed automatically.

reaction centre The site in the chloroplast which receives the energy trapped by chlorophyll and accessory pigments, and initiates the electron transfer process.

reading frame One of the three possible ways of reading a nucleotide sequence. As the genetic code is read in non-overlapping triplets (*codons*) there are three possible ways of *translating* a sequence of nucleotides into a protein, each with a different starting point. For example, given the nucleotide sequence: AGCAGCAGCAGCAGCAG-CAGC, the three reading frames are: AGC AGC AGC AGC AGC AGC / GCA GCA GCA GCA GCA / CAG CAG CAG CAG CAG CAG.

reagin Reaginic antibodies; an outmoded term for *IgE*.

***rec* A protein** Protein (40kD) product of the *rec*A (recombination) gene, which catalyses the assimilation of a single-stranded piece of DNA into pairing with its complementary sequence, displacing a loop of single-stranded DNA (D-loop).

***rec* B protein** Protein (140kD); one subunit of nuclease that unwinds double-stranded DNA and fragments the strands sequentially; the other subunit is *rec*C (128kD).

receiver cell Cells in the photosynthetic tissues of plants into which the solutes from xylem are pumped.

receptor downregulation See *downregulation*.

receptor mediated endocytosis Endocytosis of molecules by means of a specific receptor protein which normally resides in a *coated pit*, but may enter this structure after complex formation occurs. The structure then forms a coated vesicle which delivers its contents to the endosome whence it may enter the cytoplasm or the lysosomal compartment. Many bacterial toxins and viruses enter cells by this route.

receptor potential Graded potential change in a sensory receptor initiated by the adequate (and appropriate) stimulus for the receptor.

receptors In general terms, molecules that bind to, or respond to something else more mobile (the *ligand*), with high specificity. Many receptors (eg. *acetylcholine receptor*, *photoreceptors*) are membrane-bound, though others (eg. steroid hormone receptors) are free in the cytosol.

receptosome Synonym for *endosome*.

recessive An *allele* or *mutation* that is only expressed phenotypically when it is present in the *homozygote*. In the *heterozygote* it is obscured by dominant alleles.

recombinant DNA Spliced DNA formed from two or more different sources which have been cleaved by *restriction enzymes* and joined by *ligases*.

recombination The creation, by a process of intermolecular exchange, of chromosomes combining genetic information from different sources, typically two genomes of a given species. Site-specific, homologous, transpositional and non-homologous (illegitimate) types of recombination are known.

recombination nodules Protein-containing assemblies of about 90nm diameter placed at intervals in the *synaptonemal complexes* which develop between homologous chromosomes at the zygotene stage of meiosis. Some nodules may be associated with the sites of *recombination*.

recruitment zone Region of cytoplasm in the rear third of a moving amoeba where endoplasm is recruited from ectoplasm.

red blood cell (erythrocyte) Cell specialised for oxygen transport, having a high concentration of *haemoglobin* in the cytoplasm (and little else). Biconcave, anucleate discs, ca 7µm diameter in mammals; nucleus contracted and chromatin condensed in other vertebrates.

red drop effect Experimental observation that the photosynthetic efficiency of monochromatic light is greatly reduced above 680nm, even though chlorophyll absorbs well up to 700nm. Led to the discovery of the two light reactions of photosynthesis; see *photosystems I and II*.

redox potential The reducing/oxidising

power of a system measured by the potential at a hydrogen electrode.

refractile Adjective usually used in describing granules within cells that scatter (refract) light. Not to be confused with refractory.

refractory period Most commonly used in reference to the interval (typically 1ms) after an *action potential* during which a second impulse cannot be elicited. This is caused by inactivation of the *sodium channels* after opening. The maximum frequency at which neurons can fire is thus limited to a few hundred Hertz. An analogous refractory period occurs in individuals of **Dictyostelium**, which are insensitive to extracellular *cyclic AMP* immediately after secreting a pulse of it. The term can be applied to any system where a similar insensitive period follows stimulation.

regeneration Processes of repair or replacement of missing structures.

regulatory T-cell Vague term, which should probably be avoided, for any class of T-lymphocyte not directly involved in the effector side of immunity, but involved in controlling responses and actions of other cells; especially T-helper and T-suppressor cells.

regurgitation during feeding Escape of lysosomal enzymes into the medium from an incompletely closed phagosome in which lysosome–phagosome fusion is already occurring: more common when the particle is large.

rejection Usually used of grafts. Any process leading to the destruction or detachment of a graft or other specified structure.

release factor A component of the specialised transport system involved in the transport of cobalamin (vitamin B_{12}) across the wall of the intestine. Dissociates the complex between cobalamin and the extracellular cobalamin-binding glycoprotein known as *intrinsic factor*.

releasing hormone General term for a hormone that triggers the release of another hormone.

renal Associated with the kidney.

renaturation The conversion of denatured protein or DNA to its native configuration. This is rare for proteins. However, if DNA is denatured by heating the two strands separate, and if the heat-denatured DNA is

then cooled slowly the double-stranded helix reforms. This renaturation is also termed re-annealing.

renin An acid protease released from the walls of afferent arterioles in the kidney when blood flow is reduced, plasma sodium levels drop, or plasma volume diminishes. Catalyzes splitting of *angiotensin* I from *angiotensinogen*, an α_2-globulin of plasma.

Reoviridae Viruses with a segmented double-stranded RNA genome (Class III); there are about 8–10 segments each coding for a different polypeptide and only one strand of the RNA (minus strand) acts as template for mRNA (plus strand). Icosahedral capsid, and the *virion* includes all the enzymes needed to synthesise mRNA. The viruses originally included in this group do not seem to cause any disease in humans, though they have been isolated from the respiratory tract and gut of patients with a variety of diseases; the name is derived from "Respiratory, Enteric, Orphan viruses". Several pathogenic viruses are now classed as reoviruses including Orbivirus, a tick-borne virus that causes Colorado tick fever, and *Rotavirus*.

repair nucleases Class of enzymes involved in *DNA repair*. It includes *endonucleases* which recognise a site of damage or an incorrect base pairing and cut it out, and *exonucleases* which remove neighbouring nucleotides on one strand.

repetitive DNA Nucleotide sequences in DNA which are present in the *genome* as numerous copies. Originally identified by the $C_0t_{1/2}$ value derived from kinetic studies of DNA *renaturation*. These sequences are not thought to code for polypeptides. One class of repetitive DNA, termed highly-repetitive DNA, is found as short sequences, 5–100 nucleotides, repeated thousands of times in a single long stretch. It typically comprises 3–10% of the genomic DNA and is predominantly *satellite DNA*. Another class, which comprises 25–40% of the DNA and is termed moderately-repetitive DNA, usually consists of sequences about 150–300 nucleotides in length dispersed evenly throughout the genome, and includes **Alu** sequences and *transposons*.

replica methods Methods in the preparation of specimens for transmission electron microscopy. The specimen (for example, a piece of *freeze fractured* tissue) is shadowed with metal and coated with carbon and then the tissue is digested away. The replica is then picked up on a grid and it is the replica that is examined in the microscope.

replica plating Technique for testing the genetic characteristics of bacterial or fungal colonies. A dilute suspension of bacteria is first spread, in a petri dish, on agar containing a medium expected to support the growth of all bacteria, the master plate. Each bacterial cell in the suspension is expected to give rise to a colony. A sterile velvet pad, the same size as the petri dish, is then pressed onto it, picking up a sample of each colony. The bacteria can then be "stamped" onto new sterile petri dishes, plates, in the identical arrangement. The media in the new plates can be made up to lack specific nutritional requirements or to contain antibiotics. Thus colonies can be identified which cannot grow without specific nutrients or which are antibiotic-resistant, and cells with mutations in particular genes can be isolated.

replicase Generic (and rather unhelpful) term for an enzyme that duplicates a polynucleotide sequence (either RNA or DNA). The term is more usefully restricted to the enzyme involved in the replication of certain viral RNA molecules.

replication Copying, but usually the production of new molecules of nucleic acid from the parental template.

replicative intermediate Intermediate stage(s) in the replication of a nucleic acid, but most commonly applied to the replication of an RNA virus: a copy of the original RNA strand, or of a single-strand copy of the first replicative intermediate. Essentially an amplification strategy.

replicons Tandem regions of replication in a chromosome, each about 30μm long.

repressor protein A protein which binds to an *operator* of a gene preventing the *transcription* of the gene. The binding affinity of repressors for the operator may be affected by other molecules. Inducers bind to repressors and decrease their binding to the operator, while co-repressors increase the binding. The paradigm of repressor proteins is the lactose repressor protein which acts on the **lac** *operon* and for which the inducers are β-galactosides such as *lactose*; it is a polypeptide of 360 amino acids which is active as a tetramer. Other examples are the *lambda*

repressor protein of λ bacteriophage which prevents the transcription of the genes required for the lytic cycle leading to *lysogeny*, and the *cro* protein, also of λ, which represses the transcription of the *lambda* repressor protein establishing the lytic cycle. Both of these are active as dimers and have a common structural feature, the helix–turn–helix motif which is thought to bind to DNA with the helices fitting into adjacent major grooves.

reproduction Propagation of organisms. The act of producing new organisms. May be asexual or sexual.

resealed ghosts Membrane shells formed by lysis of erythrocytes resealed by adjusting the cation composition of the medium. Relatively impermeable, although more permeable than the original membrane.

residual body 1. *Secondary lysosomes* containing material that cannot be digested. 2. The surplus cytoplasm shed by spermatids during their differentiation to spermatozoa. Usually the cytoplasm from several spermatids connected by *cytoplasmic bridges*. 3. Surplus cytoplasm containing pigment and left over after production of merozoites during schizogony of *malaria* parasites.

resilin Amorphous rubber-like protein found in insect tissues: similar to elastin, but there is no fibre formation.

resolution Complete return to normal structure and function: used, for example, of an inflammatory lesion, or of a disease. See also *resolving power*.

resolving power 1. The resolution of an optical system defines the closest proximity of two objects which can be seen as two distinct regions of the image. This limit depends upon the **Numerical Aperture** (N.A.) of the optical system, the contrast step between objects and background and the shape of the objects. The often-quoted Airy limit applies only to self-luminous discs. 2. In genetics, the smallest map distance measurable by an experiment involving a certain number of classified recombinant progeny.

respiration Term used by physiologists to describe the process of breathing and by biochemists to describe the intracellular oxidation of substrates coupled with production of ATP and oxidised coenzymes (NAD^+ and FAD). This form of respiration may be

Table R1. Restriction endonucleases

Table R1. Recognition sequences of various type II restriction endonucleases

Enzyme	Bacterium from which enzyme is derived	Recognition sequence[a]
Aha III	*Aphanothece halophytica*	▼ TTT|AAA
Alu I	*Arthrobacter luteus*	▼ AG|CT
Ava I	*Anabaena variabilis*	▼ CPy C|GPuG
Bam HI	*Bacillus amyloliquefaciens* H	▼ m GGA|TCC
Bst EII	*Bacillus stearothermophilus* ET	▼ GGTNACC
Cla I	*Caryphanon latum*	▼ ATC|GAT
Dde I	*Desulfovibrio desulfuricans*	▼ CTNAG
Eco RI	*Escherichia coli*	▼ m GAA|TTC
Eco RII	*Escherichia coli*	▼ m CC(^A_T)TT
Eco RV	*Escherichia coli*	▼ GAT|ATC
Hae II	*Haemophilus aegyptius*	▼ PuGC|CGPy
Hae III	*Haemophilus aegyptius*	▼m GG|CC
Hha I	*Haemophilus haemolyticus*	m ▼ GC|GC
Hin dIII	*Haemophilus influenzae* R_d	m▼ AAG|CTT
Hin fI	*Haemophilus influenzae* R_f	▼ GANTC
Hpa I	*Haemophilus parainfluenzae*	▼ GTT|AAC
Hpa II	*Haemophilus parainfluenzae*	▼ m CC|GG
Kpn I	*Klebsiella pneumoniae* OK8	▼ GGT|ACC
Mbo I	*Moraxella bovis*	▼ GA|TC
Msp I	*Moraxella* species	▼ CC|GG
Mst I	*Microcoelus* species	▼ TGC|GCA
Pst I	*Providencia stuartii*	▼ CTG|CAG
Pvu I	*Proteus vulgaris*	▼ CGA|TCG
Sac I	*Streptomyces achromogenes*	▼ GAG|CTC

(Continued)

Table R1. Restriction endonucleases (Continued)

Enzyme	Bacterium from which enzyme is derived	Recognition sequence[a]
Sma I	Serratia marcescens	▼ CCC|GGG
Xba I	Xanthomonas badrii	▼ TCT|AGA
Xho I	Xanthomonas holcicola	▼ CTC|GAG

▼   m
[a]5'-XXXX|XXXX-3'
(▼: CLEAVAGE SITE |: AXIS OF SYMMETRY)
Pu = Purine i.e. A or G are recognised
Py = Pyrimidine i.e. C or T are recognised
N = Any base
m
X = base methylated by corresponding methylase, where known, to give
N^6-methyladenosine or 5-methylcytosine

anaerobic as in *glycolysis* or aerobic for oxidations operating via the *tricarboxylic acid cycle* and the *electron transport chain*.

respiratory burst See *metabolic burst*.

respiratory chain The mitochondrial electron transport chain.

respiratory enzyme complexes The enzymes that make up the respiratory chain: NADH-Q reductase, succinate-Q reductase, cytochrome reductase, cytochrome c and cytochrome oxidase.

resting potential The electrical potential of the inside of a cell, relative to its surroundings. Almost all animal cells are negative inside; resting potentials are in the range -20 to -100mV, -70mV typical. Resting potentials reflect the action of the *sodium pump* only indirectly; they are mainly caused by the subsequent diffusion of potassium out of the cell through potassium leak channels. The resting potential is thus close to that derived by the *Nernst equation* for potassium. See *action potential*.

restriction endonucleases Class of bacterial enzymes that cut DNA at specific sites. In bacteria their function is to destroy foreign DNA, such as that of *bacteriophages* (host DNA is specifically modified by methylation at these sites). Type I restriction endonucleases occur as a complex with the methylase and a polypeptide which binds to the recognition site on DNA; they are often not very specific and cut at a remote site. Type II restriction endonucleases are the classic experimental tools. They have very specific recognition and cutting sites. The recognition sites are short, 4–8 nucleotides, and are usually *palindromic sequences*. Because both strands have the same sequence running in opposite directions the enzymes make double-stranded breaks, which, if the site of cleavage is off-centre, generates fragments with short single-stranded tails; these can hybridise to the tails of other fragments and are called "sticky ends". They are generally named according to the bacterium from which they were isolated (the first letter of genus name and the first two letters of the specific name). The bacterial strain is identified and multiple enzymes from the same strain are given Roman numerals. For example the two enzymes isolated from the R strain of *E. coli* are designated *Eco* RI and *Eco* RII. The more commonly used restriction endonucleases are shown in Table R1.

restriction enzyme See *restriction endonuclease*.

restriction fragment length polymorphism If polymorphism exists in a population (as will normally be the case) then the fragments of DNA generated by *restriction enzymes* will also show polymorphism: this is the basis of testing for degree of relationship between individuals, and for so-called genetic fingerprinting.

restriction fragments The fragments of DNA generated by digesting DNA with a specific *restriction endonuclease*.

restriction map Map of DNA showing the

position of sites recognised and cut by various *restriction endonucleases*.

restriction nucleases See *restriction endonucleases*.

restriction point, cell cycle A point, late in *G1*, after which the cell must, normally, proceed through to division at its standard rate.

reticular fibres Fine fibres (of *reticulin*) found in extracellular matrix, particularly in lymph nodes, spleen, liver, kidneys and muscles.

reticulin Constituent protein of *reticular fibres*: collagen Type III.

reticulocyte Immature red blood cell found in the bone marrow, and in very small numbers in the circulation. Contains detectable cytoplasmic RNA, and is a useful marker for haemolytic anaemias, in which blood reticulocyte counts are high.

reticulocyte lysate Cell lysate produced from *reticulocytes*; used as an *in vitro* translation system.

reticuloendothelial system The phagocytic system of the body, including the fixed macrophages of tissues, liver and spleen. Rather old-fashioned term which has never been abandoned; "mononuclear phagocyte system" is probably better when only phagocytes are meant. The cells involved are neither reticular nor endothelial.

reticulum cells Cells of the reticuloendothelial system, found particularly in lymph nodes, bone marrow, and spleen. The ones in lymph nodes are stromal cells and probably are not reticuloendothelial cells in the current sense of that term.

retina Light-sensitive layer of the eye. In vertebrates, looking from outside, there are four major cell layers: (i) the outer neural retina, which contains neurons (ganglion cells, *amacrine cells*, *bipolar cells*) as well as blood vessels; (ii) the photoreceptor layer, a single layer of rods and cones; (iii) the *pigmented retinal epithelium* (PRE or RPE); (iv) the choroid, composed of connective tissue, fibroblasts, and including a well-vascularised layer, the chorio capillaris, underlying the basal lamina of the PRE. Behind the choroid is the sclera, a thick organ capsule. See *retinal rods*, *retinal cones*, *rhodopsin*. In molluscs (especially cephalopods such as the squid) the retina has the light-sensitive cells as the outer layer with the neural and supporting tissues below.

retinal Aldehyde of retinoic acid (vitamin A); complexed with opsin forms *rhodopsin*. Photosensitive component of all known visual systems. Absorption of light causes retinal to shift from the 11-*cis* form to the all-*trans* configuration, and through a complex cascade of reactions excites activity in the neurons synapsed with the rod cell.

retinal cone One of the two light-sensitive cell types of the retina, which, unlike the *retinal rod*, is differentially sensitive to particular wavelengths of light, and is important for colour vision. There are three types of cones, each type sensitive to red, green or blue. Present in large numbers in the fovea.

retinal ganglion cell See *ganglion cell*.

retinal pigmented epithelial cell See *pigmented retinal epithelium*.

retinal rod Major photoreceptor cell of vertebrate retina (about 125 million in a human eye). Columnar cells (about 40µm long, 1µm diameter) having three distinct regions: a region adjacent to, and synapsed with, the neural layer of the *retina* contains the nucleus and other cytoplasmic organelles; below this is the inner segment, rich in mitochondria, which is connected through a thin "neck" (in which is located a *ciliary body*) to the outer segment. The outer segment largely consists of a stack of discs (membrane infoldings which are incompletely separated in cones) which are continually replenished near the inner segment and which are shed from the distal end and phagocytosed by the pigmented epithelium. The membranes of the discs are rich in *rhodopsin*, the pigment which absorbs light.

retino-tectal connection A problem that has exercised developmental biologists is the way in which nerve fibres from the developing retina are "mapped" onto the tectum of the brain. There seems to be a good positioning system in operation, and a variety of mechanisms probably operate, including control of the fasciculation of fibres in the optic nerve, and some specific recognition of the correct target area by the nerve growth cone.

retinoblastoma Malignant tumour of the retina, usually arising in the inner nuclear layer of the neural retina. Retinoblastoma is unusual in being caused by an autosomal dominant mutation in an anti-oncogene in some cases (about 6%), in which case it may be bilateral.

retinoic acid See *retinal*.

retraction fibres Thin projections from crawling cells associated with areas where the cell body is becoming detached from the substratum, but *focal adhesions* persist. Usually contain a bundle of microfilaments which are under tension.

retrograde axonal transport The transport of vesicles from the synaptic region of an axon terminal towards the cell body: involves the interaction of *MAP1C* with microtubules.

Retroviridae *Viruses* with a single-stranded RNA genome (Class VI). On infecting a cell the virus generates a DNA replica by action of its virally coded *reverse transcriptase*. *Oncovirinae* are one of three subclasses of retroviruses, the others being *Lentivirinae* and Spumavirinae.

reverse passive haemagglutination If antibodies are bonded to the surface of red blood cells haemagglutination will occur if the appropriate bi- or multivalent antigen is added in soluble or microparticulate form. Used as a test for eg. *hepatitis B* virus in the serum.

reverse transcriptase RNA-directed DNA polymerase; enzyme first discovered in *Retroviridae*, which can construct double-stranded DNA molecules from the single-stranded RNA *templates* of their genomes. Can be inhibited by the drug AZT. Reverse transcription now appears also to be involved in movement of certain mobile genetic elements such as the *Ty* plasmid in yeast, in the replication of other viruses such as *hepatitis B*, and possibly in the generation of mammalian *pseudogenes*.

reversion Reversion of a mutation occurs when a second mutation restores the function which was lost as a result of the first mutation. The second mutation causes a change in the DNA that either reverses the original alteration or compensates for it.

Reynold's number A constant without dimensions which relates the inertial and viscous drag acting to hinder a body moving through fluid medium. For cells the Reynold's number is very small; viscous drag is dominant, and inertial resistance can be neglected.

Rhabdoviridae Class V viruses with a single negative-strand RNA genome and an associated virus-specific RNA polymerase. The capsid is bullet-shaped and enveloped by a membrane which is formed when the virus buds out of the plasma membrane of infected cells. The budded membrane contains host lipids but only glycoproteins coded for by the virus, of which there are usually 1–3 species. In the electron microscope these appear as regularly arranged spikes about 10nm long and are called spike glycoproteins. This group includes *rabies virus*, *vesicular stomatitis virus* and a number of plant viruses.

rhamnogalacturonan Plant cell-wall polysaccharide consisting principally of rhamnose and galacturonic acid. Present as a major part of the pectin of the primary cell wall. Two types known: rhamnogalacturonan I (RG-I), the major component, which contains rhamnose, galacturonic acid, arabinose and galactose, and rhamnogalacturonan II (RG-II), containing at least four different sugars in addition to galacturonic acid and rhamnose.

rhamnose (6-deoxy-L-mannose) A sugar found in plant glycosides.

rheotaxis Tactic response (*taxis*) to the direction of flow of a fluid.

Rhesus blood group Human blood group system with allelic red cell antigens C, D and E. The D antigen is the strongest. Red cells from a Rhesus positive foetus cross the placenta and can sensitise a Rhesus negative mother, especially at parturition. The mother's antibody may then, in a subsequent pregnancy, cause haemolytic disease of the newborn if the foetus is Rhesus positive. The disease can be prevented by giving anti-D IgG during the first 72 hours after parturition to mop up D+ red cells in the maternal circulation.

rheumatic fever Disease involving inflammation of joints and damage to heart valves which follows streptococcal infection and is believed to be due to *autoimmunity*, ie. antibodies to streptococcal components cross-react with host tissue antigens.

rheumatoid arthritis Chronic inflammatory disease in which there is destruction of joints. Considered by some to be an autoimmune disorder in which immune complexes are formed in joints and excite an inflammatory response (complex-mediated *hypersensitivity*). Cell-mediated (Type IV) hypersensitivity also occurs, and macrophages accumulate. This in turn leads to the destruction of the synovial lining (see *pannus*).

rheumatoid factor Complex of IgG and anti-IgG formed in a variety of locations, particularly in joints in *rheumatoid arthritis*. Serum rheumatoid factors are more usually formed from IgM antibodies directed against IgG.

Rhinoviruses *Picornaviridae* that largely infect the upper respiratory tract. Include the common cold virus and foot-and-mouth disease virus.

Rhizobium Gram negative bacterium which fixes nitrogen in association with roots of some higher plants, notably legumes. Forms root nodules, in which it is converted to the nitrogen-fixing *bacteroid* form.

rhizoid Portion of a cell or organism which serves as a basal anchor to the substratum.

rhizoplast Striated contractile structure attached to the basal region of the cilium in a variety of ciliates and flagellates. May regulate the flagellar beat pattern, and is sensitive to calcium concentration. Composed of a 20kD protein rather similar to *spasmin*.

rhodamines A group of triphenylmethane-derived dyes are referred to as rhodamines, lissamines etc. Many are fluorescent and are used as fluorochromes in labelling proteins and membrane probes.

Rhodophyta (red algae) Division of algae, many of which have branching filamentous forms and red coloration. The latter is due to the presence of *phycoerythrin*. The food reserve is floridean (starch), found outside the plastid. The walls contain sulphated galactans such as *carageenan* and *agar*.

rhodopsin (visual purple) Light-sensitive pigment formed from *retinal* linked through a Schiff's base to opsin: rhodopsin is an integral membrane protein found in the discs of *retinal rods* and *retinal cones*, comprising some 40% of the membrane. Vertebrate opsins are proteins of 38kD. See also *bacteriorhodopsin*.

Rhodospirillium rubrum A purple non-sulphur bacterium with a spiral shape; contains the pigment *bacteriochlorophyll* and under anaerobic conditions *photosynthesises* using organic compounds as electron donors for the reduction of carbon dioxide. The purple colour results from the presence of *carotenoids*, though the bacteria are often more red or brown.

rho factor Protein factors found in prokaryotes, especially *E. coli*, involved in the termination of transcription. Mutations in *rho* may cause the RNA polymerase to read through from one *operon* to the next.

rhoptries A pair of electron-dense organelles found in **Plasmodium** merozoites and other related protozoa which are obligate intracellular parasites. Function unclear.

riboflavin (vitamin B_2) Ribose attached to a flavin moiety which becomes part of flavine adenine dinucleotide and flavine mononucleotide.

ribonucleases Phosphodiesterases which degrade RNA. Very widely distributed.

ribonucleic acid See *RNA*.

ribonucleoprotein (RNP) Complexes of RNA and protein involved in a wide range of cellular processes. Besides *ribosomes* (with which RNP was originally almost synonymous), in eukaryotic cells both initial RNA transcripts in the nucleus (*hnRNA*) and cytoplasmic *messenger RNAs* exist as complexes with specific sets of proteins. Processing (splicing) of the former is carried out by small nuclear RNPs (snRNPs). Other examples are the *signal recognition particle* responsible for targetting proteins to endoplasmic reticulum and a complex involved in termination of transcription.

ribophorin Glycoproteins of the endoplasmic reticulum which interact with ribosomes whilst co-translational insertion of membrane or secreted proteins is taking place. Ribophorins may form a pore through which the nascent polypeptide chain passes.

ribose (D-ribose) A monosaccharide pentose of widespread occurrence in biological molecules, eg. RNA.

ribose binding protein Periplasmic binding protein of bacteria which interacts either with the ribose transport system or with the methyl-accepting chemotaxis protein, MCP III (*trg*).

ribosomal protein Protein present within the ribosomal subunits. In prokaryotes there are 31–34 proteins in the large subunit and 21 in the small subunit. Eukaryotic subunits have 45–50 (large subunit) and 33 (small subunit) proteins.

ribosomal RNA (rRNA) Structural RNA components of the ribosome. Prokaryotes have 5S and 23S species in the large subunit and a 16S species in the small subunit. Eukaryotes have a 5S, 5.8S and 28S species in the large subunit and an 18S species in the small subunit.

ribosome A small particulate organelle found in prokaryotes and eukaryotes and also within mitochondria and chloroplasts, but differing in size and composition. Made of two subunits, each being an RNA-protein complex. Ribosomes are responsible for the translation of *messenger RNA*, which may occur in the cytoplasm (see *polyribosomes*) or on *rough endoplasmic reticulum*.

ribulose 1,5-bisphosphate An intermediate in the *Calvin–Benson cycle* of photosynthesis.

ribulose bisphosphate carboxylase/oxidase (RUBISCO) Enzyme responsible for CO_2 fixation in photosynthesis. Carbon dioxide is combined with ribulose diphosphate to give two molecules of 3-phosphoglycerate, as part of the *Calvin–Benson cycle*. It is the sole CO_2-fixing enzyme in *C3 plants*, and collaborates with *PEP carboxylase* in CO_2 fixation in *C4 plants*. In the presence of oxygen the products of the reaction are one molecule of phosphoglyceric acid and one molecule of phosphoglycolic acid. The latter is the initial substrate for photorespiration and this oxygenase function occurs in C3 plants where the enzyme is not protected from ambient oxygen; in C4 plants the enzyme acts exclusively as a carboxylase since it is protected from oxygen. Also known as Fraction 1 protein, the major protein of leaves.

ribulose diphosphate carboxylase (RuDPC; RuDP carboxylase) See *ribulose bisphosphate carboxylase/oxidase*.

ricin Highly toxic *lectin* (66kD) from seeds of the castor bean, *Ricinus communis*. Has toxic A subunit (32kD), carbohydrate-binding B subunit (34kD). Toxic subunit inactivates *ribosomes*, and the binding subunit is specific for β-galactosyl residues.

Ricinus communis agglutinin (RCA) *Lectin* (120kD) from castor bean, with specificity similar to *ricin*, but much less toxic.

rifampicin Semi-synthetic member of the *rifamycin* group of antibiotics.

rifamycin Antibiotic produced by *Streptomyces mediterranei*, which acts by inhibiting prokaryotic but not eukaryotic DNA-dependent RNA synthesis. Blocks initiation but not elongation of transcripts.

rigidity An important property of the environment for a moving cell. Two-dimensional rigidity (surface viscosity) determines whether a cell will be able to crawl over a surface, particularly for cells such as fibroblasts which exert considerable traction forces. The rigidity of a matrix will affect the resistance offered to the passage of an invading cell.

rigor Stiffening of muscle as a result of high calcium levels and ATP depletion, so that actin–myosin links are made, but not broken.

ristocetin Mixture of ristocetins A and B: isolated from the actinomycete, *Nocardia lurida*. Induces platelet aggregation.

RNA (ribonucleic acid) This molecular species has an informational role, a structural role and an enzymic role and is thus used in a more versatile way than either DNA or proteins. Considered by many to be the earliest macromolecule of living systems. The structure is of ribose units joined in the 3′ and 5′ positions through a phosphodiester linkage with a purine or pyrimidine base attached to the 1′ position. Almost all RNA species are synthesised by transcription of DNA sequences, but there may be post-transcriptional modification.

RNA polymerases Enzymes that polymerise ribonucleotides in accordance with the information present in DNA. Eukaryotes have type I which synthesises all *ribosomal RNA* (rRNA) except the 5S component, type II which synthesises *messenger RNA* and *hnRNA* and type III which synthesises *transfer RNA* and the 5S component of rRNA. Prokaryotes have a single enzyme for the three RNA types which is subject to stringent regulatory mechanisms.

RNA primase An RNA polymerase which synthesises a short RNA primer sequence to initiate DNA replication.

RNA primer The primer sequence synthesised by RNA primase.

RNA processing Modifications of primary RNA transcripts including splicing, cleavage, base modification, capping and the addition of *poly-A tails*.

RNA splicing The removal of *introns* from primary RNA transcripts.

RNA tumour virus See *Oncovirinae*.

RNP See *ribonucleoprotein*.

rod cell See *retinal rod*.

rod outer segment See *retinal rod*.

root cap Tissue found at the apex of roots, overlying the root apical meristem and protecting it from friction as the root grows through the soil. Secretes a glycoprotein mucilage as a lubricant.

root hair cell Root epidermal cell, part of which projects from the root surface as a thin tube, thus increasing the root surface area and promoting absorption of water and ions.

root nodule Globular structure formed on the roots of certain plants, notably legumes and alder, by symbiotic association between the plant and a nitrogen-fixing microorganism (*Rhizobium* in the case of legumes and *Frankia* in the case of alder and a variety of other plants).

rootlet system Microtubules associated with the base of the flagellum in ciliates and flagellates. Also associated with this region is the *rhizoplast*.

Rotavirus Genus of the *Reoviridae* having a double-layered capsid and 11 double-stranded RNA molecules in the genome. They have a wheel-like appearance in the electron microscope, and cause acute diarrhoeal disease in their mammalian and avian hosts.

rotenone An inhibitor of electron transport which blocks transfer of reducing equivalents from NADH dehydrogenase to coenzyme Q. A very potent poison for fish and for insects.

rough endoplasmic reticulum (RER) Membrane organelle of eukaryotes which forms sheets and tubules. Contains the receptor for the signal receptor particle and binds ribosomes engaged in translating mRNA for secreted proteins and the majority of transmembrane proteins. Also a site of membrane lipid synthesis. The membrane is very similar to the nuclear outer membrane. The lumen contains a number of proteins which possess the C-terminal signal KDEL.

rough microsome Small vesicles obtained by sonicating cells and which are derived from the rough endoplasmic reticulum. Contain bound ribosomes and can be used to study protein synthesis.

rouleaux Cylindrical masses of red blood cells. Horse blood will spontaneously form rouleaux, in other species they can be induced by reducing the repulsion forces between erythrocytes by the addition of high-molecular weight molecules such as dextran. The erythrocyte sedimentation rate is widely used as a marker for human diseases in which an increased immunoglobulin concentration causes rapid sedimentation through rouleaux formation.

Rous sarcoma virus (RSV) The virus responsible for the classical first cell-free transmission of a solid tumour, the chicken *sarcoma*, first reported by Rous in 1911. An avian C-type member of the *Oncovirinae*, original source of the *src* oncogene.

rRNA See *ribosomal RNA*.

RSV See *Rous sarcoma virus*.

RUBISCO See *ribulose bisphosphate carboxylase/oxidase*.

RuDP carboxylase Ribulose diphosphate carboxylase; see *ribulose bisphosphate carboxylase/oxidase*.

ruffles Projections at the leading edge of a crawling cell. In time-lapse films the active edge appears to "ruffle". The protrusions are apparently supported by a microfilament meshwork, and can move centripetally over the dorsal surface of a cell in culture.

ruthenium red A stain used in electron microscopy for acid mucopolysaccharides on the outer surfaces of cells.

S

S1 Soluble fragment (102kD) of heavy meromyosin (*HMM*) which is produced by papain cleavage: retains the ATPase and actin-binding activity, and can be used to decorate actin filaments for identification by electron microscopy.

S2 Fibrous fragment of heavy meromyosin (*HMM*). Links the *S1* head to the *LMM* region which lies in the body of the thick filament and acts as a flexible hinge.

S180 Sarcoma 180: highly malignant mouse sarcoma cells, often passaged in *ascites* form. Used in some of the classical studies on contact inhibition of locomotion.

S antigens Soluble heat-stable antigens (195kD) on the surface of **Plasmodium** *falciparum* that are responsible for antigenic heterogeneity.

S gene complex Genes coding for molecular components of the pollen–stigma recognition system in the cabbage genus (*Brassica*). The gene products govern the *self-incompatibility* response and include a glycoprotein found on the stigma surface and a lectin on the pollen grain surface which binds to the stigma glycoprotein.

S phase The phase of the *cell cycle* during which DNA replication takes place.

S region The non-MHC gene in the midst of the H-2 *major histocompatibility complex* of the mouse genome which codes for complement component C4. Sometimes confusingly known as the gene for the type III MHC product in mice.

S value Svedberg Unit. See *sedimentation coefficient*.

S-ring The static part of the bacterial motor: a ring of 15 or 17 subunits (one less or one more than the number of subunits in the *M-ring*), anchored to the inner surface of the cell wall.

Saccharomyces Genus of *Ascomycetes*; yeasts. Normally haploid unicellular fungi that reproduce asexually by budding. Also have a sexual cycle in which cells of different mating-types fuse to form a diploid *zygote*. Economically important in brewing and baking, and are also suitable eukaryotic cells for the processes of genetic engineering and for the analysis of, for example, cell division cycle control by selecting for mutants (see **cdc** *genes*). *S. cerevisiae* is baker's yeast; *S. carlsbergensis* is now the major brewer's yeast.

Salmonella Genus of Enterobacteriaceae; motile, Gram negative. Enteric organisms which, if invasive, cause enteric fever (eg. typhoid, due to *S. typhi*); if non-invasive, may cause food-poisoning (usually *S. typhimurium* or *S. enteritidis*, the latter notorious for contamination of poultry).

saltatory conduction A method of neuronal transmission in vertebrate myelinated axons, where only specialised *nodes of Ranvier* participate in excitation. This reduces the capacitance of the neuron, allowing much faster transmission. See *myelin*, *Schwann cell*.

saltatory movements Abrupt jumping movements of the sort shown by some intracellular particles. Mechanism unclear.

saltatory replication The sudden amplification of a DNA sequence laterally to generate many copies in a tandem arrangement. Possible mechanism for the origin of *satellite DNA*.

Sanfillipo syndrome *Lysosomal disease* in which either keratan sulphate sulphatase or N-acetyl α-D-glucosaminidase is defective: cross correction (complementation) of co-cultured fibroblasts from apparently clinically-identical patients can therefore occur if a different enzyme is missing in each.

saprophyte Organism which feeds on complex organic materials, often the dead and decaying bodies of other organisms. Many fungi are saprophytic.

sarcoidosis Disease of unknown aetiology in which there are chronic inflammatory granulomatous lesions in lymph nodes and other organs.

sarcolemma Plasma membrane of a striated muscle fibre.

sarcoma cells Cells of a malignant tumour derived from connective tissue. Often given a prefix denoting tissue of origin, eg. osteosarcoma (from bone).

sarcoma growth factor Group of polypep-tides released by *sarcoma cells* which pro-mote the growth of cells by binding to a cell surface receptor; the sarcoma cell is therefore self-sufficient and independent of normal growth control. See *growth factors*. The name is not now commonly used, a more specific name being given now that the growth factors are better characterised.

sarcoma virus Virus which causes tumours originating from cells of connective tissue such as *fibroblasts*. See *Rous sarcoma virus*, src *gene*.

sarcomere Repeating subunit from which the *myofibrils* of striated muscle are built. Has *A-* and *I-bands*, the I-band being subdi-vided by the *Z-disc*, and the A-band being split by the *M-line* and the H-zone.

sarcoplasm Cytoplasm of striated muscle fibre.

sarcoplasmic reticulum Endoplasmic reticulum of striated muscle, specialised for the sequestration of calcium ions, which are released upon receipt of a signal relayed by the *T tubules* from the neuromuscular junc-tion, ie. by an *action potential* propagating along the T tubule membrane.

satellite cells Sparse population of mononu-cleate cells found in close contact with mus-cle fibres in vertebrate skeletal muscle. Seem normally to be inactive, but may be important in regeneration after damage. May be considered *quiescent stem cells*.

satellite DNA Eukaryotic DNA which can be isolated as a discrete band when the DNA is centrifuged to isopycnic equilibrium on a *caesium chloride* density gradient. The buoyant density of DNA is largely determined by its G–C content (usually about 40%) but may be higher or lower in satellite DNA. In mice 8% of the DNA forms a satellite band with 30% G–C. Satel-lite DNA is highly repetitive with many short sequences repeated many times, is mostly found in *heterochromatin*, is not usually tran-scribed and is of unknown function.

saturated fatty acids In eukaryotic membranes refers to stearic, palmitic and myristic acids, which are linear aliphatic chains with no double bonds. Prokaryotes have numerous branched-chain saturated fatty acids.

saturation density The density of cells in a culture of, for example, fibroblasts, at which further increase in density (cell number per area) ceases. Further proliferation may occur, but one of the two daughter cells will fail to become attached and will be lost. Thought to be a consequence of the inability of the crowded cells to gain adequate access to growth factors in the medium. *Transfor-med cells* characteristically grow to higher saturation density than their normal coun-terparts. See *anchorage dependence*, *density dependent inhibition of growth*.

saturation of receptors Saturation, the state in which receptors are effectively occu-pied all of the time, can be said to occur in a simple binding equilibrium when the con-centration of ligand is more than 5 times the K_d value, although strictly this will only be true at infinite ligand concentration.

sauvagine Peptide (40 amino acids) origi-nally isolated from the skin of the frog, *Phyllomedusa sauvagei*, and which is closely related to *corticotrophin releasing factor* and to urotensin I.

saxitoxin (STX) *Neurotoxin* produced by the "red tide" dinoflagellates, *Gonyaulax catenella* and *G. tamarensis*. It binds to the *sodium channel*, blocking the passage of *action potentials*. Its action closely resembles that of *tetrodotoxin*. The toxin was originally isolated from the clam, *Saxidomus gigan-teus*.

scanning electron microscopy (SEM) Technique of electron microscopy in which the specimen is coated with heavy metal, and then scanned by an electron beam. The image is built up on a monitor screen (in the same way as the raster builds a conventional television image). The resolution is not so great as with transmission electron micro-scopy, but preparation is easier (often by fixation followed by critical point drying), the depth of focus is relatively enormous, the surface of a specimen can be seen (though not the interior unless the specimen is cracked open) and the image is aesthetically pleasing.

Scatchard plot A method for analysing data for freely reversible ligand/receptor binding interactions. The plot is:- (Bound ligand/ Free ligand) against (Bound ligand); the slope gives the negative reciprocal of the binding affinity, the intercept on the x-axis the number of receptors (Bound/Free becomes zero at infinite ligand concent-ration). The Scatchard plot is preferable to the *Eadie–Hoffstee plot* for binding data because it is more dependent upon the values at high ligand concentration (which will be the most reliable values). A non-linear Scatchard plot is often taken to indicate heterogeneity of receptors, although this is not the only explanation possible.

Scheie syndrome Mucopolysaccharidosis (*lysosomal disease*) in which there is a defect in α-L-iduronidase. Fibroblasts from Scheie syndrome patients do not cross-correct fibroblasts from *Hurler's disease*, although the two conditions are clinically distinct.

schistocytes Fragments of red blood cells found in the circulation.

schistosomiasis Disease (bilharzia) caused by digenetic trematode worms of the genus *Schistosoma*, the adults of which live in the urinary or mesenteric blood vessels. Eggs shed by the female worms pass to the outside in the urine or faeces, but many also lodge in and obstruct the blood flow in the liver. *Eosinophils* seem to be particularly important in the killing of the invasive larval stage (schistosomulum). Evasion of the host's immune response by adult schistosomes seems to involve the acquisition of a coat of host cell-surface material by the parasite.

schistosomulum See *schistosomiasis*.

Schwann cell A specialised glial cell which wraps around vertebrate *axons* providing extremely good electrical insulation. Separated by *nodes of Ranvier* about once every millimetre, at which the axon surface is exposed to the environment. See *saltatory conduction*, *myelin*.

scintillation counting Technique for measuring quantity of a radioactive isotope present in a sample. In biology, liquid scintillation counting is mainly used for low-energy beta emitters such as ^{14}C and ^{35}S, and particularly for the very low-energy beta emission of ^{3}H. Gamma emissions are often measured by counting the scintillations that they cause in a crystal. Autoradiographic images can be enhanced by using a screen of scintillant behind the film.

sclereid Type of *sclerenchyma* cell which differs from the *fibre cell* by not being greatly elongated. Often occurs singly (an idioblast) or in small groups, giving rise to a gritty texture in, for instance, the pear fruit, where

it is known as a "stone cell". May also occur in layers, eg. in hard seed coats.

sclerenchyma Plant cell type with thick lignified walls, normally dead at maturity and specialised for structural strength. Includes *fibre cells*, which are greatly elongated, and *sclereids*, which are more isodiametric. Intermediate types exist.

scrapie A chronic disease of sheep in which neurological defects appear. One theory is that the disease is caused by a *prion*, but more generally thought to be a *slow virus*.

scruin Actin-binding protein found associated with the acrosomal process of the horseshoe crab, *Limulus polyphemus*. Scruin holds the microfilaments of the core process in a strained configuration so that the process is coiled. Experimentally, the myosin binding sites on the microfilaments are blocked, indicating the unusual packing conformation; when the scruin–actin binding is released the process straightens (and myosin binding is possible).

scurvy Disease caused by vitamin C deficiency. The effects are due to a failure of the hydroxylation of proline residues in collagen synthesis, and the consequent failure of fibroblasts to produce mature collagen. See *hydroxyproline*.

scutellum Part of the embryo in seeds of the Poaceae (grasses). Can be considered equivalent to the cotyledon of other monocotyledenous seeds. During germination, absorbs degraded storage material from the endosperm and transfers it to the growing axis.

SDS (sodium dodecyl sulphate; sodium lauryl sulphate) Anionic detergent which at millimolar concentrations will bind to and denature proteins, forming an SDS–protein complex. The amount of SDS bound is proportional to the molecular weight of the protein, and each SDS molecule, bound by its hydrophobic domain, contributes one negative charge to the protein thus swamping its intrinsic charge. This property is exploited in the separation of proteins by *SDS-PAGE*.

SDS-PAGE *Polyacrylamide gel electrophoresis* (PAGE) in which the charge on the proteins results from their binding of *SDS*. Since the charge is proportional to the surface area of the protein, and the resistance to movement proportional to diameter, small proteins migrate further.

second messenger In many hormone-sensitive systems the systemic hormone does not enter the target cell but binds to a receptor on the cell surface and indirectly affects the production of another molecule within the cell; this diffuses intracellularly to the target enzymes to produce the response. This intracellular mediator is called the second messenger. Examples include cyclic AMP, cyclic GMP, inositol trisphosphate (see *phosphatidyl inositol*) and diacylglycerol.

secondary active transport Accumulation of a substance across a membrane against a net electrochemical gradient, without direct linkage to metabolism (typically ATP hydrolysis). This is generally accomplished by a co-transport or counter-transport mechanism, linking the "uphill" movement of the substrate to the "downhill" movement of another solute, whose gradient had been established by another primary active transport process. Many cellular secondary active transport systems (eg. the sodium/hydrogen *antiport*, the sodium/glucose *symport*) are driven indirectly by the sodium gradient established by the *sodium-potassium ATPase*.

secondary immune response The response of the immune system to the second or subsequent occasion on which it encounters a specific antigen. See *primary immune response*.

secondary lysosome Term used to describe the intracellular vacuole formed by the fusion of a lysosome with organelles (*autosomes*) or with primary *phagosomes*. *Residual bodies* are the remnants of secondary lysosomes containing indigestible material.

secondary product End-product of plant cell metabolism, which accumulates in, or is secreted from, the cell. Includes anthocyanins, alkaloids, etc. Some are of major economic importance, eg. as drugs. In contrast to a primary product which is involved in the vital metabolism of the plant.

secondary structure Structure produced in polypeptide chains involving interactions between amino acids within the chain, especially an α-*helical* structure or a *beta-pleated sheet*. Also applied to a nucleic acid structure which folds back on itself, eg. the clover-leaf structure of transfer RNA.

secondary wall (plants) That part of the plant cell wall which is laid down on top of the *primary cell wall* after the wall has ceased to increase in surface area. Only occurs in certain cell types, eg. tracheids, vessel elements and sclerenchyma. Differs from the primary wall both in composition and structure, and is often diagnostic for a particular cell type.

secretin Peptide hormone of gastrointestinal tract (27 residues) found in the mucosal cells of duodenum. Stimulates pancreatic, pepsin and bile secretion, inhibits gastric acid secretion. Considerable homology with *gastric inhibitory polypeptide*, *vasoactive intestinal peptide* and *glucagon*.

secretion Release of synthesised product from cells. Release may be of membrane-bounded vesicles (merocrine secretion) or of vesicle content following fusion of the vesicle with the plasma membrane (apocrine secretion). In holocrine secretion whole cells are released.

secretory cells Cells specialised for secretion, usually epithelial. Those which secrete proteins characteristically have well-developed *rough endoplasmic reticulum*, whereas conspicuous *smooth endoplasmic reticulum* is typical of cells that secrete lipid or lipid-derived products (eg. *steroids*).

secretory component of IgA A polypeptide chain of about 60kD which aids secretion of the IgA; a portion of the IgA receptor on the plasmalemma of the inner side of the epithelial cells lining the gut, which is proteolysed when the IgA–receptor complex has travelled through the cell after receptor-mediated endocytosis at the inner face, to the outer (luminal) face.

secretory granule See *secretory vesicle*.

secretory proteins In eukaryotes, proteins synthesised on *rough endoplasmic reticulum* and destined for export. Nearly all proteins secreted from cells are glycosylated (in the *Golgi apparatus*), although there are exceptions (*albumin*). In prokaryotes, secreted proteins may be synthesised on ribosomes associated with the plasma membrane or exported post-translation.

secretory vesicle Membrane-bounded vesicle derived from the *Golgi apparatus* and containing material which is to be released from the cell. The contents may be densely packed, often in an inactive precursor form (*zymogen*).

sedimentation The settling of one or more components of a mixture under natural or artificial gravitational fields so that there is separation into two or more phases or zones.

sedimentation coefficient The ratio of the velocity of sedimentation of a molecule to the centrifugal force required to produce this sedimentation. It is a constant for a particular species of molecule, and the value is given in Svedberg units (S) which, it should be noted, are non-additive.

sedoheptulose Seven-carbon sugar, whose phosphate derivatives are involved in the *pentose phosphate pathway* and the *Calvin–Benson cycle*.

segment long spacing collagen See *SLS-collagen*.

segregation of chromosomes The separation of homologous pairs of chromosomes that occurs at meiosis so that only one chromosome from each pair is present in any single gamete.

selenocysteine An unusual amino acid of proteins, the selenium analogue of *cysteine*, in which a selenium atom replaces sulphur. Involved in the catalytic mechanism of seleno-enzymes such as formate dehydrogenase of *E. coli*, and mammalian glutathione peroxidase. May be co-translationally coded by a special *opal suppressor* tRNA which recognises certain UGA nonsense codons.

self assembly The property of forming structures from subunits (protomers) without any external source of information about the structure to be formed such as priming structure or template.

self-antigens The antigens of an organism's own cells and cell products are self-antigens to its own immune system. Clones of immune cells reactive with self-antigens are normally eliminated.

self-incompatibility A mechanism to ensure that self-pollination does not result in self-fertilisation. Usually the pollen tube is rejected by the stylar tissues. Ensures out-breeding within some plant species, eg. in the case of the *S gene complex* in Brassicas.

self-replicating Literally, replication of a system by itself without outside intervention. In practice often taken to refer to

systems which replicate without the contribution of any information from outside the system.

semiautonomous Of systems or processes which are not wholly independent of other systems or processes.

semiconservative replication The system of replication of DNA found in all cells in which each daughter cell receives one old strand of DNA and one strand newly synthesised at the preceding *S phase*. The existence of semiconservative replication was demonstrated by the Meselson–Stahl experiment and implies the two- or multi-strandedness of DNA.

semipermeable membrane A membrane which is selectively permeable to only one (or a few) solutes. The potential developed across a membrane permeable to only one ionic species is given by the *Nernst equation* for the species: this is the basis for the operation of *ion-selective electrodes*.

Semliki forest virus Enveloped virus of the alphavirus group of *Togaviridae*. First isolated from mosquitoes in the Semliki Forest in Uganda; not known to cause any illness in humans. The synthesis and export of its three spike glycoproteins, *via* the endoplasmic reticulum and Golgi complex, have been used as a model for the synthesis and export of plasma membrane proteins.

Sénarmont compensation In interference microscopy, compensation for the phase difference introduced by the object, measured by introducing a quarter-wavelength plate and rotating the analyser: the angle of rotation is proportional to the optical path difference.

Sendai virus Parainfluenza virus type 1 (Paramyxoviridae). Can cause fatal pneumonia in mice, and may cause respiratory diseases in humans. The ability of ultraviolet-inactivated virus to fuse mammalian cells has been extensively used in the study of *heterokaryons* and *hybrid cells*.

senescent cell antigen An antigen (62kD) which appears on the surface of senescent erythrocytes and is immunologically cross-reactive with isolated *band III*. Seems to be recognised by an autoantibody, and the immunoglobulin-coated erythrocyte is then removed from circulation by cells such as *Kupffer cells* of the liver which have *Fc-receptors*. Intracellular cleavage of intact band III by a calcium-activated protease, *calpain*, may reveal the antigen *in situ*.

sensitisation A state of heightened responsiveness, usually referring to the state of an animal after primary challenge with an antigen. The term is frequently used in the context of *hypersensitivity*.

sensory neuron 1. A *neuron* which receives input from sensory cells. 2. Sensory cells such as cutaneous mechanoreceptors and muscle receptors.

Sephacryl Trade-name for a covalently cross-linked allyl dextrose gel formed into beads. Used in *gel filtration* columns for separating molecules in the size range 5kD to 1.5 million D.

Sephadex Trade-name for a cross-linked dextran gel in bead form used for *gel filtration* columns: by varying the degree of cross-linking the effective fractionation range of the gel can be altered.

Sepharose Trade-name for a gel of agarose in bead form from which charged polysaccharides have been removed. Used in *gel filtration* columns.

septate junction An intercellular junction found in invertebrate epithelia that is characterised by a ladder-like appearance in electron micrographs. Thought to provide structural strength, and to provide a barrier to diffusion of solutes through the intercellular space. Occurs widely in transporting epithelia, and is controversially considered analogous to a *zonula occludens*.

septic shock Condition of clinical shock caused by *endotoxin* in the blood. A serious complication of severe burns and abdominal wounds, frequently fatal. Part of the problem seems to be due to increased leucocyte adhesiveness, which leads to massive sequestration of neutrophils in the lung, increased vascular permeability, a fall in blood pressure and acute respiratory distress syndrome.

septicaemia See *bacteraemia*.

sequence homology Strictly, refers to the situation where nucleic acid or protein sequences are similar because they have a common evolutionary origin. Often used loosely to indicate that sequences are very similar. Sequence similarity is observable; sequence homology can be no more than an hypothesis derived from the observed similarity.

SER See *smooth endoplasmic reticulum*.

serine (Ser; S; MW 105) One of the amino acids found in proteins which can be phosphorylated. See Table A2.

serine protease One of a group of endoproteases from both animal and bacterial sources which share a common reaction mechanism based on formation of an acyl-enzyme intermediate on a specific active serine residue. Serine proteases are all irreversibly inactivated by a series of organophosphorus esters, such as di-isopropyl-fluorophosphate (DFP) and by naturally-occurring inhibitors (*serpins*). Examples are *trypsin*, *chymotrypsin* and the bacterial *subtilisins*.

serotonin (5-hydroxytryptamine; 5-HT) A *neurotransmitter* and *hormone* (MW 176), found in vertebrates, invertebrates and plants.

serotype The genotype of a unicellular organism as defined by antisera against antigenic determinants expressed on the surface.

serous gland An exocrine gland which produces a watery, protein-rich secretion, as opposed to a carbohydrate-rich mucous secretion.

serpins Superfamily of proteins, mostly serine protease inhibitors, that includes *ovalbumin*, α_1-*protease inhibitor*, *antithrombin*.

Serratia marcescens A Gram negative bacterium found in soil and water; most strains produce a characteristic pigment, prodigiosin. Opportunistic human pathogens, infecting mainly hospital patients.

Sertoli cells Tall columnar cells found in the mammalian testis closely associated with developing spermatocytes and spermatids. Probably provide appropriate microenvironment for sperm differentiation and phagocytose degenerate sperm.

serum Fluid which is left when blood clots; the cells are enmeshed in *fibrin* and the clot retracts because of the contraction of platelets. It differs from plasma in having lost various proteins involved in clot formation (*fibrinogen*, *prothrombin*, various blood-clotting factors such as *Hagemann factor*,

Factor VIII etc.) and in containing various platelet-released factors, notably *platelet-derived growth factor*. For this reason serum is a better supplement for cell culture medium than defibrinated plasma (plasma-derived serum).

serum amyloid In secondary amyloidosis the fibrils deposited in tissues are unrelated to immunoglobulin light chains (in contrast to the situation in primary amyloidosis) and are made of amyloid A protein (AA protein). This is derived from serum amyloid A (SAA) which is the apolipoprotein of a high-density lipoprotein and an *acute phase protein*. Partial proteolysis converts SAA into the *beta-pleated sheet* configuration of the amyloid fibrils. Amyloid P protein is also found as a minor component of the fibrils (in both primary and secondary amyloidosis) and is derived from serum amyloid P which has similarity to *C-reactive protein*. The physiological role remains obscure.

serum hepatitis See *hepatitis B*.

serum requirement The amount of serum that must be added to culture medium to permit growth of an animal cell in culture. Transformed cells frequently have less stringent serum requirements than their normal counterparts.

serum sickness A *hypersensitivity* response (Type III) to the injection of large amounts of antigen, as might happen when large amounts of antiserum are given in a passive immunisation. The effects are caused by the presence of soluble immune complexes in the tissues.

severin Calcium-dependent F-actin cleaving protein isolated from *Dictyostelium discoideum*, that binds irreversibly to the barbed ends of the microfilament; not, apparently, essential for movement.

sex chromatin Condensed chromatin of the inactivated X chromosome in female mammals (*Barr body*).

sex chromosome Chromosome which determines the sex of an animal. In humans, where the two sex chromosomes (X and Y) are dissimilar, the female has two X chromosomes, and the male is heterogametic (XY). In birds, the opposite is the case, the male being XX and the female XY; in many organisms, there is only one sex chromosome, and one sex is XX, the other X0. A portion of the X and Y chromosomes is

similar and is known as the pseudo-autosomal region.

sex hormone Hormone which is secreted by gonads, or which influences gonadal development. Examples are *oestrogen*, *testosterone*, *gonadotrophins*.

sex pili Fine filamentous projections (*pili*) on the surface of a bacterium which are important in conjugation. Often seem to be coded for by plasmids which confer conjugative potential on the host; in the case of the F-plasmid, the F-pili are 8–9nm diameter and several microns long, composed of *pilin*. Whether the pili merely serve to establish and maintain adhesive contact between the partners in conjugation, or whether DNA is actually transferred through the central core of the pilus is still unresolved, although a simple adhesion role is more generally accepted.

sex-duction The transfer of genes from one bacterium to another by the process of conjugation. May involve one bacterium with an F-plasmid, in which case the process is called F-duction.

sex-linkage Mode of inheritance due to a gene situated in the unpaired portions of the sex chromosomes. Comprises X-linkage and Y-linkage, but very few Y-linked genes have been identified, none in humans.

sex-linked disorder A genetic defect, usually due to a gene on the unpaired portion of the X chromosome. Recessive X-linked alleles are fully expressed in the heterogametic sex because they can have only one copy of the gene. Thus X-linked mutant disorders are more common in human males than in females.

shadowing Procedure much used in electron microscopy, in which a thin layer of material, usually heavy metal or carbon, is deposited onto a surface from one side, in such a way as to cast "shadows". Deposition is usually done by vapourising the metal on an electrode under vacuum.

Shigella Genus of non-motile Gram negative enterobacteria: cause dysentery.

shikimic acid pathway Metabolic pathway in plants and micro-organisms, by which the aromatic amino acids (phenylalanine, tyrosine and tryptophan) are formed from phosphoenolpyruvate and erythrose-4-phosphate *via* shikimic acid. The aromatic amino acids in turn serve as precursors for the formation of lignin and other phenolic compounds in plants. Inhibitors of this pathway (eg. glyphosphate) are used as herbicides.

Shine–Dalgarno region A poly-*purine* sequence found in bacterial mRNA about 7 nucleotides in front of the *initiation codon*, AUG. The complete sequence is 5'-AGGAGG-3' and almost all messengers contain at least half of this sequence. Complementary to a highly-conserved sequence at the 3' end of 16s *ribosomal RNA*, 3'-UCCUCC-5', and thought to be involved in the binding of the mRNA to the ribosome.

shingles Disease caused by reactivation of latent *Varicella-zoster* virus (Herpetoviridae), which in children causes chicken pox. The disease may be associated with a decline in cell-mediated immunity.

shock Condition associated with circulatory collapse; a result either of blood loss, bacteraemia, an anaphylactic reaction, or emotional stress.

Shope fibroma virus Poxvirus associated with the production of benign skin tumours in rabbits.

Shope papilloma virus *Papovavirus* which produces *papillomas* (warts) in rabbits.

shuttle flow See *cytoplasmic streaming*.

shuttle vector Cloning *vector* which replicates in cells of more than one organism, eg. *E. coli* and yeast. This combination allows DNA from yeast to be grown in *E. coli* and tested directly for *complementation* in yeast. Shuttle vectors are constructed so that they have the origins of replication of the various hosts.

Shwartzman reaction Reaction which occurs when two injections of *endotoxin* are given to the same animal, particularly rabbits, 24h apart. In the local Shwartzman reaction the first injection is given intradermally, the second intravenously, and a haemorrhagic reaction develops at the dermal site. If both injections are intravenous the result is a generalised Shwartzman reaction, often accompanied by *disseminated intravascular coagulation*. The reaction depends upon the response of platelets and neutrophils to endotoxin.

sialic acid See *neuraminic acid*.

sialidase See *neuraminidase*.

sialoglycoprotein Glycoprotein of which the N- or O-glycan chains include residues of *neuraminic acid*.

sickle cell anaemia Disease common in races of people from areas in which malaria is endemic. The cause is a point mutation in haemoglobin (valine instead of glutamic acid at position 6), and the altered haemoglobin (HbS) crystallises readily at low oxygen tension. In consequence, erythrocytes from homozygotes change from the normal discoid shape to a sickled shape when the oxygen tension is low, and these sickled cells become trapped in capillaries or damaged in transit, leading to severe anaemia. In heterozygotes, the disadvantages of the abnormal haemoglobin are apparently outweighed by increased resistance to *Plasmodium falciparum malaria*, probably because parasitised cells tend to sickle and are then removed from circulation.

sideramines Naturally-occurring iron-binding compounds, hydroxamic acids.

sideroblasts Red blood cells containing Pappenheimer bodies; small, deeply basophilic granules which contain ferric iron.

sideromycins Non-chelating antibiotic analogues produced by some enteric bacteria; interfere with the uptake of sideramine–ferric ion complexes.

siderophores Natural iron-binding compounds that chelate ferric ions (which form insoluble colloidal hydroxides at neutral pH and are then inaccessible) and are then taken up together with the metal ion. See *sideramines*.

sieve plate Perforated end walls separating the component cells (sieve elements) which make up the phloem *sieve tubes* in vascular plants. The perforations permit the flow of water and dissolved organic solutes along the tube, and are lined with *callose*. The plates are readily blocked by further deposition of callose when the sieve tube is stressed or damaged.

sieve tube The structure within the *phloem* of higher plants which is responsible for transporting organic material (sucrose, raffinose, amino acids, etc.) from the photosynthetic tissues (eg. leaves) to other parts of the plant. Made up of a column of cells (sieve elements) connected by *sieve plates*.

sigma factor (σ factor) *Initiation factor*

(86kD) which binds to *E. coli* DNA-dependent **RNA polymerase** and promotes attachment to specific initiation sites on DNA. Following attachment, the sigma factor is released.

signal peptidase See *signal peptide*.

signal peptide A peptide present on eukaryotic proteins that are destined either to be secreted or to be membrane components. It is usually at the N-terminus and normally absent from the mature protein. Normally refers to the sequence (ca 20 amino acids) which interacts with *signal recognition particle* and directs the ribosome to the *endoplasmic reticulum* where co-translational insertion takes place. Could also refer to sequences which direct post-translational uptake by organelles such as peroxisomes or the nucleus. Signal peptides are highly hydrophobic but with some positively-charged residues. The signal sequence is normally removed from the growing peptide chain by signal peptidase, a specific protease located on the cisternal face of the endoplasmic reticulum.

signal recognition particle A complex between a 7S RNA species and 6 proteins. It binds to a class of signal peptides and may halt translation. The complex then interacts with a receptor protein on the *rough endoplasmic reticulum* to resume translation (if halted) and cause insertion of the protein into, or through, the membrane.

signal–response coupling The mechanisms whereby a signal at the outer surface of a cell membrane causes a response to take place within the cell. Frequently involves a cascade process with *GTP-binding protein* activation and *second messenger* production.

silent mutation Mutation that has no effect on *phenotype* because it does not affect the activity of the product of the gene, usually because of codon ambiguity.

silicosis Inflammation of the lung caused by foreign bodies (inhaled particles of silica): leads to fibrosis but unlike *asbestosis* does not predispose to neoplasia.

Simian virus 40 See *SV40*.

Sindbis virus Enveloped virus of the alphavirus group of *Togaviridae*. It is thought to be an infection of birds spread by fleas, and there is little evidence that it causes any serious infection in humans. The synthesis and export of the spike proteins, *via* the endoplasmic reticulum and Golgi complex, have been used as a model for the synthesis and export of plasma membrane proteins.

single-stranded DNA (ssDNA) DNA that consists of only one chain of nucleotides rather than the two *base-pairing* strands found in DNA in the double-helix form. *Parvoviridae* have a single-stranded DNA genome. Single-stranded DNA can be produced experimentally by rapidly cooling heat-denatured DNA. Heating causes the strands to separate and rapid cooling prevents *renaturation*.

singlet oxygen ($^1O_2$) An energised but uncharged form of oxygen which is produced in the *metabolic burst* of leucocytes and which can be toxic to cells.

sister chromatid One of the two *chromatids* making up a *chromosome*. Both are semi-conservative copies of the original chromatid.

site-specific mutagenesis (site-directed mutagenesis) An *in vitro* technique in which an alteration is made at a specific site in a DNA molecule, which is then reintroduced into a cell. Various techniques are used; for the cell biologist, a very powerful approach to determining which parts of a protein or nucleotide sequence are important.

site-specific recombination A type of *recombination* which occurs between two specific short DNA sequences present in the same or in different molecules. An example is the integration and excision of λ prophage.

situs inversus Condition in which the normal asymmetry of the body (in respect of circulatory system and intestinal coiling) is reversed. Interesting because it occurs in approximately 50% of patients with *immotile cilia syndrome*, a disorder of ciliary dynein. A similar mutation is known in mice.

skeletal muscle A rather non-specific term usually applied to the striated muscle of vertebrates which is under voluntary control. The muscle fibres are syncytial and contain myofibrils, tandem arrays of *sarcomeres*.

skyllocytosis Newly recognised phagocytic process in the pseudopodial network (reticulopods) of the marine foraminiferan, *Allogromia*.

sleeping sickness See **Trypanosoma**.

sliding filament model Generally accepted

model for the way in which contraction occurs in the *sarcomere* of striated muscle, by the sliding of the thick filaments relative to the thin filaments.

slime moulds Two distinct groups of fungi, the cellular slime moulds or *Acrasidae* which include *Dictyostelium*, and the *acellular slime moulds* or Myxomycetes which include *Physarum*.

slow muscle Striated muscle used for long-term activity (eg. postural support). Depends therefore on oxidative metabolism and has many mitochondria and abundant *myoglobin*.

slow reacting substance of anaphylaxis (SRS-A) Potent bronchoconstrictor and inflammatory agent released by mast cells; an important mediator of allergic bronchial asthma. A mixture of three *leucotrienes* (LTC4 mainly, LTD4 and LTE4).

slow virus 1. Specifically one of the *Lentivirinae*. 2. Any virus causing a disease which has a very slow onset. Diseases such as sub-acute spongiform encephalopathy, *scrapie*, *kuru* and *Creutzfeld–Jacob disease* are caused by slow viruses.

SLS collagen (segment long spacing collagen) Abnormal packing pattern of collagen molecules formed if ATP is added to acidic collagen solutions, in which lateral aggregates of molecules are produced. Each aggregate is 300nm long, and the molecules are all in register. If SLS-aggregates are overlapped with a quarter-stagger, the 67nm banding pattern of normal fibrils is reconstituted.

small cell carcinoma Common malignant neoplasm of bronchus. Cells of the tumour have endocrine-like characteristics and may secrete one or more of a wide range of hormones, especially regulatory peptides like *bombesin*.

small nuclear RNA (snRNA) Abundant class of RNA found in the nucleus of eukaryotes, usually including those RNAs with sedimentation coefficients of 7s or less. They are about 100–300 nucleotides long. Although 5s *ribosomal RNA* and *transfer RNA* are of a similar size, they are not normally regarded as snRNAs. Most are found in complexes with proteins (see *ribonucleoprotein*) and at least some have a role in processing *hnRNA*.

smooth endoplasmic reticulum (SER) An internal membrane structure of the eukaryotic cell. Biochemically similar to the rough endoplasmic reticulum (RER), but lacks the ribosome-binding function. Tends to be tubular rather than sheet-like, may be separate from the RER or may be an extension of it. Abundant in cells concerned with lipid metabolism and proliferates in hepatocytes when animals are challenged with lipophilic drugs.

smooth microsome Fraction produced by ultracentrifugation of a cellular homogenate. It consists of membrane vesicles derived largely from the *smooth endoplasmic reticulum*.

smooth muscle Muscle tissue in vertebrates made up from long tapering cells which may be anything from 20–500μm long. Smooth muscle is generally involuntary, and differs from striated muscle in the much higher actin/myosin ratio, the absence of conspicuous sarcomeres, and the ability to contract to a much smaller fraction of its resting length. Smooth muscle cells are found particularly in blood vessel walls, surrounding the intestine (especially the gizzard in birds), and in the uterus. The contractile system and its control resemble those of motile tissue cells (eg. fibroblasts, leucocytes), and antibodies against smooth muscle myosin will cross-react with myosin from tissue cells, whereas antibodies against skeletal muscle myosin will not. See also *dense bodies*.

sodium channel (sodium gate) The protein responsible for electrical excitability of *neurons*. A transmembrane *ion channel*, containing an aqueous pore around 0.4nm diameter, with a negatively-charged region internally (the "selectivity filter") to block passage of anions. The channel is *voltage gated*: it opens in response to a small *depolarisation* of the cell (usually caused by an approaching action potential), by a multi-step process. Around 1000 sodium ions pass in the next millisecond, before the channel spontaneously closes (an event with single-step kinetics). The channel is then refractory to further depolarisations until returned to near the *resting potential*. There are around 100 channels/μm^2 in unmyelinated axons; in myelinated axons, they are concentrated at the *nodes of Ranvier*. The sodium channel is the target of many of the deadliest *neurotoxins*.

sodium dodecyl sulphate See *SDS*.

sodium pump See *sodium–potassium ATPase*.

sodium–potassium ATPase A major transport protein of the plasma membrane. A multi-unit enzyme, it moves 3 sodium ions out of the cell, and 2 potassium ions in, for each ATP hydrolysed. The sodium gradient established is used for several purposes (see *facilitated diffusion, action potential*), while the potassium gradient is dissipated through the potassium leak channel. Must not be confused with a *sodium channel*.

soft agar Semi-solid agar used to gelate medium for culture of animal cells. Placed in such a medium, over a denser agar layer, the cells are denied access to a solid substratum on which to spread, so that only anchorage-independent (usually transformed) cells are able to grow.

sol-gel transformation Transition between more fluid cytoplasm (*endoplasm*; plasmasol) and stiffer gel-like *ectoplasm* (plasmagel) proposed as a mechanism for amoeboid locomotion: since the endoplasm cannot really be considerd a simple fluid and has visco-elastic properties like a gel, the term is misleading.

somatic cell Usually any cell of a multicellular organism which will not contribute to the production of gametes, ie. most cells of which an organism is made: not a *germ cell*. Notice, however, the alternative use in *somatic mesoderm*.

somatic cell genetics Method for identifying the chromosomal location of a particular gene without sexual crossing. Unstable *heterokaryons* are made between the cell of interest and another cell with identifiably different characteristics (or without the gene in question), and a series of clones isolated. By correlating retention of gene expression with the remaining chromosomes, it is possible to deduce which chromosome must carry the gene. Human–mouse heterokaryons have been extensively used in this sort of work.

somatic hybrid *Heterokaryon* formed between two somatic cells, usually from different species. See *somatic cell genetics*.

somatic mesoderm That portion of the embryonic mesoderm which is associated with the body wall, and is divided from the splanchnic (visceral) mesoderm by the coelomic cavity.

somatic mutation *Mutation* which occurs in the somatic tissues of an organism, and which will not, therefore, be heritable, since it is not present in the *germ line*. Some neoplasia is due to somatic mutation; a more conspicuous example is the reversion of some branches of variegated shrubs to the wild-type (completely green) phenotype. Somatic mutation is probably also important in generating diversity in V-gene regions of immunoglobulins.

somatic recombination One of the mechanisms used to generate diversity in antibody production is to rearrange the DNA in B-cells during their differentiation, a process that involves cutting and splicing the immunoglobulin genes. Somatic recombination *via* homologous crossing-over occurs at a low frequency in *Aspergillus, Drosophila* and *Saccharomyces* and in mammalian cells in culture. It may be detected through the production of homozygous patches or sectors after mitosis of cells heterozygous for suitable marker genes.

somatocrinin Peptide (44 residues) with high growth hormone releasing activity. Can be isolated from rat hypothalamus and some human pancreatic tumours. Acts on adenylate cyclase.

somatomedin Peptide hormone (4kD) which is produced mainly in the liver and is released in response to somatotropin. Somatomedin stimulates the growth of bone and muscle, and also influences calcium, phosphate, carbohydrate and lipid metabolism. See *insulin-like growth factor*.

somatostatin Gastrointestinal and hypothalamic peptide hormone (two forms: 14 and 28 residues); found in gastric mucosa, pancreatic Islets, nerves of the gastrointestinal tract, in posterior pituitary and in the central nervous system. Inhibits gastric secretion and motility: in hypothalamus/pituitary inhibits somatotropin release.

somatotropin See *growth hormone*.

somites Segmentally arranged blocks of mesoderm lying on either side of the notochord and neural tube during development of the vertebrate embryo. Somites are formed sequentially, starting at the head. Each somite will give rise to muscle (from the myotome region), spinal column (from the sclerotome) and dermis (from dermatome).

sorbitol (glucitol) The polyol (polyhydric alcohol) corresponding to glucose. Occurs naturally in some plants, is used as a growth substrate in some tests for bacteria, and is sometimes used to maintain the tonicity of low ionic strength media.

sorbose A monosaccharide hexose: L-sorbose is an intermediate in the commercial synthesis of *ascorbic acid*.

sorocarp Fruiting body formed by some cellular slime moulds; has both stalk and spore-mass.

sorting out Phenomenon observed to occur when mixed aggregates of dissimilar embryonic cell types are formed *in vitro*. The original aggregate sorts out so that similar cells come together into homotypic domains, usually with one cell type sorting out to form a central mass which is surrounded by the other cell type. Much controversy has arisen over the years as to the underlying mechanism, whether there is specificity in the adhesive interactions (which would imply tissue-specific receptor–ligand interactions), or whether it is sufficient to suppose that there are quantitative differences in homotypic and *heterotypic* adhesion (the *differential adhesion* hypothesis). With the exception perhaps of the main protagonists, most cell biologists consider that there are probably elements both of tissue specificity (*CAMs*) and of quantitative adhesive differences involved.

sorus A group of *sporangia* or spore cases, eg. on the underside of fern leaves.

Southern blot An electroblotting method in which DNA from an electrophoretic gel separation is transferred to a thin rigid sheet (usually nitrocellulose) and detected by hybridisation to a ^{32}P-labelled single-stranded DNA or RNA probe. The purpose of the blot is to make more of the sample available for hybridising with the probe than could be achieved with the original gel. The technique is named after its inventor, E. M. Southern, and was the first of the blot-hybridisation techniques to be developed. See *blots*.

soybean trypsin inhibitor (STI; SBTI) Single polypeptide (21kD; 181 amino acids) which forms a stable, stoichiometric, enzymically-inactive complex with trypsin.

spacer DNA The DNA sequence between genes. In bacteria, only a few nucleotides long. In eukaryotes, can be extensive and include *repetitive DNA*, comprising the majority of the DNA of the *genome*. The term is used particularly for the spacer DNA between the many tandemly-repeated copies of the ribosomal RNA genes.

sparsomycin Antibiotic that inhibits peptidyl transferase in both prokaryotes and eukaryotes.

spasmin Protein (20kD) which forms the *spasmoneme*. Thought to change its shape when the calcium ion concentration rises, and to revert when the calcium concentration falls: the reversible shape change is used as a motor mechanism. Contraction does not require ATP, relaxation does, probably to pump calcium ions back into the smooth endoplasmic reticulum.

spasmoneme Contractile organelle found in **Vorticella** and related ciliate protozoans. Capable of shortening faster than any actin–myosin system, and of expanding actively. See *spasmin*.

spatial sensing Mechanism of sensing a gradient in which the signal is compared at different points on the cell surface and cell movement directed accordingly. Translocation of all or part of the cell is not required. See *temporal* and *pseudospatial gradient sensing*.

spectinomycin Aminocyclitol antibiotic: acts on ribosomes, but is bacteriostatic rather than bactericidal.

spectrin Membrane-associated dimeric protein (240 and 220kD) of erythrocytes. Forms a complex with *ankyrin*, actin and probably other components of the "membrane cytoskeleton", so that there is a meshwork of proteins underlying the plasma membrane, potentially restricting the lateral mobility of integral proteins. Isoforms have been described from other tissues (*fodrin*, *TW-240/260*), where they are assumed to play a similar role.

spectrophotometry Quantitative measurements of concentrations of reagents made by measuring the absorption of visible, ultraviolet or infrared light.

speract Sperm-activating peptide from the jelly coat of the eggs of the sea-urchins *Strongylocentrotus purpuratus* and *Hemicentrotus pulcherrinus*.

spermatids The haploid products of the

second meiotic division in spermatogenesis. Differentiate into mature spermatozoa.

spermatocytes Cells of the male reproductive system which undergo two meiotic divisions to give haploid spermatids.

spermatogenesis The process whereby primordial germ cells form mature spermatazoa.

Spermatophyte Division of the plant kingdom, consisting of plants which reproduce by means of seeds.

spermatozoon The mature sperm cell (male gamete).

spermidine See *spermine*.

spermine (N,N'bis(3-aminopropyl)-1,4-butanediamine; spermidine) A polybasic amine. Found in human sperm, in ribosomes and in some viruses. Involved in nucleic acid packaging. Synthesis is regulated by ornithine decarboxylase which plays a key role in control of DNA replication.

spherical aberration The distortion of an image by a lens; the production of a curved image of a flat object. There may also be magnification differences between the central and the peripheral portions of the image. "Plan" lenses are ones which have been corrected for this defect. See also *chromatic aberration*.

spherocytosis A condition in which erythrocytes lose their biconcave shape and become spherical. It occurs as cells age, and is also found in individuals with abnormal cytoskeletal proteins (hereditary spherocytosis, a disorder which leads to haemolytic anaemia).

spheroplast Bacterium from which the cell wall has been removed but which has not lysed.

spherosome Lysosome-like compartment in plants which derives from the endoplasmic reticulum and is a site for lipid storage.

sphingolipid Structural lipid of which the parent structure is sphingosine rather than glycerol. Synthesised in the Golgi complex.

sphingomyelin A sphingolipid in which the head group is phosphoryl choline. A close analogue of phosphatidyl choline. In many cells the concentrations of sphingomyelin and phosphatidyl choline in the plasma membrane seem to bear a reciprocal relationship.

sphingosine Long-chain amino alcohol which bears an approximate similarity to glycerol with a hydrophobic chain attached to carbon 3. Forms the class of sphingolipids when it carries an acyl group joined by an amide link to the nitrogen. Forms sphingomyelin when phosphoryl choline is attached to the 1-hydroxyl group. Gives rise to the cerebroside and ganglioside classes of glycolipids when oligosaccharides are attached to the 1-hydroxyl group. Not found in the free form.

spinal ganglion (dorsal root ganglion) Enlargement of the dorsal root of the spinal cord containing cell bodies of afferent spinal neurons. Neural outgrowth from dorsal root ganglia has been studied extensively *in vitro*.

spindle See *mitosis*.

spindle fibres Microtubules of the spindle which interdigitate at the equatorial plane with microtubules of the opposite polarity derived from the opposite pole *microtubule organising centre*. Usually distinguished from kinetochore fibres which are microtubules that link the poles with the kinetochore, although these could be included in a broader use of the term.

spiral cleavage Pattern of early cleavage found in molluscs and annelids (both with *mosaic eggs*). The animal pole blastomeres are rotated with respect to those of the vegetal pole. The handedness of the spiral twist is *maternally inherited*.

spirochaete (USA spirochete) An elongated, spirally-shaped bacterium, eg. the organism responsible for syphilis.

spironolactone *Aldosterone* antagonist: diuretic; used to treat low-*renin* hypertension and Conn's syndrome (in which there is overproduction of aldosterone).

splenocytes Vague term which should be avoided; sometimes applied to phagocytic cells (macrophages) of the spleen.

split gene See *intron*.

split ratio The fraction of the cells in a fully grown culture of animal cells which should be used to start a subsequent culture. Minimum may be dictated by medium inadequacies which result in poor growth of some cells at high dilution.

spokein Constituent protein of the *radial spokes* of the ciliary *axoneme*. Since a number of complementary spoke mutants are known to occur in *Chlamydomonas*, and one mutant lacks 17 proteins, it seems likely that spokein is a complex mixture.

spongioblast Cell found in developing nervous system: gives rise to *astrocytes* and *oligodendrocytes*.

spongiocytes Lipid droplet-rich cells from the middle region of the cortex of the adrenal gland.

spongy parenchyma Tissue usually found in the lower part of the leaf *mesophyll*. Consists of irregularly-shaped, photosynthetic parenchyma cells, separated by large air spaces.

spontaneous transformation *Transformation* of a cultured cell which occurs without the deliberate addition of a transforming agent. Cells from some species, especially rodents, are particularly prone to such spontaneous transformation.

sporangium A specialised structure in which spores are produced.

spore Highly resistant dehydrated form of reproductive cell adapted to withstand conditions of environmental stress. Usually has very resistant cell wall (integument) and low metabolic rate until activated. Bacterial spores may survive quite extraordinary extremes of temperature, dehydration or chemical insult. Gives rise to a new individual without fusion with another cell.

sporophyte Spore-producing plant generation. The dominant generation in *Pteridophyta* and higher plants, and alternates with the gametophyte generation.

sporopollenin Polymer of carotenoids, found in the exine of the pollen wall. Extremely resistant to chemical or enzymic degradation.

spot desmosome Macula adhaerens: see *desmosome*.

squames Flat, keratinised, dead cells shed from the outermost layer of a stratified *squamous epithelium*.

squamous epithelium An epithelium in which the cells are flattened. May be simple (eg. *endothelium*) or stratified (eg. *epidermis*).

src **gene** The transforming (sarcoma-inducing) gene of *Rous sarcoma virus*. Protein product is pp60^{ysrc}, a cytoplasmic protein with tyrosine-specific *protein kinase*

activity, which associates with the cytoplasmic face of the plasma membrane.

stachyose Digalactosyl-sucrose, a compound involved in carbohydrate transport in the phloem of many plants, and also in carbohydrate storage in some seeds.

staphylococcins *Bacteriocins* produced by staphylococci.

Staphylococcus Genus of non-motile Gram positive bacteria whose cells occur in clusters, and which produce important exotoxins (see Table E1). *Staphylococcus aureus* is pyogenic, an opportunistic pathogen, and responsible for a range of infections. It has *protein A* on the surface of the cell wall. Coagulase production correlates with virulence: hyaluronidase, lipase and *staphylokinase* are released in addition to the toxins.

staphylokinase Enzyme released by *Staphylococcus aureus* which acts as a *plasminogen activator*.

starch Storage carbohydrate of plants, consisting of *amylose* (a linear α(1–4)-glucan) and *amylopectin* (an α(1–4)-glucan with α(1–6) branch points). Present as starch grains in plastids, especially in *amyloplasts* and *chloroplasts*.

start codon See *initiation codon*.

statocyst An organ for the perception of gravity and thus body orientation, found in many invertebrate animals; a cavity lined with sensory cells and containing a *statolith*.

statocyte A root-tip cell containing one or more *statoliths*, involved in the detection of gravity.

statolith 1. *Bot*. A type of *amyloplast* found in root-tip cells of higher plants. It can sediment within the cell under the influence of gravity, and is thought to be involved in the detection of gravity. 2. *Zool*. A sand-grain or a structure of calcium carbonate or other hard secreted substance, found in the cavity of a *statocyst*. It stimulates sensory cells lining the cavity with which it comes in contact under the influence of gravity.

stearic acid (n-octadecanoic acid) See *fatty acids*.

steatoblasts Cells which give rise to fat cells (adipocytes).

stem cell 1. Cell which gives rise to a lineage of cells. 2. More commonly used of a cell which, upon division, produces dissimilar daughters, one replacing the original stem cell, the other differentiating further (eg. stem cells in basal layers of skin, in haemato-poietic tissue and in meristems).

stereocilium Microfilament bundle-suppor-ted projection, several microns long, from the apical surface of sensory epithelial cells (*hair cells*) in inner ear: like a *microvillus*, but larger. It is stiff and may act as a transducer directly, or merely restrict the movement of the sensory cilium (which does have an axon-eme). Also described on cells of pseudostrati-fied epithelium of the epididymal duct. Recently, stereocilia have been referred to as stereovilli, a much better and less confus-ing name.

steroid hormones A group of structurally related hormones, based on the cholesterol molecule. They control sex and growth characteristics, are highly lipophilic and are unique in that their receptors are in the nucleus, rather than on the plasma mem-brane. Examples: *testosterone*, *oestrogen*.

sterols Molecules that have a 17-carbon steroid structure, but with additional alcohol groups and side chains. Commonest example is *cholesterol*.

"sticky" ends The short stretches of single-stranded DNA produced by cutting DNA with *restriction endonucleases* whose site of cleavage is not at the axis of symmetry. The cut generates two complementary sequences which will hybridise (stick) to one another or to the sequences on other DNA fragments produced by the same restriction endonu-clease.

stimulus-secretion coupling A term used to describe the events which link receipt of a stimulus with the release of materials from membrane-bounded vesicles (the analogy is with excitation-contraction coupling in the control of muscle contraction). A classical example is the link between membrane depolarisation at the presynaptic terminal and the release of neurotransmitter into the synaptic cleft. Although it has long been known that intracellular calcium is involved (as with muscle, hence the analogy) the details remain obscure.

sting cells *Nematocysts* of coelenterates.

stipe A stalk, especially of fungal fruiting bodies or of large brown algae.

stoma (*pl.* stomata) Pore in the epidermis of leaves and some stems, which permits gas exchange through the epidermis. Can be open or closed, depending upon the physio-logical state of the plant. Flanked by stoma-tal *guard cells*.

stone cell See *sclereid*.

stop codon See *termination codon*.

storage diseases Another name for *lysoso-mal diseases*.

storage granules 1. Membrane-bounded vesicles containing condensed secretory materials (often in an inactive, zymogen, form). Otherwise known as zymogen gran-ules or condensing vacuoles. 2. Granules found in plastids, or in cytoplasm; assumed to be "food reserves", often of glycogen or other carbohydrate polymer.

strain birefringence See *birefringence*.

stratified epithelium An epithelium composed of multiple layers of cells, only the basal layer being in contact with the *basal lamina*. The basal layer is of *stem cells* which divide to produce the cells of the upper layers; in skin, these become heavily keratin-ised before dying and being shed as squames. Stratified epithelia usually have a mechanical/protective role.

streptococcal toxins Group of haemolytic *exotoxins* released by Streptococci. α-hae-molysin: 26–39kD (four types), forms ring-like structures in membranes (see *streptoly-sin O*); lipid target unclear. β-haemolysin: a hot–cold *haemolysin* with sphingomyelinase C activity. γ-haemolysin: complex of two proteins (29 and 26kD) which act synergisti-cally; rabbit erythrocytes particularly sensi-tive. δ-toxin: heat-stable peptide (5kD) with high proportion of hydrophobic amino acids; seems to act in a detergent-like man-ner (cf *subtilysin*), but may form hydrophilic transmembrane pores by cooperative inter-action with other δ-toxin molecules. Leuco-cidin (Panton–Valentine leucocidin): two components, F (fast migration on CM-cellulose column: 32kD), and S (slow: 38kD); mode of action contentious. See also **Streptococcus**, *streptolysins O and S*, *erythro-genic toxin*.

Streptococcus Genus of Gram positive cocci which grow in chains. Some species (*S. pyogenes* in particular) are responsible for important diseases in humans (pharyngitis, scarlet fever, rheumatic fever): *S. pneumoniae* is the main culprit in lobar- and broncho-pneumonia. Streptococci have anti-phagocytic components (hyaluronic acid-rich capsule and *M-protein*), and release various toxins (*streptolysins O and S, erythrogenic toxin*) and enzymes (*streptokinase, streptodornase*, hyaluronidase and proteinase). α-haemolytic streptococci (viridans streptococci) produce limited haemolysis on blood agar; include *S. mutans*, *S. salivarius*, *S. pneumoniae*. β-haemolytic streptococci, of which *S. pyogenes* is the only species, though there are many serotypes, produce a broad zone of almost complete haemolysis on blood agar as a result of *streptolysin O and S* release. γ-streptococci are non-haemolytic (eg. *S. faecalis*).

streptococcins *Bacteriocins* released by streptococci.

streptodornase Mixture of four DNAases released by streptococci. By digesting DNA released from dead cells the enzyme reduces the viscosity of pus and allows the organism greater motility.

streptokinase *Plasminogen activator* released by *Streptococcus pyogenes*. Occurs in two forms, A and B.

streptolydigin Antibiotic which blocks peptide chain elongation by binding to the polymerase.

streptolysin O Oxygen-labile thiol-activated haemolysin (68kD) from streptococci. Haemolysis is inhibited by cholesterol, and only cells with cholesterol in their membranes are susceptible. Toxin aggregates are linked to cholesterol to form a channel 30nm diameter in the membrane, and non-osmotic lysis follows. Markedly inhibits *neutrophil* movement and stimulates secretion, but has little effect on *monocytes*.

streptolysin S Thought to be a peptide toxin of 28 residues: causes zone of β-haemolysis around streptococcal colonies on blood agar. Mechanism of haemolysis unclear. Toxic to leucocytes, platelets and several cell lines.

Streptomyces Genus of Gram positive spore-forming bacteria which grow slowly in soil or water as a branching filamentous mycelium similar to that of fungi. Important as the source of many antibiotics, eg. *streptomycin, tetracycline, chloramphenicol, macrolides*.

streptomycin Commonly used antibiotic in cell culture media: acts only on prokaryotes, and blocks transition from *initiation complex* to chain-elongating ribosome. Isolated originally from a soil streptomycete.

streptovaricins Antibiotics which block initiation of transcription in prokaryotes. (cf *rifamycins* and *rifampicin*).

streptozotocin Methyl nitroso-urea with a 2-substituted glucose, used as an antibiotic (effective against growing Gram positive and Gram negative organisms), and also to induce a form of diabetes in experimental animals (rapidly induces pancreatic *B cell* necrosis if given in high dose). By using multiple low doses in a particular strain of mice, it is possible to produce insulitis followed later by diabetes, a model for juvenile-onset diabetes in humans. Also used to treat B cell carcinoma of the pancreas.

stress-fibres Bundles of microfilaments and other proteins found in fibroblasts, particularly slow-moving fibroblasts cultured on rigid substrata. Shown to be contractile; have a periodicity reminiscent of the *sarcomere*. Anchored at one end to a *focal adhesion*, although sometimes seem to stretch between two focal adhesions.

stress-induced proteins Alternative and preferable name for *heat shock proteins* of eukaryotic cells, which emphasises that the same small group of proteins is induced both by heat and various other stresses.

striated border Obsolete term for the apical surface of an epithelium with microvilli.

striated muscle Muscle in which the repeating units (*sarcomeres*) of the contractile *myofibrils* are arranged in registry throughout the cell, resulting in transverse or oblique striations observable at the level of the light microscope, eg. the voluntary (*skeletal*) and *cardiac muscle* of vertebrates.

stroma The soluble, aqueous phase within the chloroplast, containing water-soluble enzymes such as those of the *Calvin–Benson cycle*. The site of the *dark reaction* of photosynthesis.

structural gene A gene which codes for a product (eg. an enzyme, structural protein, tRNA), as opposed to a gene which serves a regulatory role.

structure–activity analysis Study in

which systematic variation in the structure of a compound is correlated with its activity, in an attempt to determine the characteristics of the (receptor) site at which it acts.

STX See *saxitoxin*.

Stylonychia mytilus Large ciliate protozoan of the Order Hypotrichida, that has compound cilia (cirri) which can be used for walking or swimming, and membranelles used for feeding.

subacute Description of a disease that progresses more rapidly than a chronic disease and more slowly than an acute one.

subacute sclerosing panencephalitis (SSPE) Chronic progressive illness seen in children a few years after measles infection, and involving demyelination of the cerebral cortex. Measles virus apparently persists in brain cells.

suberin Fatty substance, containing long-chain fatty acids and fatty esters, found in the cell walls of cork cells (phellem) in higher plants. Also found in the *Casparian band*. Renders the cell wall impervious to water.

subfragment 1 of myosin See *S1*.

submitochondrial particles Formed by sonicating mitochondria. Small vesicles in which the inner mitochondrial membrane is inverted to expose the innermost surface.

substance P A *vasoactive intestinal peptide* (MW 1348) found in the brain, spinal ganglia and intestine of vertebrates. Induces vasodilation, salivation and increases capillary permeability. Sequence: Arg-Pro-Lys-Pro-Gln-Phe- Phe-Gly-Leu-Met-NH$_2$.

substrate Substance which is acted upon by an enzyme: one can also speak of a suitable chemical substrate for maintaining a species of bacterium — the compound is one that can support cell growth.

substratum The solid surface over which a cell moves, or upon which a cell grows: should be used in this sense in preference to *substrate*, to avoid confusion.

subtilisin Extracellular alkaline serine protease produced by *Bacillus* spp.

subtilysin Haemolytic surfactant produced by *Bacillus subtilis*: cyclic heptapeptide linked to a long-chain hydroxy fatty acid.

subunits Components from which a structure is built; thus myosin has six subunits,

microtubules are built of tubulin subunits. In some cases it may be more informative to speak of *protomers*.

succinate (ethane dicarboxylic acid) Intermediate of the *tricarboxylic acid cycle* and *glyoxylate cycle*.

succinyl CoA An intermediate product in the *tricarboxylic acid cycle*.

succinylcholine Cholinergic antagonist and therefore a skeletal muscle relaxant.

sucrose (table sugar) Non-reducing disaccharide, α-D-glucopyranosyl- β-D-fructofuranose.

Sudan stains Histochemical stains used for lipids.

sugars See separate entries, and Table S1.

sulphatase An *esterase* in which one of the substituents of the substrate is a sulphate group.

sulphinpyrazone Pyrazole compound related to phenylbutazone, but without anti-inflammatory activity. Has no effect on platelet aggregation *in vitro*, but inhibits platelet adhesion and release reactions. Inhibits uric acid resorption in the proximal convoluted tubule of the kidney and is therefore uricosuric.

sulpholipids Lipids in which the polar head group contains sulphate species. Synthesised in the Golgi complex. Compounds that are gaining increasing recognition.

sulphonamides Group of drugs derived from sulphanilamide: act by blocking folic acid synthesis from p-aminobenzoic acid (PABA), because they are competitive analogues.

sulphydryl reagents Compounds which bind to SH groups. Include p-chlormercuribenzoate, N-ethyl maleimide, iodoacetamide. Very important in studies of protein structure.

supercoiling In circular DNA or closed loops of DNA, twisting of the DNA about its own axis changes the number of turns of the double helix. If twisting is in the opposite direction to the turns of the double helix, ie. anticlockwise, the DNA strands will either have to unwind or the whole structure will twist or supercoil — termed negative supercoiling. If twisting is in the same direction as the helix, clockwise, which winds the DNA up more tightly, positive supercoiling

Table S1. Sugars

Table S1. Sugars. The list includes only the most common compounds found in metabolic pathways and in structural molecules. The structures are presented as Haworth models and it should be noted the configuration at the carbon which carries the carbonyl oxygen is not determined unless the hydroxyl-group takes part in a glycosidic linkage, which it always does in higher oligomers. The convention for depicting glycosidic linkages is:

glycosyl carbon → acceptor hydroxyl. Configuration not defined in free molecule.

Monosaccharides

PENTOSES.

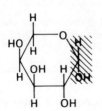

L-arabinose

D-ribose

D-xylose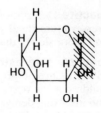

2-deoxy-D-ribose MW 134.1

HEXOSES

D-fructose

D-mannose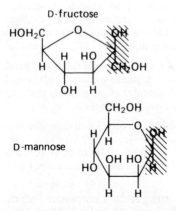

D-galactose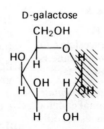

D-glucose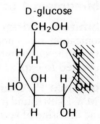

Free amino-sugars are not found in structural oligosaccharides but N-acetyl aminohexoses are widely distributed. Most common are:

N-acetylgalactosamine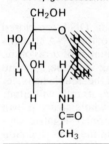

N-acetylglucosamine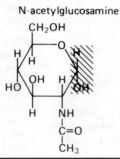

(Continued)

Table S1. Sugars (Continued)

Monosaccharides (continued)

Other common components of structural oligosaccharides are:

fucose

sialic acids
(N-acetyl neuraminic acid)

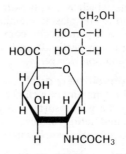

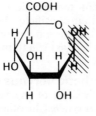

Hexose derivatives found in proteoglycans also include:

D-glucuronic acid

muramic acid

Sulphated derivatives of N-acetyl aminohexoses are also widespread and include the 4- and 6- sulphate esters of N-acetyl glucosamine and N-acetyl galactosamine.

Disaccharides and Polysaccharides

These are fully specified by the residue names, sequence, bond-direction, and the position numbers of the carbon atoms giving rise to the linkage. The configuration around the glycosidic carbon is also specified as alpha or beta.

---

is generated. DNA which shows no supercoiling is said to be relaxed. Supercoiling in circular DNA can be detected by electrophoresis because supercoiled DNA migrates faster than relaxed DNA. Circular DNA is commonly negatively supercoiled and the DNA of eukaryotes largely exists as supercoils associated with protein in the *nucleosome*. The degree of supercoiling can be altered by *topoisomerases*.

superoxide (superoxide radical) Term used interchangeably for the superoxide anion O_2^-, or the weak acid HO_2. Superoxide is generated both by prokaryotes and eukaryotes, and is an important product of the *metabolic burst* of neutrophil leucocytes. A very active oxygen species, it can cause substantial damage, and is thought to be responsible for the inactivation of plasma antiproteases which contributes to the pathogenesis of emphysema.

superoxide dismutase (SOD; EC 1.15.1.1) Any of a range of metalloenzymes which catalyzes the formation of hydrogen peroxide and oxygen from superoxide, and thus protects against superoxide-induced damage. Usually has either iron or manganese as the metal cation in prokaryotes, copper or zinc in eukaryotes.

suppressor factor 1. Factor released by T-suppressor cell. 2. See *suppressor mutation* and *ochre suppressor*, *opal suppressor*.

suppressor mutation Mutation which alleviates the effect of a primary mutation at a different locus. May be through almost any mechanism which can give a primary mutation, but perhaps the most interesting class is the *amber* and *ochre suppressors*, where the anticodon of the tRNA is altered so that it misreads the termination codon and inserts an amino acid, preventing premature termination of the peptide chain.

suppressor T-cell See *T-cells*.

surface active compound Usually, in biological systems, means a detergent-like molecule which is amphipathic and which will bind to the plasma membrane, or to a surface with which cells come in contact, altering its properties from hydrophobic to hydrophilic, or *vice versa*.

surface envelope model A way of treating the hydrodynamics of a ciliary field, by considering the whole surface of the ciliate to have an undulating surface. The undulations arise because of *metachronism*.

surfactant A *surface active compound*, the best known example of which is the lung surfactant which renders the alveolar surfaces hydrophobic and prevents the lung filling with water by capillary action. The lung surfactant is produced just at parturition, and it has often been speculated that deficiencies in surfactant metabolism might cause cot death.

suspensor cell Plant cell linking the growing embryo to the wall of the embryo sac in developing seeds.

suxamethonium A depolarising neuromuscular blocking agent, which resembles acetylcholine in structure and binds to acetylcholine receptors, acting as an agonist. When used as a drug, it diffuses slowly to the motor end plate and persists for long enough to cause the loss of electrical excitability.

SV3T3 *Swiss 3T3 cells* transformed with *SV40*.

SV40 Simian virus 40; a small DNA *tumour virus*, member of the *Papovaviridae*. Isolated from monkey cells which were being used for the preparation of *poliovirus* vaccine, and originally named vacuolating agent owing to a cytopathic effect observed in infected cells. Found to induce tumours in new-born hamsters. Transforms in culture the cells of many non- and semi-permissive species, including mouse and human.

swainsonine Fungal alkaloid that inhibits the mannosidase in the Golgi apparatus which is involved in processing the oligosaccharide chains of glycoproteins.

Swiss 3T3 cells An immortal line of fibroblast-like cells established from whole trypsinised embryos of Swiss mice (not an inbred stock), under conditions which favour establishment of cells with low saturation density in culture.

switch regions Regions within immunoglobulin heavy-chain genes which recombine with each other to delete intervening DNA, and thus alter the antibody class synthesised by a cell from IgM or D to IgG, E or A. See *isotype switching*.

symbiont One of the partners in a symbiotic relationship.

symbiotic algae Algae (often *Chlorella* spp.) which live intracellularly in animal cells (eg. endoderm of the cnidarian *Hydra viridis*). The relationship is complex, because lysosomes do not fuse with the vacuoles containing the algae, and the growth rates of both cells are regulated to maintain the symbiosis. There is considerable strain-specificity. The term is imprecise, since there are many other symbiotic algae (as in lichens) where the relationship is different.

sympathetic nervous system One of the two divisions of the vertebrate *autonomic nervous system* (the other being the *parasympathetic nervous system*). The sympathetic preganglionic neurons have their cell bodies in the thoracic and lumbar regions of the spinal cord, and connect to the paravertebral chain of sympathetic ganglia. Innervate heart and blood vessels, sweat glands, viscera and the adrenal medulla. Most sympathetic neurons, but not all, use noradrenaline as a postganglionic neurotransmitter.

symplast The intracellular compartment of plants, consisting of the cytosol of a large number of cells connected by *plasmodesmata*.

symplectic metachronism See *metachronism*.

symport A mechanism of transport across a membrane in which two different molecules move in the same direction. Often, one molecule can move up an electrochemical gradient because the movement of the other molecule is more favourable (see *facilitated diffusion*); example: the sodium/glucose co-transport. See *antiport*, *uniport*.

synapse A connection between *excitable cells*, by which an excitation is conveyed from one to the other. Chemical synapse: one in which an *action potential* causes the exocytosis of *neurotransmitter* from the presynaptic cell, which diffuses across the *synaptic cleft* and binds to *ligand-gated ion channels* on the postsynaptic cell. These ion channels then affect the resting potential of the postsynaptic cell. Electrical synapse: one in which electrical connection is made directly through the cytoplasm, *via gap junctions*. Rectifying synapse: one in which action potentials can only pass across the synapse in one direction (all chemical and some electrical synapses). Excitatory synapse: one in which the firing of the presynaptic cell increases the probability of firing of the postsynaptic cell. Inhibitory synapse: one in which the firing of the presynaptic cell reduces the probability of firing of the postsynaptic cell.

synapsin I Protein associated with synaptic vesicles, an analogue of the erythrocyte membrane-associated protein Band 4.1.

synapsis The gene–by–gene pairing of the chromatids of homologous chromosomes during *prophase* I of meiosis. It allows *crossing over* to take place.

synaptic cleft The narrow space between the *presynaptic cell* and the *postsynaptic cell* in a *chemical synapse*, across which the *neurotransmitter* diffuses.

synaptic transmission The process of propagating a signal from one cell to another *via* a *synapse*.

synaptic vesicles Intracellular vesicles found in the presynaptic terminals of *chemical synapses*, which contain *neurotransmitter*.

synaptogenesis Formation of a *synapse*.

synaptonemal complex Structure, identified by electron microscopy, lying between chromosomes during *synapsis*; consists of two lateral plates closely apposed to the chromosomes and connected to a central plate by filaments. It appears to act as a scaffold, and is essential for *crossing over*.

synaptosome A subcellular fraction prepared from tissues rich in *chemical synapses*, used in biochemical studies. Consists mainly of *synaptic vesicles* from presynaptic terminals.

synchronous cell population A culture of cells which all divide in synchrony. Particularly useful for certain studies of the cell cycle, cells can be made synchronous by depriving them of essential molecules, which are then restored. Synchronisation breaks down after a few cycles, however, as individual cells have unique division rates.

syncytium An *epithelium* or tissue in which there is cytoplasmic continuity between the constituent cells.

synemin Protein (230kD) which cross-links *vimentin* filaments and binds to filaments in a periodic fashion.

syngeneic Having identical genotypes. Syngeneic organisms may be identical twins or may be the product of extensive inbreeding to produce identical genotypes.

synkaryon A somatic hybrid cell (*heterokaryon*) in which chromosomes from two different parental cells are enveloped in a single nucleus.

synthetases Enzymes of Class 6 in the *E classification*; catalyze synthesis of molecules, their activity often being coupled to the breakdown of a nucleotide triphosphate.

syringyl alcohol (sinapyl alcohol) A phenylpropanoid alcohol, one of the three precursors of lignin.

systemic lupus erythematosus (SLE) Disease of humans, probably autoimmune with antinuclear and other antibodies in plasma. Immune complex deposition in the glomerular capillaries is a particular problem.

T

T tubules Invaginations of the plasma membrane (sarcolemma) of striated muscle cells; depolarise following neuronal stimulation at the motor end plate and thus disseminate the activation through the muscle. Very closely associated with elements of the sarcoplasmic reticulum (forming a "triad"), but the exact nature of the link between the T tubule and the sarcoplasmic reticulum is still obscure.

t-antigen The small *T-antigen* of polyoma virus.

T-antigens Proteins coded by viral genes which are expressed early in the replication cycle of papovaviruses, such as SV40 and polyoma. Essential for normal viral replication, they are also expressed in non-permissive cells transformed by these viruses. Originally detected as Tumour-antigens by immunofluorescence with anti-sera from tumour-bearing animals. SV40 has two, large T and small t; polyoma has three, large, middle and small. Appear to be collectively responsible for transformation by these viruses.

T-cells A class of lymphocytes, so called because they are of thymic origin and have been through thymic processing. Involved primarily in cell-mediated immune reactions and in the control of B-cell development. They bear T-cell antigen receptors; the T-cell receptor is always found in association with the trimeric CD3 (see *CD nomenclature*) molecule, and the latter is a widely-used marker for T-cells. Within the T-cell series are major subsets CD4+ (mainly helper cells) and CD8+ (mainly cytotoxic or suppressor cells).

T-cell leukaemia/lymphoma viruses See *HTLV-I, HTLV-II*.

T-cell receptor (TCR) The antigen-recognising receptor on the surface of a *T-cell*. Heterodimeric (disulphide-linked), one of the immunoglobulin superfamily of proteins; binds antigen in association with the *major histocompatibility complex* (MHC), leading to the activation of the cell. There

are two subunits (α and β, 42–44kD in mouse, 50–40kD in humans), each with variable and constant regions, which are associated non-covalently with CD3 (20–30kD). A second heterodimer on CD3+ cells with γ (35kD in mice, 55kD in humans) and δ (45kD in mice, 40kD in humans) chains is thought to represent a second T-cell antigen receptor which is not **MHC-restricted**. The $\gamma\delta$TCR are found on very early T-cells in the thymus and are later replaced by $\alpha\beta$TCR.

T-helper cells Set of T-lymphocytes which is involved specifically in the differentiation of B-lymphocytes into antibody-secreting cells. Most, but not all, T-helper cells are CD4+ and associated with Class II MHC accessory cells bearing antigen. Loss of T-helper cells, as in *HIV* infection, leads to immunodeficiency. There may also be T-cell help of T-cell function.

T-loop of RNA (thymine pseudo-uracil loop; TΨC loop) The T-loop of tRNA is the region of the molecule which is responsible for ribosome recognition.

T-lymphocyte See *T-cells*.

T-suppressor cells Set of T-lymphocytes, usually CD8+, specifically involved in suppressing B-cell differentiation into antibody-secreting cells. There may also be T-suppressors of T-cell functions such as *CTL* killing.

Tacaribe virus Species of *Arenaviridae* isolated from S.American bat.

talin Protein (215kD) which binds to *vinculin*, but not to actin, and is associated with the subplasmalemmal cytoskeleton.

Tamiami virus Species of *Arenaviridae* of the Tacaribe complex.

tandem repeats Copies of genes repeated one after another along a chromosome: for example the 40S rRNA genes in somatic cells of toads, of which there are about 500 copies.

tannic acid Penta-(m-digalloyl)-glucose, or any soluble tannin; used in electron microscopy to enhance the contrast. Addition of tannic acid to fixatives greatly improves, for example, the image obtained of tubulin subunits in the microtubule, or the *HMM* decoration of microfilaments.

tannins Complex phenolic compounds found in the vacuoles of certain plant cells, eg. in bark. They are strongly astringent, and are used in tanning and dyeing.

tapetum 1. Layer of reflective tissue just behind the pigmented retinal epithelium of many vertebrate eyes. May consist either of a layer of guanine crystals, or a layer of connective tissue. In bovine eyes reflects a blue-green iridescent colour. 2. Layer of cells in the sporangium of a vascular plant which nourishes the developing spores.

TATA box (Goldberg–Hogness box) A consensus sequence found in the *promoter* region of most genes transcribed by eukaryotic *RNA polymerase* II. Found about 25 nucleotides before the site of initiation of transcription and has the consensus sequence: 5′-TATAAAA-3′. This sequence seems to be important in determining accurately the position at which transcription is initiated.

tau (*tau* factor; *tau* protein) Protein (60–70kD) which co-purifies with *tubulin* through cycles of assembly and disassembly, and which is important in the assembly process.

taxis A response in which the direction of movement is affected by an environmental cue. Should be clearly distinguished from a *kinesis*.

taxol Drug isolated from yew (*Taxus brevifolis*) which stabilises microtubules: analogous in this respect to *phalloidin* which stabilises microfilaments.

Tay–Sachs disease *Lysosomal disease* (*lipidosis*) in which hexosaminidase A, an enzyme which degrades *ganglioside* GM2, is absent. A lethal autosomal recessive; mostly affects brain, where ganglion cells become swollen and die.

TCA cycle See *tricarboxylic acid cycle*.

teichoic acid Acidic polymer (glycerol or ribitol linked by phosphodiester bridges) found in cell wall of Gram positive bacteria. May constitute 10–50% of wall dry weight and is cross-linked to peptidoglycan. Related to *lipoteichoic acid*.

telocentric chromosome Chromosome with the centromere located at one end.

telomere The end of a chromosome.

telophase The final stage of mitosis or meiosis, when chromosome separation is completed.

temperature-sensitive (*ts*) mutation A type of conditional mutation in organism, somatic cell or virus which makes it possible to study genes whose total inactivation would be lethal. Such *ts* mutations can also make possible studies of the effect of reversible switching (by temperature changes) in expression of the mutated gene. The usual mechanism of temperature sensitivity is that the mutated gene codes for a protein with a temperature-dependent conformational instability, so that it possesses normal activity at one temperature (the permissive temperature), but is inactive at a second (non-permissive) temperature.

template A structure which in some direct physical process can cause the patterning of a second structure, usually complementary to it in some sense. In current biology almost exclusively used to refer to a nucleotide sequence that directs the synthesis of a sequence complementary to it by the rules of Watson–Crick base-pairing.

temporal sensing Mechanism of gradient sensing in which the value of some environmental property is compared with the value at some previous time, the cell having moved position between the two samplings. Initial movement is random; until the second observation is made the gradient cannot be detected. See *spatial* and *pseudospatial gradient sensing* mechanisms. Bacterial chemotaxis (so called) is based on this mechanism.

tenascin (myotendinous antigen) Protein of the extracellular matrix (240kD subunit: usually as a hexabrachion, a six-armed hexamer of more than 1000kD) selectively present in mesenchyme surrounding foetal (but not adult) rat mammary glands, hair follicles and teeth. Found in the matrix surrounding mammary tumours of rat. Tenascin contaminates cell-surface fibronectin and accounts for most of the haemagglutinating activity of extracellular matrix protein.

teratocarcinoma Malignant tumour, thought to originate from primordial germ cells or misplaced blastomeres, which contains tissues derived from all three embryonic layers, eg. bone, muscle, cartilage, nerve, tooth-buds and various glands. Accompanied by undifferentiated, pluripotent epithelial cells known as embryonal carcinoma cells.

teratogen Agent capable of causing malformations in embryos. Notorious example is *thalidomide*.

teratoma See *teratocarcinoma*.

terminal bar Obsolete name for *zonula adhaerens*.

terminal cisternae Regions of the *sarcoplasmic reticulum* adjacent to *T tubules*, and from which calcium is released when striated muscle is activated.

terminal web The cytoplasmic region at the base of microvilli in intestinal epithelial cells, a region rich in microfilaments from the microvillar core and from *adhaerens junctions*, in myosin, and in other proteins characteristic of an actomyosin motor system.

termination codon The three *codons*, UAA known as ochre, UAG as amber and UGA as opal, that do not code for an amino acid but act as signals for the *termination* of protein synthesis. They are not represented by any *transfer RNA* and termination is catalyzed by protein release factors. There are two release factors in *E. coli*; RF1 recognises UAA and UAG, RF2 recognises UAA and UGA. Eukaryotes have a single *GTP*-requiring factor, eRF.

terminator DNA sequence at the end of a *transcription unit* that causes *RNA polymerase* to stop transcription.

terpene Lipid species, very abundant in plants. In principle terpenes are polymers of isoprene units. In animals *dolichol*, an important carrier species in the formation of glycoproteins, is a terpenoid. Similarly, squalene, an intermediate in the synthesis of cholesterol, is a terpene.

tertiary structure The third level of structural organisation in a macromolecule. The primary structure of a protein (for example) is the amino acid sequence, the secondary structure is the folding of the peptide chain (α-helical or beta-pleated), the tertiary structure is the way in which the helices or sheets are folded or arranged to give the three-dimensional structure of the protein. Quarternary structure refers to the arrangement of protomers in a multimeric protein.

testa (seed-coat) That part of the plant seed coat derived from the integument of the ovary.

testicular feminisation If genetic males lack receptors for testosterone they develop as females and are unresponsive to male hormones.

testosterone Male sex hormone (androgen) secreted by the interstitial cells of the testis of mammals and responsible for triggering the development of sperm and of many secondary sexual characteristics.

tetanolysin *Thiol-activated haemolysin* released by the bacterium *Clostridium tetani*.

tetanospasmin See *tetanus toxin*.

tetanus (lock-jaw) Disease caused by the bacterium *Clostridium tetani*, spores of which persist in soil but can proliferate anaerobically in an infected wound. Disease entirely due to the *tetanus toxin*, released by bacterial autolysis.

tetanus toxin (tetanospasmin) Neurotoxin released by *Clostridium tetani*; becomes active when peptide cleaved proteolytically to heavy (100kD) and light (50kD) chains held together by disulphide bond. Heavy chain binds to disialogangliosides (GD2 and GD1b), and part of the peptide (the amino-terminal B-fragment) forms a pore: light chain essential to toxicity but mechanism unclear. Toxin transported from motoneuron endings through axon to spinal cord, where it acts presynaptically to inhibit release of glycine (an inhibitory transmitter) and thus causes hyperactivity of motoneurons.

tetracaine (amethocaine) Potent local anaesthetic.

tetracycline Broad-spectrum antibiotic which blocks binding of aminoacyl-tRNA to the ribosomes of both Gram positive and Gram negative organisms (and those of organelles). Produced by *Streptomyces aureofasciens*.

tetraethylammonium ion (TEA) A monovalent cation widely used in neurophysiology as a specific blocker of potassium channels. It is similar in size to the hydrated potassium ion, and gets stuck (reversibly) in the channels.

Tetrahymena Genus of ciliate protozoan, frequently used in studies on ciliary axonemes, self-splicing RNA and telomere replication.

tetraploid Nucleus, cell or organism that has

four copies of the normal *haploid* chromosome set.

tetrodotoxin (TTX) A potent *neurotoxin* (MW 319) from the Japanese puffer fish. It binds to the *sodium channel*, blocking the passage of *action potentials*. Its activity closely resembles that of *saxitoxin*.

tetrose General term for a monosaccharide with 4 carbon atoms.

TFP See *trifluoperazine*.

thalassemia Hereditary blood disease in which there is abnormality of the globin portion of haemoglobin. Widespread in Mediterranean countries.

thalidomide Sedative drug, which when taken between 3rd and 5th week of pregnancy produces a range of malformations of the foetus, in severe cases complete absence of limbs (amelia), or much reduced limb development (phocomelia). A *teratogen*.

theophylline (1,3-dimethylxanthine) Inhibits cyclic AMP phosphodiesterase, and is often used in conjunction with exogenous dibutyryl cyclic AMP to raise cellular cyclic AMP levels. Other, less potent, methylxanthines are caffeine, theobromine and aminophylline.

thermodynamics The study of energy and energy flow in closed and open systems.

thermolysin Heat-stable metalloproteinase (EC 3.4.24.4.) produced by a strain of *Bacillus stearothermophilus*. Retains 50% of its activity after 1h at 80°C.

thermophile An organism which thrives at high temperature. Typical examples are *cyanobacteria* from hot-springs which have optima of 50–55°C, and will tolerate temperatures of 90°C.

thermotaxis A directed motile response to temperature. The grex of *Dictyostelium discoideum* shows a positive thermotaxis.

thiamine pyrophosphatase (TPP; carboxylase) The coenzyme form of vitamin B_1 (thiamine), deficiency of which causes beri-beri. Forms the prosthetic group of pyruvate dehydrogenase, α-ketoglutarate dehydrogenase and transketolase, in which it is involved in transfer of a 2-carbon unit.

thiamine pyrophosphate (co-carboxylase) A cofactor which has an unusually acidic carbon atom able to form carbon-

carbon bonds. Found in pyruvate dehydrogenase and transketolase.

thick filaments Bipolar *myosin* filaments (12–14nm diameter, 1.6μm long) of striated muscle. Myosin filaments elsewhere are often referred to as "thick filaments", although their length may be considerably less. The myosin heads project from the thick filament in a regular fashion. There is a central "bare" zone without projecting heads, the core being formed from antiparallel arrays of *LMM* regions of the myosin heavy chains. Self-assembly will occur.

thin filaments Filaments 7–9nm diameter attached to the *Z-discs* of striated muscle, have opposite polarity in each half-*sarcomere*. Built of F-actin with associated *tropomyosin* and *troponin*.

thin layer chromatography (TLC) Chromatography using a thin layer of powdered medium on an inert sheet to support the stationary phase. Faster than paper chromatography, gives higher resolution and requires smaller samples.

thiol endopeptidases Proteases which have an active thiol group. Include papain and ficin.

thiol proteinases See *thiol endopeptidases*.

thiol-activated haemolysins (oxygen-labile haemolysins) Cytolytic bacterial exotoxins which act by binding to cholesterol in cell membranes and forming ring-like complexes which act as pores. SH-groups of these toxins must be in the reduced state for the toxin to function. Oxidation (to disulphide bridges) inactivates the toxin. Examples: *tetanolysin*, *streptolysin O*, *cereolysin*.

Thomson's disease Glycogen storage disease in which the missing enzyme is phosphoglucomutase. See *lysosomal diseases*.

thoracic duct The major efferent lymph duct into which lymph from most of the peripheral lymph nodes drains. Recirculating lymphocytes which have left the circulation in the lymph node return to the blood through the thoracic duct.

threonine (Thr; T; MW 119) The hydroxylated polar amino acid. See Table A2.

threose The 4-carbon sugar, CHO.CH.OH-.CH.OH.CH$_2$OH. The two central hydroxyl groups are in *trans* orientation (*cis* in erythrose).

thrombasthenia Condition in which there is defective platelet aggregation, though adherence is normal. See *Glanzmann's thrombasthenia*.

thrombin Protease (34kD) generated in blood clotting which acts on *fibrinogen* to produce *fibrin*. Consists of two chains, A and B, linked by a disulphide bond. B-chain has sequence homology with pancreatic serine proteases: cleaves at Arg-Gly. Thrombin is produced from prothrombin by the action either of the extrinsic system (tissue factor plus phospholipid) or, more importantly, the intrinsic system (contact of blood with a foreign surface or connective tissue). Both extrinsic and intrinsic systems activate plasma Factor X to form Factor Xa which then, in conjunction with phospholipid (tissue derived or *platelet Factor III*) and Factor V, catalyzes the conversion.

thrombocyte Archaic name for a blood *platelet*.

thrombocytopenia Gross deficiency in *platelet* number, consequently a tendency to bleeding.

thrombocytopenic purpura In severe *thrombocytopenia*, bleeding into skin leads to small petechial haemorrhages. In primary thrombocytopenic purpura an autoimmune mechanism seems to cause platelet destruction; secondary thrombocytopenic purpura may be a result of drug-induced Type II *hypersensitivity* in which platelets coated with antibody to the drug (which is acting as a *hapten*) are destroyed in a complement-mediated reaction.

thrombomodulin Specific endothelial cell receptor (100kD: luminal surface only) which forms a 1:1 complex with thrombin. This complex then converts protein C to Ca, which in turn acts on Factors Va and VIIIa. Structurally similar to *coated pit* receptors.

thromboplastin Traditional name for substance in plasma which converts prothrombin to *thrombin*. Now known not to be a single substance. (See *thrombin*).

thrombosis Formation of a solid mass (a *thrombus*) in the lumen of a blood vessel or the heart.

thrombospondin Homotrimeric glycoprotein (450kD) from α granules of platelets, and synthesised by various cell types in culture. Also found in extracellular matrix of cultured endothelial, smooth muscle and fibroblastic cells. May have autocrine growth-regulatory properties: involved in platelet aggregation.

thrombosthenin Obsolete name for platelet contractile protein: now known to be actomyosin (which makes up 15–20% of the total platelet protein).

thromboxanes Arachidonic acid metabolites produced by the action of thromboxane synthetase on prostaglandin cyclic endoperoxides. Thromboxane A_2 (TxA_2) is a potent inducer of platelet aggregation and release, and although unstable, the activation of platelets leads to the further production of TxA_2. Also causes arteriolar constriction. Another endoperoxide product, *prostacyclin*, has the opposite effects.

thrombus Solid mass which forms in a blood vessel, usually as a result of damage to the wall. The first aggregate is of platelets and fibrin, but the thrombus may propagate by clotting in the stagnant downstream blood.

Thy-1 glycoprotein A molecule of 19kD found on the surface of lymphoid and brain tissues of some mammals (not humans). One of the immunoglobulin superfamily with only one domain; phosphoinositol-linked. Used as a marker for T-lymphocytes in those species displaying it. Two allelic forms are known, Thy-1a, Thy-1b, determining Thy-1.1 and Thy-1.2 molecules respectively, formerly known as θ-AKR and θ-C3H. Functions unclear.

thylakoids Membranous cisternae of the chloroplast, found as part of the *grana* and also as single cisternae interconnecting the grana. Contain the photosynthetic pigments, reaction centres and electron-transport chain. Each thylakoid consists of a flattened sac of membrane enclosing a narrow intra-thylakoid space.

thymectomy The excision of the thymus by operation, radiation or chemical means.

thymic aplasia (hypoplasia) A lack of T-lymphocytes, due to failure of the thymus to develop, resulting in very reduced cell-mediated immunity though serum immunoglobulin levels may be normal. See also *DiGeorge syndrome*.

thymidine The nucleoside thymine deoxyriboside; not the riboside which naming of the other nucleosides might lead one to expect.

thymidine kinase (TK) Enzyme of pyrimidine salvage, catalyzing phosphorylation of thymine deoxyriboside to form its 5′ phosphate, the nucleotide thymidylate. Animal cells lacking this enzyme can be selected by lethal synthesis, eg. by resistance to bromodeoxyuridine, and can be used as parentals in somatic hybridisation, since they are unable to grow in *HAT medium*.

thymine (2,6-dihydroxy,5-methylpyrimidine; 5-methyluracil) Pyrimidine base found in DNA (in place of uracil of RNA).

thymine dimer Dimer which can be formed in DNA by covalent linkage between two adjacent (*cis*) thymidine residues, in response to ultraviolet irradiation. Occurrence potentially mutagenic, although repair enzymes exist which can excise thymine dimers. See *xeroderma pigmentosum*.

thymocyte Lymphocyte within the thymus; term usually applied to an immature lymphocyte.

thymoma A tumour of thymus origin.

thymosins Group of peptides (28 amino acids in the case of thymosine α_1) which are believed to regulate the development of T-cells. Possibly thymic hormones.

thymus The lymphoid organ in which T-lymphocytes are formed, composed of stroma (thymic epithelium) and lymphocytes, almost entirely of the T-cell lineage, and situated in mammals just anterior to the heart within the ribcage; in other vertebrates in variably distributed regions of the neck or within the gill chamber in teleost fish. The thymus regresses as the animal matures.

thymus-derived lymphocyte See *T-lymphocyte*.

thyroglobulin The 650kD protein of the thyroid gland which binds thyroxine.

thyroid hormones Thyroxine and tri-iodothyronine are hormones secreted by the thyroid gland in vertebrates. These iodinated aromatic amino acid compounds influence growth and metabolism and, in amphibia, metamorphosis. The hormone *calcitonin* which has hypocalcaemic effects is also of thyroid origin but is not usually classed with thyroxine and tri-iodothyronine as a thyroid hormone.

thyroid-stimulating antibodies Long-acting thyroid stimulator is an autoantibody found in many cases of primary thyrotoxicosis which causes hyperplasia of the thyroid by undetermined mechanisms. Human thyroid-stimulating immunoglobulins are different antibodies found in nearly all cases of primary thyrotoxicosis and act in some cases by binding to the thyrotropin (TSH) receptor site, causing increased synthesis of thyroglobulin.

thyroid-stimulating hormone (TSH; thyrotropin) Polypeptide hormone (28kD), secreted by the anterior pituitary gland, which activates cyclic AMP production in thyroid cells leading to production and release of the *thyroid hormones*.

thyroiditis Disease of the thyroid, especially Hashimoto's disease, in which autoimmune destruction of the thyroid takes place.

thyroliberin See *thyrotropic-releasing hormone*.

thyrotropic-releasing hormone (protirelin; TRH; thyroliberin) Tripeptide (pyroGlu-His-Pro-NH$_2$) which releases thyrotropin from the anterior pituitary by stimulating adenyl cyclase. May also have neurotransmitter and paracrine functions.

thyrotropin See *thyroid-stimulating hormone*.

thyroxine (T4; tetra-iodothyronine) *Thyroid hormone*.

Ti plasmid Plasmid of **Agrobacterium tumefaciens**, transferred to higher plant cells in crown gall disease, carrying the T-DNA which is incorporated into the plant cell genome. Used as a vector to introduce foreign DNA into plant cells.

tight junction See *zonula occludens*.

time-lapse Technique applied to speed up the action in a film or videotape sequence. In filming by taking frames every few seconds, and projecting at conventional speed (16 or 24 frames per second), the movements of cells can be greatly speeded up, and then become conspicuous. With videotape, the recording is made at slow tape-speed and replayed at full speed. The opposite of slow-motion.

tissue culture Originally the maintenance and growth of pieces of explanted tissue (plant or animal) in culture away from the source organism. Now usually refers to the (much more frequently used) technique of cell culture, using cells dispersed from tissues, or distant descendants of such cells.

tissue culture plastic Polystyrene which has been rendered wettable by oxidation, a treatment which increases its adhesiveness for cells from animal tissues, and without which *anchorage dependent* cells will not grow. Commercially achieved by treatment known as glow discharge.

tissue-typing The process of determining the allelic types of the antigens of the *major histocompatibility complex* (MHC) which determine whether a tissue graft will be accepted or rejected. At present carried out either by use of polyclonal or monoclonal antibodies against MHC antigens, or less usually by tests of MHC-restricted cell function or skin grafting (the latter not in humans).

TL antigens The mouse antigens coded for by the *TLa complex*; in normal animals only found on intrathymic lymphocytes, but also seen on leukaemic cells (hence, thymus leukaemia antigen) in certain forms of the disease in mice. The molecules have structures similar in some ways to Class I MHC products but are disulphide-bonded tetramers of two 45kD chains and two 12kD chains of β_2-*microglobulin* type.

TLa complex Genes coding for and controlling *TL antigens*; the complex is situated close to the H-2 complex on mouse chromosome 17 and resembles H-2 in several ways.

TMB-8 Inhibitor of the release of calcium from intracellular stores.

TMV See *Tobacco Mosaic Virus*.

TNF See *tumour necrosis factor*.

tobacco mosaic virus Plant RNA virus, the first to be isolated. Consists of a single helical strand of RNA (6500 nucleotides) on which is based a coat of 2130 identical capsomeres which, in the absence of the RNA, will self-assemble into a cylinder similar to the normal virus but of indeterminate length. Causes mottling of the leaves of the tobacco plant.

α-tocopherol (vitamin E) Protects unsaturated membrane lipids from oxidation, and may prevent free-radical damage. See Table V1.

Togaviridae Class IV viruses with a single positive-strand RNA genome. Probable icosahedral capsid, enveloped by a membrane formed from the host cell plasma membrane; the budded membrane contains host lipids and viral ("spike") glycoproteins. The group can be divided into two main groups: alphaviruses, which include *Semliki Forest virus* and *Sindbis virus*, and *flaviviruses*, which include yellow fever virus and rubella (German measles) virus. Many are transmitted by insects and were previously classified as *arboviruses*.

tolerance The development of specific non-reactivity to an antigen. See *immunological tolerance*.

toluidine blue (CI Basic Blue 17) A thiazin dye related to methylene blue and Azure A in structure; often used for staining thick resin sections and typically exhibits metachromasia.

tonic See *adaptation*.

tonofilaments Cytoplasmic filaments (10nm diameter: *intermediate filaments*) inserted into *desmosomes*.

tonoplast Membrane which surrounds the vacuole in a plant cell.

tophus Mass of urate crystals surrounded by a chronic inflammatory reaction: characteristic of gout.

topoinhibition Term used to describe the inhibition of cell proliferation as the cells become closely packed on a culture dish: generally superseded by the term *density dependent inhibition*.

topoisomerases Enzymes which change the degree of *supercoiling* in DNA by cutting one or both strands. Type I topoisomerases cut only one strand of DNA; type I topoisomerase of *E. coli* (omega protein) relaxes negatively supercoiled DNA and does not act on positively supercoiled DNA. Type II topoisomerases cut both strands of DNA; type II topoisomerase of *E. coli* (DNA gyrase) increases the degree of negative supercoiling in DNA and requires ATP. It is inhibited by several antibiotics, including nalidixic acid and ovobiocin.

Torres body Intranuclear inclusion body in liver cells infected with yellow fever virus (*Togaviridae*).

torus Structure found at the centre of a bordered *pit*, especially in conifers, forming a thickened region of the pit membrane. When subjected to a pressure gradient, it seals the pit by pressing against the pit border.

totipotent Capable of giving rise to all types of differentiated cell found in that organism. A single totipotent cell could, by division, reproduce the whole organism.

toxigenicity The ability of a pathogenic organism to produce injurious substances which damage the host.

toxin A naturally-produced poisonous substance which will damage or kill other cells. Bacterial toxins are frequently the major cause of the pathogenicity of the organism in question. See *endotoxins* and *exotoxins*.

toxoid Non-toxic derivative of a bacterial exotoxin produced by formaldehyde or other chemical treatment: useful as a vaccine because it retains most antigenic properties of the toxin.

tracheid Water-conducting cell forming part of the plant *xylem*. Contains thick, lignified secondary cell walls, with no protoplast at maturity. Interconnects with neighbouring tracheids through pits in the end-walls.

transacylase An enzyme which transfers an acyl group; eg. transacetylase which transfers an acetyl group from acetyl-lipoamide to coenzyme A.

transaldolase Together with transketolase, links the *pentose phosphate pathway* with glycolysis by converting pentoses to hexoses.

transaminases Enzymes which convert amino acids to keto acids in a reversible process using pyridoxal phosphate as cofactor; eg. aspartate amino transferase catalyses the reaction: aspartate + α-keto-glutarate = oxaloacetate + glutamate.

transcriptase See *reverse transcriptase*.

transcription Synthesis of RNA by RNA polymerases using a DNA template.

transcription factor Protein required for recognition by RNA polymerases of specific stimulatory sequences in eukaryotic genes. Several are known which activate transcription by RNA polymerase II when bound to *upstream* promoters. Transcription of the 5S RNA gene in *Xenopus* by RNA polymerase III is dependent on a 40kD protein TFIIIA which binds to a regulatory site in the centre of the gene, and was the first protein found to exhibit the metal-binding domains known as zinc fingers.

transcription unit A region of DNA that is transcribed to produce a single primary RNA transcript, ie. a newly synthesised RNA molecule that has not been processed. Transcription units can be mapped by kinetic studies of RNA synthesis, and in some instances directly visualised by electron microscopy.

transcriptional control Control of gene expression by controlling the number of RNA transcripts of a region of DNA. A major regulatory mechanism for differential control of protein synthesis in both prokaryotic and eukaryotic cells.

transcytotic vesicle Membrane-bounded vesicle which shuttles fluid from one side of the endothelium to the other. There is some controversy as to whether the vesicles form pores (ie. continuous channels from one side to the other) at some stages.

transcytosis Process of transport of material across an epithelium by uptake on one face into a coated vesicle, which may then be sorted through the *trans-Golgi network* and transported to the opposite face in another set of vesicles.

transdetermination Change in determined state observed in experiments on *Drosophila imaginal discs*. These can be cultured for many generations in the abdomen of an adult, where their cells proliferate but do not differentiate. If transplanted into a larva, the cells differentiate after pupation according to the disc from which they were derived; they maintain their determination. Occasionally the disc will differentiate into a structure appropriate to another disc. This is termed transdetermination. It is a rare event, involves a population of cells and certain changes are more common than others; eg. leg to wing is more frequent than wing to leg.

transdifferentiation Change of a cell or tissue from one differentiated state to another. Rare, and has mainly been observed with cultured cells. In newts the pigmented cells of the iris transdifferentiate to form lens cells if the existing lens is removed.

transducin A *GTP-binding protein* found in the disc membrane of *retinal rods and cones*; part of the cascade involved in transduction of light to a nervous impulse. A complex of three subunits; α (39kD), β (36kD) and γ (8kD). Photoexcited rhodopsin interacts with transducin and promotes the exchange of *GTP* for GDP on the α subunit. The GTP-α subunit dissociates from the complex and activates a cyclic GMP-*phosphodiesterase* by removing an inhibitory subunit. The α subunit of transducin can be ADP-ribosylated by cholera toxin and pertussis toxin.

transduction 1. The transfer of a gene from one bacterium to another by a *bacteriophage*. In generalised transduction any gene may be transferred as a result of accidental incorporation during phage packaging. In specialised transduction only specific genes can be transferred, as a result of improper recombination out of the host chromosome of the *prophage* of a lysogenic phage. Transduction is an infrequent event but transducing phages have proved useful in the genetic analysis of bacteria. 2. The conversion of a signal from one form to another. For example, various types of sensory cells convert or transduce light, pressure, chemicals, etc. into *nerve impulses* and the binding of many hormones to receptors at the cell surface is transduced into an increase in *cyclic AMP* within the cell.

transfection The introduction of DNA into a recipient eukaryote cell and its subsequent integration into the recipient cell's chromosomal DNA. Usually accomplished using DNA precipitated with calcium ions. Only about 1% of cultured cells are normally transfected. Transfection is analogous to bacterial transformation but in eukaryotes *transformation* is used to describe the changes in cultured cells caused by *tumour viruses*.

transfer cell Parenchyma cell specialised for transfer of water-soluble material to or from a neighbouring cell, usually a phloem sieve tube or a xylem tracheid. Elaborate wall ingrowths greatly increase the area of plasma membrane at the cell face across which transfer occurs.

transfer factor A dialysable factor obtained from sensitised T-cells by freezing and thawing, which may possibly enhance immune responses. The transfer of specific immunity from one animal to another has been claimed.

transfer RNA (s-RNA; 4S RNA; tRNA) The low-molecular weight RNAs which specifically bind amino acids by amino-acylation and which possess a special nucleotide triplet, the anticodon, sometimes containing the base inosine, by which they recognise codons on mRNA. By this recognition the appropriate tRNAs are brought into alignment in turn in the ribosome during protein synthesis (translation), there being at least one species of tRNA for each amino acid. In practice most cells possess about 30 types of tRNA. The amino acids are bound at the 3′ terminus which is always 3′-ACC. The anticodon is around 34–38 nucleotides from the 5′ end and the total length of the various tRNAs is 70–80 bases.

transferase A suffix to the name of an enzyme indicating that it transfers a specific grouping from one molecule to another; eg. acyl transferases transfer acyl groups.

transferrin The iron storage protein (80kD) found in mammalian serum; a β-globulin. Binds ferric iron with a K_{ass} of 21.3 at pH 7.4, 18.1 at pH 6.6. An important constituent of growth media. Transferrin receptors on the cell surface bind transferrin as part of the transport route of iron into cells.

transformasome Membranous extension responsible for binding and uptake of DNA; found on the surface of transformation-competent *Haemophilus influenzae* bacteria.

transformation Any alteration in the properties of a cell which is stably inherited by its progeny. Classical example was the transformation of *Diplococcus pneumoniae* to virulence by DNA, achieved in 1944 by Avery, MacLeod and McCarty. Currently usually refers to malignant transformation, but is used in other senses also, such as blast transformation of lymphocytes, which can be distinguished only by context. Malignant transformation is a change in animal cells in culture which usually greatly increases their ability to cause tumours when injected into animals. (It is assumed that parallel changes occur during carcinogenesis *in vivo*). Transformation can be recognised by changes in growth characteristics, particularly in requirements for macromolecular growth factors, and often also by changes in morphology.

transformed cell See *transformation*.

transforming genes Genes, originally of tumour viruses, responsible for their ability to transform cells. The term now serves as an operational definition of *oncogenes*.

transforming growth factors (TGF) Proteins secreted by transformed cells that can stimulate growth of normal cells. Unfortunate misnomer, since TGFs induce aspects of transformed phenotype, such as growth in semi-solid agar, but do not actually transform. TGF-α, a 50 amino acid polypeptide originally isolated from viral-transformed rodent cells, has sequence similarity to *epidermal growth factor* (EGF) and binds to the EGF receptor. Stimulates growth of microvascular endothelial cells, ie. is angiogenic. TGF-β, homodimer of two 112 amino acid chains secreted by many different cell types, stimulates wound healing but *in vitro* is also a growth inhibitor for certain cell types.

transforming virus Virus capable of inducing malignant transformation of animal cells in culture. Among the *Oncovirinae*, nondefective viruses which lack oncogenes can induce tumours such as leukaemias in animals, but cannot transform *in vitro*. On acquisition of oncogenes they become (acute) transforming viruses.

transglycosylation Transfer of a glycosidically bound sugar to another hydroxyl group.

trans-Golgi network A term, which has replaced *GERL*, for the complex of membranous tubules and vesicles near the *trans*-face of the Golgi; thought to be a major intersection for intracellular traffic of vesicles and the region where proteins to be secreted are segregated from lysosomal enzymes.

transhydrogenases Enzymes in mitochondria from liver or heart, for example, which catalyze (reversibly) the transfer of a hydrogen atom from NADH to $NADP^+$.

transin Protease secreted by carcinoma cells: carboxy-terminal domain has distant homology to haemopexin (haem binding protein), and the N-terminal domain has the proteolytic activity. May be involved in digestion of extracellular matrix.

transition probability model A model to account for the apparently random variation in *cell cycle* time between individual animal tissue cells in culture, which postulates that transition from G1 to S phase is probabilistic. Contrasts with hypotheses that require

the accumulation of critical levels of particular proteins.

transition temperature The temperature at which there is a transition in the organisation of, for example, the phospholipids of a membrane where the transition temperature marks the shift from fluid to more crystalline. Usually determined by using an *Arrhenius plot* of activity against the reciprocal of absolute temperature, the transition temperature being that temperature at which there is an abrupt change in the slope of the plot. In membranes such phase-transitions tend to be inhibited by the presence of cholesterol.

transitional elements Regions at the boundary of the *rough endoplasmic reticulum* (RER) and the *Golgi apparatus*. *Transport vesicles* are responsible for the transfer of secretory proteins from this part of the RER to the Golgi system.

transitional endoplasmic reticulum See *transitional elements*.

transketolase See *transaldolase*.

translation The process that occurs at the ribosome whereby the information in mRNA is used to specify the sequence of amino acids in a polypeptide chain.

translational control Control of protein synthesis by regulation of the translation step, for example by selective usage of pre-formed mRNA or instability of the mRNA.

translocase (elongation factor G; EF-G) The enzyme that causes peptidyl-tRNA to move from the A site to the P site in the *ribosome* and the mRNA to move so that the next codon is in position for usage.

translocation Rearrangement of a chromosome in which a segment is moved from one location to another, either within the same chromosome or to another chromosome. This is sometimes reciprocal, when one fragment is exchanged for another.

transmembrane protein A protein subunit in which the polypeptide chain is exposed on both sides of the membrane. The term does not apply when different subunits of a protein complex are exposed at opposite surfaces. Most integral membrane proteins are also transmembrane proteins.

transmembrane transducer A system which transmits a chemical or electrical signal across a membrane. Usually involves a transmembrane receptor protein which is thought to undergo a conformation change that is expressed on the inner surface of the membrane. Many such transducing species are dimeric and the conformation change may involve interaction between the two components.

transmissible mink encephalopathy "Unconventional" type of slow virus infection, similar to *kuru*, *scrapie*, and *Creutzfeld–Jacob disease*.

transmission electron microscopy (TEM) Those forms of electron microscopy in which electrons are transmitted through the object to be imaged, suffering energy loss by diffraction and to a small extent by absorption.

transpiration Loss of water-vapour from land-plants into the atmosphere, causing movement of water through the plant from the soil to the atmosphere *via* roots, shoot and leaves. Occurs mainly through the *stomata*.

transplantation antigen Any antigen that is antigenically active in graft rejection. In practice the *major histocompatibility complex* and the H-Y antigens, and to a lesser extent minor histocompatibility antigens.

transplantation reaction The set of cellular phenomena observed after an allogeneic (mismatched) graft is made to an organism, which leads to destruction, detachment or isolation of the graft. In mammals this includes the invasion and destruction of the graft by cytotoxic lymphocytes or activated macrophages, inhibition of *angiogenesis* and other processes.

transport diseases Single-gene defect diseases in which there is an inability to transport particular small molecules across membranes. Examples are aminoacidurias such as cystinuria, *Fanconi syndrome*, *Hartnup disease*, *iminoglycinuria*.

transport protein A class of transmembrane protein which allows substances to cross plasma membranes much faster than would be possible by diffusion alone. A major class of transport proteins expend energy to move substances (*active transport*); these are transport ATPases. See *facilitated diffusion*, *symport*, *antiport*.

transport vesicle Vesicle which tranfers material from the rough endoplasmic reticulum (RER) to the receiving face of the Golgi. How it is targeted remains unclear at present.

transportase See *transport protein*.

transporter See *transport protein*.

transposable element *Transposon*.

transposon Small, mobile DNA sequence which can replicate and insert copies at random sites within chromosomes. Transposons have nearly identical sequences at each end, oppositely oriented (inverted) repeats, and code for the enzyme, transposase, which catalyzes their insertion. Bacteria have two types of transposon; simple transposons which have only the genes needed for insertion, and complex transposons which contain genes in addition to those needed for insertion. Eukaryotes contain two classes of mobile genetic elements; the first are like bacterial transposons in that DNA sequences move directly. The second class (retrotransposons) move by producing RNA which is transcribed, by reverse transcriptase, into DNA which is then inserted at a new site.

transudate Plasma-derived fluid which accumulates in tissue and causes *oedema*. A result of increased venous and capillary pressure, rather than altered vascular permeability (which leads to exudate formation).

transverse tubule See *T tubule*.

transversions *Point mutation* in which a purine is substituted by a pyrimidine or *vice versa*.

treadmilling Name given to the process in microtubules in which there is continual addition of subunits at one end, and disassembly at the other, so that the tubule stays of constant length, but individual subunits move along. Could in principle be used as a transport mechanism, although this is not currently favoured as a possibility. Has also been suggested for microfilaments.

trehalose A disaccharide sugar (MW 342) found widely in invertebrates, bacteria, algae, plants and fungi, formed by the dimerisation of glucose. Yields glucose on acid hydrolysis.

trephone Substance supposedly released at a wound which stimulate mitosis: the opposite of *chalone*.

Treponema Genus of bacteria of the spirochaete family (Spirochaetaceae). *T. pallidum* causes syphilis. Cells are corkscrew-like, 6–15μm long, 0.1–0.2μm wide, motile, anaerobic and with a *peptidoglycan* cell wall and a capsule of *glycosaminoglycans* similar to *hyaluronic acid* and *chondroitin sulphate* in composition. Membrane has *cardiolipin*.

triacyl glycerols See *triglycerides*.

triad (triad junction) The junction between the T tubules and the sarcoplasmic reticulum in striated muscle.

tricarboxylic acid cycle (TCA cycle; citric acid cycle; Krebs cycle) The central feature of oxidative metabolism. Cyclic reactions whereby acetyl CoA is oxidised to carbon dioxide providing reducing equivalents (NADH or $FADH_2$) to power the *electron transport chain*. Also provides intermediates for biosynthetic processes.

trichocyst Membrane-bounded organelle lying below the pellicle of many ciliates and some flagellates. Fusion of the trichocyst membrane with the plasma membrane results in discharge and three-dimensional reorganisation of the trichocyst in milliseconds. Discharge occurs at a predictable site which can therefore be examined for membrane specialisation.

trifluoperazine (TFP; trifluperazine; Stellazine) Antipsychotic drug which inhibits *calmodulin* at levels just below those at which it kills cells.

trigger protein See *U protein*.

triglycerides Storage fats of animal adipose tissue where they are largely glycerol esters of saturated fatty acids. In plants they tend to be esters of unsaturated fatty acids (vegetable oils). Present as a minor component of cell membranes. Important energy supply in heart muscle.

triiodobenzoic acid (TIBA) An inhibitor of basipetal *auxin* transport in plants.

trimethoprim A drug which inhibits the reduction of dihydrofolate (DHF) to tetrahydrofolate (a later step than that inhibited by *sulphonamides*). Selective for certain bacterial DHF reductases, therefore usually given together with sulphonamides to achieve a synergistic antibacterial action.

triple response The vascular changes in the skin in response to mild mechanical injury, an outward-spreading zone of reddening (flare) followed rapidly by a weal (swelling) at the site of injury. Redness, heat and swelling, three of the "cardinal signs" of inflammation, are present.

triploid Having three times the haploid number of chromosomes.

triskelion A three-legged structure assumed by *clathrin* isolated from coated vesicles. A trimer of clathrin with three light chains. It is probably the physiological subunit of clathrin coats in coated vesicles.

tritium Long-lived radioactive isotope of hydrogen (^3H; half-life 12.26 years). Weak beta-emitter, very suitable for autoradiography, and relatively easy to incorporate into complex molecules.

Triton X-100 Non-ionic detergent used for cell lysis and in isolating membrane proteins: the detergent replaces the phospholipids which normally surround such a protein. Other detergents of the Triton group are occasionally used so the full name should be quoted.

tRNA See *transfer RNA*.

trophectoderm The extra-embryonic part of the ectoderm of mammalian embryos at the blastocyst stage before the mesoderm becomes associated with the ectoderm.

trophoblast Extra-embryonic layer of epithelium which develops around the mammalian blastocyst, and attaches the embryo to the uterus wall. Forms the outer layer of the chorion, and together with maternal tissue will form the placenta.

trophozoite The feeding stage of a protozoan (as distinct from reproductive or encysted stages).

tropocollagen Subunit from which *collagen* fibrils self-assemble: generated from *procollagen* by proteolytic cleavage of the extension peptides.

tropomyosin Protein (66kD) associated with actin filaments both in cytoplasm and (in association with *troponin*) in the thin filament of striated muscle. Composed of two elongated α-helical chains (each about 33kD), 40nm long, 2nm diameter. Each chain has six or seven similar domains and interacts with as many G-actin molecules as there are domains. Not only does the binding of tropomyosin stabilise the F-actin, but the association with troponin in striated muscle is important in control by calcium ions.

troponin Complex of three proteins, troponins C, I and T, associated with *tropomyosin* and actin on the thin filament of striated muscle, upon which it confers calcium sensitivity. There is one troponin complex per tropomyosin. Troponin C (18kD) binds calcium ions reversibly, and has marked homology with *calmodulin*. It is the least variable of the subunits and binds TnI and TnC, but not actin. Troponin I (23kD) binds to actin and at 1:1 stoichiometry can inhibit the actin-myosin interaction on its own. Troponin T (37kD) binds strongly to tropomyosin.

Trypan blue A blue azo-dye (MW 960) often used to test for viability of cells; live cells exclude the dye, whereas dead cells stain blue.

Trypanosoma Genus of Protozoa which causes serious infections in humans and domestic animals. African trypanosomes, of the *brucei* group, are carried by tsetse flies and, when they enter the bloodstream of the mammalian host, go through a complex series of stages. Perhaps the most interesting feature is that there are recurrent bouts of parasitaemia as the parasite alters its surface antigens to evade the immune response of the host (see *antigenic variation*). The repertoire of variable antigens is considerable. The S. American trypanosome, *T. cruzi*, is carried by reduviid bugs and cause a chronic and incurable disease. Other interesting features of trypanosomes are the kinetoplast DNA and glycosomes (organelles containing enzymes of the glycolytic chain).

trypanosomiasis Disease caused by **Trypanosoma**.

trypsin (EC 3.4.21.4; 23kD) Serine protease from the pancreas of vertebrates. Cleaves peptide bonds involving the amino groups of lysine or arginine.

tryptophan (Trp; W; MW 204) One of the 20 amino acids found in proteins. Essential dietary component in humans. Precursor of nicotinamide. See Table A2.

TSH See *thyroid stimulating hormone*.

TSH releasing factor (*thyroid stimulating hormone* releasing factor) A tripeptide produced by the hypothalamus which stimulates the anterior pituitary to release TSH.

TTX See *tetrodotoxin*.

tubercle Chronic inflammatory focus, a *granuloma*, caused by **Mycobacterium tuberculosis**.

tuberculin skin test See *Mantoux test*, *Heaf test*.

tubulin Abundant cytoplasmic protein (55kD), found in two forms, α and β. A tubulin *heterodimer* (one α, one β) constitutes the *protomer* for microtubule assembly. Multiple copies of tubulin genes are present (and are expressed) in most eukaryotic cells studied so far. The different tubulin isoforms seem, however, to be functionally equivalent.

tularemia Disease of rodents and rabbits caused by *Pasteurella tularense*. Can infect humans (either transmitted by the deer-fly, or by direct contact with the bacterium).

tumour (USA tumor) Strictly, any abnormal swelling, most often applied to a mass of neoplastic cells.

tumour angiogenesis factor (TAF) Substance(s) released from a tumour which promotes vascularisation of the mass of neoplastic cells. Once a tumour becomes vascularised, it will grow more rapidly and is more likely to *metastasise*. TAF is almost certainly more than one substance. See *angiogenin*.

tumour cell Cell derived from a tumour in an animal. Refers to a tumour-causing malignant cell, and not an adventitious normal cell. Loosely, a transformed cell able to give rise to tumours.

tumour necrosis factor (TNF) TNFα is the same as *cachectin*, a tumour-inhibiting factor originally found in the blood of animals exposed to bacterial *lipopolysaccaride* or *Bacillus Calmette-Guerin* (BCG). Preferentially kills tumour cells *in vivo* and *in vitro*, causes necrosis of certain transplanted tumours in mice and inhibits experimental metastases. Human TNFα is a protein of 157 amino acids.

tumour promoter Agent which in classical studies of carcinogenesis in rodent skin was able to increase the sensitivity of tumour formation by a previously applied primary carcinogen, but was unable to induce tumours when used alone. Important example was croton oil, active ingredients of which are now believed to be phorbol esters.

Thought to act as analogues of diacylglycerols, and may activate protein kinase C. Strictly speaking, not the same as co-carcinogen, which is defined as being active when administered at the same time. Tumour promoters generally are carcinogens when tested more stringently.

tumour specific antigen (tumour specific transplantation antigen; TSTA) Antigen on tumour cells detected by cell-mediated immunity. For virus-transformed cells TSTA (unlike T-antigen) is found to differ for different individual tumours induced by the same virus. May consist of fragments of T-antigens exposed at the cell surface.

tumour virus Virus capable of inducing tumours.

tumourigenic Capable of causing tumours. Can refer either to a carcinogenic substance or agent such as radiation which affects cells, or to transformed cells themselves.

tunicamycin Nucleoside antibiotic from *Streptomyces lysosuperificus* which acts in eukaryotic cells to inhibit N-glycosylation. Tunicamycin inhibits the first step in synthesis of the dolichol-linked oligosaccharide, by preventing the addition of N-acetyl glucosamine to dolichol phosphate.

turgor The pressure within a cell, especially a plant cell, derived from osmotic pressure differences between the inside and outside of the cell, giving rise to mechanical rigidity of the cell. Turgor drives cell expansion and certain movements such as the closing or opening of stomata.

Turner's syndrome Genetic defect in humans in which there is only one X chromosome (affected individuals are therefore phenotypically female), probably as a result of meiotic non-disjunction.

turnover number Equivalent to V_{max}, being the number of substrate molecules converted to product by one molecule of enzyme in unit time, when the substrate is saturating.

TW-240/260 Protein (240/260kD) found in the *terminal web* of intestinal epithelial cells. Probably an isoform of *spectrin* and *fodrin*.

twitch-muscle Striated muscle innervated by a single motoneuron and having an electrically excitable membrane which exhibits an all–or–none response (cf *tonic* muscle): in mammals almost all skeletal muscles are twitch muscles. Physiologists often divide muscles into fast- and slow-twitch types, the fast-twitch muscles being associated with fast motor units.

two-dimensional gel electrophoresis A high resolution separation technique in which protein samples are separated by isoelectric focusing in one dimension and then laid on an SDS gel for size-determined separation in the second dimension. Can resolve hundreds of components on a single gel.

TxA$_2$ See *thromboxanes*.

tylose A parenchyma-cell outgrowth which wholly or partly blocks a *xylem* vessel. It grows out from an axial or ray parenchyma cell through a pit in the vessel wall.

tyrosinase A copper-containing protein (a monoxygenase) which catalyzes the oxidation of tyrosine, and sets in train spontaneous reactions which yield melanin, the black pigment of skin, hair and eyes. The first intermediate is 3,4-dihydroxyphenylalanine (DOPA). Lack of tyrosinase activity is responsible for albinism.

tyrosine (Tyr; Y; MW 181) One of the twenty directly coded amino acids in proteins. Nonessential in humans since can be synthesised from phenylalanine. See Table A2.

tyrosine phosphorylation Enzymically catalyzed phosphorylation of specific tyrosine residues in proteins. This activity is possessed by receptors for some *growth factors*, such as EGF and PDGF, by the insulin receptor, and by the protein products of several oncogenes, some of which are truncated growth factor receptors. Phosphotyrosine is present even in transformed cells as a very small proportion of total protein phosphate, but is believed to be of crucial importance in growth control, by a mechanism that is not understood at present.

U

U protein Hypothetical protein thought to regulate the transition of cells from G0 to G1 phase of the *cell cycle*, and thus inevitably into S phase. The idea would be that the concentration of this unstable (U) protein would have to exceed a threshold level for triggering progression through the cycle, and that this would only happen if the cell had adequate access to growth factors or to nutrients. Also known as trigger protein.

ubiquinone (coenzyme Q) Small molecule with a hydrocarbon chain (usually of several isoprene units) which serves as an important electron carrier in the respiratory chain. The acquisition of an electron and a proton by ubiquinone produces ubisemiquinone (a free radical); a second proton and electron convert this to dihydroubiquinone. Plastoquinone, which is almost identical to ubiquinone, is the plant form.

ubiquitin A protein (8.5kD) found in all eukaryote cells. Can be linked to the lysine side chains of proteins by formation of an amide bond to its C terminal glycine in an ATP-requiring process. The protein/ubiquitin complex is often subject to rapid proteolysis. Ubiquitin also has a role in the heat-shock response.

ubisemiquinone See *ubiquinone*.

UDP-galactose (uridine diphosphate-galactose) Sugar nucleotide, active form of galactose for galactosyl transfer reactions.

UDP-glucose (uridine diphosphate-glucose) Sugar nucleotide, active form of glucose for glucosyl transfer reactions.

ulcer Inflamed area where the epithelium and underlying tissue is eroded.

ulcerative colitis Inflammation of the colon and rectum: cause unclear, although there are often antibodies to colonic epithelium and *E. coli* strain 0119 B14.

ultracentrifugation Centrifugation at very high *g*-forces: used to separate molecules eg. mitochondrial from nuclear DNA on a *caesium chloride* gradient.

ultrafiltration Filtration under pressure. In the kidney, an ultrafiltrate is formed from plasma because the blood is at higher pressure than the lumen of the glomerulus. Also used experimentally to fractionate and concentrate solutions in the laboratory using selectively permeable artificial membranes.

ultrastructure General term to describe the level of organisation that is below the level of resolution of the light microscope. In practice, a shorthand term for "structure observed using the electron microscope", although other techniques could give information about structure in the submicrometre range.

ultraviolet (UV) Continuous spectrum beyond the violet end of the visible spectrum (wavelength less than 400nm), and above the X-ray wavelengths (greater than 5 nm). Glass absorbs UV, so optical systems at these wavelengths have to be of quartz. Nucleic acids absorb UV most strongly at around 260nm, and this is the wavelength most likely to cause mutational damage (by the formation of thymine dimers). It is the UV component of sunlight that causes actinic *keratoses* to form in skin, but which is also required for vitamin D synthesis.

uncouplers Agents which uncouple electron transport from oxidative phosphorylation. Ionophores can do this by discharging the ion gradient across the mitochondrial membrane which is generated by electron transport. In general the term applies to any agent capable of dissociating two linked processes.

underlapping Possible outcome of collision between two cells in culture, particularly head–side collision: one cell crawls underneath the other, retaining contact with the substratum, and obtaining traction from contact with the rigid substratum (unlike *overlapping*, where traction must be gained on the dorsal surface of the other cell).

unequal crossing over *Crossing over* between *homologous chromosomes* that are not precisely paired, resulting in nonreciprocal exchange of material and chromosomes of unequal length. Favoured in regions containing tandemly repeated sequences.

unineme theory Theory which proposes that each chromosome (before S phase) consists of a single strand of DNA. Now generally accepted and, being non-controversial, the term is falling into disuse.

uniport A class of transmembrane *transport protein* which conveys a single species across the plasma membrane.

unit membrane The three-ply membrane structure found in electron micrographs of all cells. The unit membrane concept carries with it the presumption that all biological membranes have basically the same structure.

upstream Refers to nucleotide sequences which precede the codons specifying the mRNA or which precede (are on the 5' side of) the protein-coding sequence. Also used of the early events in any process which involves sequential reactions.

uracil (2,6-dihydroxypyrimidine) The pyrimidine base from which uridine is derived.

uranyl acetate Uranium salt which is very electron-dense, and which is used as a stain in electron microscopy, usually for staining nucleic acid-containing structures in sections.

urea The final nitrogenous excretion product of many organisms.

uric acid The final product of nitrogenous excretion in animals that require to conserve water, such as terrestrial insects, or have limited storage space, such as birds and their eggs. Uric acid has very low water-solubility, and crystals may be deposited in, for example, butterflies' wings to impart iridescence. See also *tophus*.

uridine The ribonucleoside formed by the combination of ribose and uracil.

urogastrone A peptide isolated from human urine which inhibits gastric acid secretion. Now known to be identical to epithelial *growth factor*.

uroid Tail region of a moving amoeba.

urokinase (uPA) *Serine protease* from kidney which is a *plasminogen activator*.

uroporphyrinogen I synthetase An enzyme of haem biosynthesis, which is defective in the inherited (autosomal dominant) disease, acute intermittent porphyria. UP I is isomerised to UP III by UP III synthetase, defective in the autosomal recessive disease, congenital erythropoietic porphyria.

uvomorulin Glycoprotein (120kD) originally defined as the antigen responsible for eliciting antibodies capable of blocking compaction in early mouse embryos (at the morula stage), and inhibiting calcium-dependent aggregation of mouse teratocarcinoma cells. May be the mouse equivalent of LCAM, the chick *cell-adhesion molecule*.

V

V_H and V_L genes/domains V_H and V_L genes define in part the sequences of the variable heavy and light regions of immunoglobulin molecules. V_H and V_L domains are the regions of amino acid sequence so defined. J genes and, in the case of the heavy chain, a D gene (D = diversity) also define these regions. Gene rearrangement plays a role in determining the sequences in which the genes are joined as the DNA of the immunoglobulin-producing cell matures.

V8 protease Protease from *Staphylococcus aureus* strain V8. Cleaves peptide bonds on the carboxyl side of aspartic and glutamic acid residues. Used experimentally for selective cleavage of proteins for amino acid sequence determination or peptide mapping.

v-onc Generic term for retroviral transforming genes (oncogenes), used to encompass eg. *v-src*, *v-ras*, *v-myc* etc.

V-region Those regions of both the heavy and the light chains of immunoglobulins where there is considerable sequence variability from one immunoglobulin to another of the same class, in contrast to Constant sequence regions. The V regions are associated with antigen-binding areas. They contain hypervariable regions of particularly high sequence diversity.

vaccination The process of inducing immunity to a pathogenic organism by injecting either an antigenically related but non-pathogenic strain (attenuated strain) of the organism or related non-pathogenic species, or killed or chemically modified organism of low pathogenicity. In all cases the aim is to expose the human or animal being vaccinated to an antigenic stimulus which leads to immune protection against disease, without inducing appreciable pathogenesis from the injection.

vaccine An antigen preparation which when injected will elicit the expansion of one or more clones of responding lymphocytes so that immune protection is provided against a disease.

vaccinia The virus of cowpox used in vaccination against smallpox.

vacuole Membrane-bounded vesicle of eukaryotic cells. Secretory, endocytotic and phagocytotic vesicles can be termed vacuoles. Botanists tend to confine the term to the large vesicles found in plant cells which provide both storage and space-filling functions.

valine (Val; V; MW 117) An essential amino acid. See Table A2.

valinomycin A potassium *ionophore* antibiotic, produced by *Streptomyces fulvissimus*. Composed of three molecules (L-valine, D-α-hydroxyisovaleric acid, L-lactic acid) linked alternately to form a 36-membered ring, which folds to make a cage shaped like a tennis-ball seam. This wraps specifically around potassium ions, presenting them with a hydrophilic interior, and presents a lipid bilayer with a hydrophobic exterior. Potassium is thus free to diffuse through the lipid bilayer. Highly ion-specific, valinomycin is used in *ion-selective electrodes*.

van der Waals' attraction Electrodynamic forces arise between atoms, molecules and assemblies of molecules due to their vibrations giving rise to electromagnetic interactions; these are attractive when the vibrational frequencies and absorptions are identical or similar, repulsive when non-identical. Other interactions originally proposed by van der Waals were included in this name, but these are usually separated into the Coulomb force, the Keesom force and the London force. Only the last is of electrodynamic nature. Probably important in holding lipid membranes into that structure and possibly in other interactions, eg. cell adhesion. Electrodynamic forces between large-scale assemblies can be of relatively long-range nature.

vanadate The ion, VO_4, which is believed to be the extracellular form of free (pentavalent) vanadium in animals. Vanadate inhibits a large number of enzymes *in vitro*, apparently behaving as an analogue of phosphate. The microtubule-associated ATPase *dynein* is especially sensitive. Supplied extracellularly to intact animal cells however, most of these ATPases are unaffected, and selective effects sometimes resembling those of insulin or growth factors have been observed.

variable antigen Term usually applied to the surface antigens of those parasitic or pathogenic organisms which can alter their antigenic character to evade host immune responses. (See *antigenic variation*).

variable gene See *V-region*.

variable region See *V-region*.

Varicella zoster Member of the Alphaherpesvirinae; human herpes simplex virus type 3, causative agent of chickenpox and *shingles*.

variola virus Virus responsible for smallpox. Said to have been completely eradicated. Large DNA virus ("brick-like", 250–390nm × 20–260nm) with complex outer and inner membranes (not derived from plasma membrane of host cell).

vascular bundle Strand of vascular tissue in a plant; composed of xylem and phloem.

vascularisation Growth of blood vessels into a tissue with the result that the oxygen and nutrient supply is improved. Vascularisation of tumours is usually a prelude to more rapid growth and often to metastasis; excessive vascularisation of the retina in diabetic retinopathy can lead indirectly to retinal detachment. Vascularisation seems to be triggered by "angiogenesis factors" which stimulate endothelial cell proliferation and migration. See *angiogenin, tumour angiogenesis factor*).

vasculitis Inflammation of the blood vessel wall. May be caused by immune complex deposition in or on the vessel wall.

vasoactive intestinal peptide (VIP) Peptide of 28 amino acids, originally isolated from porcine intestine, but later found in the central nervous system where it acts as a neuropeptide, and is released by specific *interneurons*. May also affect behaviour of cells of the immune system.

vasopressin (antidiuretic hormone; ADH) A peptide hormone released from the posterior pituitary lobe but synthesised in the hypothalamus. There are two forms, differing only in the amino acid at position 8: arginine vasopressin is widespread, while lysine vasopressin is found in pigs. Has antidiuretic and *vasopressor* actions. Used in the treatment of *diabetes insipidus*.

vasopressor Any compound which causes constriction of blood vessels (vasoconstriction) thereby causing an increase in blood pressure.

vector 1. Mathematical term implying both magnitude and direction (as opposed to scalars which have only magnitude). 2. Something which transfers things from one place to another. For example, mosquitoes are the vectors of *malaria* parasites, *plasmids* are vectors transferring DNA from cell to cell. Directionality is implicit in both cases.

vectorial synthesis Term usually applied to the mode of synthesis of proteins destined for export from the cell. As the protein is made it moves (vectorially) through the membrane of the rough endoplasmic reticulum, to which the ribosome is attached, and into the cisternal space.

vectorial transport Transport of an ion or molecule across an epithelium in a certain direction (eg. absorption of glucose by the gut). Vectorial transport implies a nonuniform distribution of *transport proteins* on the plasma membranes of two faces of the epithelium.

vegetal pole Of vertebrate eggs. The surface of the egg opposite to the animal pole. Usually the cytoplasm in this region is incorporated into future endoderm cells.

veiled cell A cell type found in afferent lymph and defined (rather unsatisfactorily) on the basis of its morphology. Probably an accessory cell migrating from the periphery (where it is referred to as a *Langerhans cell* if in the skin) to the draining lymph node. In the lymph node known as an interdigitating cell and found in the T-dependent areas of spleen or lymph nodes, involved in antigen presentation (Class II MHC positive). Has high levels of surface Ia antigens.

vein 1. Blood vessel which returns blood from the microvasculature to the heart; walls thinner and less elastic than those of artery. 2. Thickened portion of leaf containing *vascular bundle*; the pattern, venation, is characteristic for each species.

venom A toxic secretion in animals which is actively delivered to the target organism, either to paralyse or incapacitate or else to cause pain as a defence mechanism. Commonly includes protein and peptide toxins.

verapamil A calcium-channel blocking drug (MW 454), used as a coronary vasodilator and anti-arhythmic.

vesicle A closed membrane shell, derived from membranes either by a physiological process (budding) or mechanically by soni-cation. Vesicles of dimensions in excess of 50nm are believed to be important in intracellular transport processes. See also *coated vesicle*.

vesicular stomatitis virus *Rhabdovirus* causing the disease "soremouth" in cattle. Widely used as a laboratory tool especially in studies on the spike glycoprotein as a model for the synthesis, post-translational modification and export of membrane proteins.

vessel Water-conducting system in the *xylem*, consisting of a column of cells (vessel elements) whose end-walls have been totally degraded, resulting in an uninterrupted tube.

vessel element Part of a *xylem* vessel in a higher plant, arising from a single cell. The end-walls are perforated and may completely disappear, giving rise to a continuous tube. The remaining walls are thickened and lignified, and there is no protoplast.

viability test Test to determine the proportion of living individuals, cells or organisms, in a sample. Viability tests are most commonly performed on cultured cells and usually depend on the ability of living cells to exclude a dye such as *Trypan blue* (an exclusion test), or to specifically take it up (inclusion test).

Vibrio cholerae Bacterium which causes cholera, the life-threatening aspects of which are due to the exotoxin (see *cholera toxin*). Short, slightly curved rods, highly motile (single polar flagellum), Gram negative. Adhere to intestinal epithelium (adhesion mechanism unknown), and produce enzymes (neuraminidase, proteases) which facilitate access of the bacterium to the epithelial surface.

vidarabine (adenine arabinoside; Ara-A) Nucleoside analogue with antiviral properties which has been used to treat severe herpesvirus infections.

villin Microfilament-severing and -capping protein (95kD) from microvillar core of intestinal epithelial cells. Severs at high calcium concentrations, caps at low.

vimentin *Intermediate filament* protein (58kD) found in mesodermally-derived cells (including muscle).

vinblastine Alkaloid (818D) isolated from

Vinca (periwinkle): binds to *tubulin* heterodimer and induces formation of paracrystals rather than tubules. Net result is that *microtubules* disappear as they disassemble and are not replaced. Used in tumour chemotherapy.

vinca alkaloids See *vinblastine* and *vincristine*.

vincristine Cytotoxic alkaloid which binds to *tubulin* and interferes with *microtubule* assembly. See also *vinblastine*.

vinculin Protein (130kD) isolated from muscle (cardiac and smooth), fibroblasts and epithelial cells. Associated with the cytoplasmic face of *focal adhesions*: may connect microfilaments to plasma membrane integral proteins through *talin*.

VIP See *vasoactive intestinal peptide*.

viral antigens Those antigens specified by the viral genome (often coat proteins) which can be detected by a specific immunological response. Often of diagnostic importance.

viral transformation Malignant transformation of an animal cell in culture, induced by a virus.

virgin lymphocyte A lymphocyte which has not, and whose precursors have not, encountered the antigenic determinant for which it possesses receptors.

virion A single virus particle, complete with coat.

viroids Extremely small viruses of plants. Their genome is a 240–350 nucleotide circular RNA strand, extensively base-paired with itself, so they resist RNAase attack. At one time the term was also used casually of self-replicative particles such as the *kappa particle* in *Paramecium*.

virus Viruses are obligate intracellular parasites of living but non-cellular nature, consisting of DNA or RNA and a protein coat. They range in diameter from 20–300nm. Class I viruses (Baltimore classification) have double-stranded DNA as their genome; Class II have a single-stranded DNA genome; Class III have a double-stranded RNA genome; Class IV have a positive single-stranded RNA genome, the genome itself acting as mRNA; Class V have a negative single-stranded RNA genome used as a template for mRNA synthesis, and Class VI have a positive single-stranded RNA genome but with a DNA intermediate not only in replication but also in mRNA synthesis. The majority of viruses are recognised by the diseases they cause in plants, animals and prokaryotes. Viruses of prokaryotes are known as bacteriophages.

viscoelastic Of substances or structures showing non-Newtonian viscous behaviour, ie. elastic and viscous properties are demonstrable in response to mechanical shear.

viscous–mechanical coupling Method by which adjacent *cilia* are synchronised in a field. Coupling is through the transmission of mechanical forces, rather than of a synchronising signal.

visinin Now called *calbindin-D-28k*.

Visna–maedi virus A retrovirus of sheep and goats. A member of the *Lentivirinae* related to *HIV*. First identified in Iceland when it was introduced by sheep imported from Germany, and causes two diseases; the most common, maedi, is a pulmonary infection (maedi is Icelandic for shortness of breath) and when it infects the nervous system, visna, a paralysis similar to multiple sclerosis (visna is Icelandic for wasting).

vital stain A stain which is taken up by live cells and which can be used to stain, for example, a group of cells in a developing embryo in order to try to determine a *fate map*.

vitamins Low-molecular weight organic compounds of which small amounts are essential components of the food supply for a particular animal or plant. For humans vitamin A, the B series, C, D1 and D2, E and K are required. Deficiencies of one or more vitamins in the nutrient supply result in deficiency diseases. See Table V1.

vitelline layer The membrane, usually of protein fibres, immediately outside the plasmalemma of the ovum and the earlier stages of the developing embryo. Its structure and composition vary in differing animal groups.

vitellogenin A protein, precursor of several yolk proteins, especially phosvitin and lipovitellin in the eggs of various vertebrates, synthesised in the liver cells after oestrogen stimulation. Also found in large amounts in the haemolymph of female insects, synthesised and released from the fat-body during egg-formation.

vitronectin Serum protein (70kD) also known as serum spreading factor from its activity in promoting adhesion and spreading of tissue cells in culture. Contains the cell-binding sequence Arg-Gly-Asp (RGD) first found in *fibronectin*.

vitrosin Old term for collagen isolated from embryonic chick vitreous. Synthesised by neural retina at early developmental stages, and by cells of the vitreous body later.

VLAs (very late antigens) VLA-1 and VLA-2 were originally defined as antigens appearing on the surfaces of T-lymphocytes 2–4 weeks after *in vitro* activation; they are now known to be part of the *integrin* family. Additional members of the subset are now known (VLA-3, VLA-4 and VLA-5), the β subunits all being identical. Some of the VLA proteins are receptors for collagen, laminin or fibronectin, and many are now known to be expressed on cells other than leucocytes.

V_{max} The maximum initial velocity of an enzyme-catalyzed reaction, ie. at saturating substrate levels.

voltage clamp A technique in electrophysiology, in which a *microelectrode* is inserted into a cell, and current injected through the electrode so as to hold the cell's *membrane potential* at some predefined level. The technique can be used with separate electrodes for voltage-sensing and current-passing; for small cells, the same electrode can be used for both. Voltage clamp is a powerful technique for the study of *ion channels*. See *patch clamp*.

voltage gated ion-channel A transmembrane *ion channel* whose permeability to ions is extremely sensitive to the transmembrane potential difference. These channels are essential for neuronal signal transmission, and for intracellular signal transduction. See *sodium channel*.

voltage gradient Literally, the electric field in a region, defined as the potential difference between two points divided by the distance between them. Used more loosely, the potential difference across a plasma membrane.

Volvox A genus of colonial flagellates. The colony is a hollow sphere about 0.5mm in diameter comprising about 50,000 cells embedded in a gelatinous wall and the cells are sometimes connected by cytoplasmic

Table V1. Vitamins

Vitamin	Full name	Occurrence	Action	Deficiency disease
Fat soluble				
A	Retinol (11-cis retinal)	Vegetables	Phototransduction Morphogen	Night blindness Xerophthalmia
D	1,25-dihydroxycholecalciferol	Action of sunlight on 7-dehydrocholesterol in skin	Ca^{2+} regulation Phosphate regulation	Rickets
E	α-tocopherol	Plants, esp. seeds Wheatgerm	Antioxidant	Failure to grow to maturity Infertility
K_1 K_2	Range of molecules	Higher green plants Intestinal bacteria		
Water soluble				
B_1	Thiamine Folic acid (Tetrahydrofolic acid) Nicotinic acid (niacin) Pantothenic acid (CoA)	Plants (eg. rice husk) Plants Can be made Plants and micro-organisms	Degradation of α-keto acids Purine biosynthesis	Beriberi Anaemia Pellagra
B_2 B_6	Riboflavine Pyridoxine Pyridoxal Pyridoxamine	Plants and micro-organisms Plants	Constituent of flavoproteins Transamination	Acrodynia in rats Convulsions
B_{12} C H	Cobalamine Ascorbic acid Biotin	Intestinal micro-organisms Plants, esp. citrus fruits Intestinal bacteria	Hydrogen transfer reactions Cofactor Protects against avidin toxicity Intermediate CO_2 carrier	Pernicious anaemia Scurvy

bridges. Each cell has a *chloroplast* and two *flagella*.

von Willebrand factor Plasma factor involved in platelet adhesion through an interaction with Factor VIII. See *von Willebrand's disease*.

von Willebrand's disease Autosomal dominant platelet disorder in which adhesion to collagen, but not aggregation, is reduced. Both bleeding time and coagulation are increased. Factor VIII levels are secondarily reduced.

Vorticella Genus of ciliate Protozoa. It has a bell-shaped body with a belt of cilia round the mouth of the bell to sweep food particles towards the "mouth", and a long stalk which contains the contractile *spasmoneme*.

W

warm antibodies Most IgG antibodies react better at 37°C than at lower temperatures, especially against red cell antigens. These are the warm antibodies as contrasted with *cold agglutinins*, especially IgM, which agglutinate below 28°C.

wart Benign tumour of basal cell of skin, the result of the infection of a single cell with wart virus (*Papillomavirus*). Virus is undetectable in the basal layer, but proliferates in keratinising cells of outer layers.

water potential The chemical potential (ie. free energy per mole) of water in plants. Water moves within plants from regions of high water potential to regions of lower water potential, ie. down-gradient.

weal-and-flare See *triple response*.

Weibel–Palade body Cytoplasmic organelle found in the vascular endothelial cells of some animals, though not in the endothelium of all vessels. Although a marker for endothelium, its absence does not necessarily mean the cells are not of endothelial origin.

Weismann's germ plasm theory The theory that organisms maintain genetic continuity from organism to offspring through the germ line cells (germ plasm) and that the other (somatic) cells play no part in the transmission of heritable factors.

Western blotting An electroblotting method in which proteins are transferred from a gel to a thin, rigid support (nitrocellulose) and detected by binding of radioactive antibody. See *blots*.

wheat germ The embryonic plant at the tip of the seed of wheat. Wheat germ has been used as the starting material for a cell-free translation system and is also the source of *wheat germ agglutinin*.

wheat germ agglutinin (WGA) Lectin from wheat germ which binds to N-acetyl glucosaminyl and neuraminic acid residues. See *lectins*.

white blood cells See *leucocytes*, and specific classes (*basophils, eosinophils, haemocytes, lymphocytes, monocytes, neutrophils*).

whole cell patch A variant of *patch clamp* technique, in which the patch electrode seals against the cell, with direct communication between the interior of the electrode and the cytoplasm.

Wiskott–Aldrich syndrome *Thrombocytopenia* with severe immunodeficiency (both cell-mediated and IgM production). Associated with increased incidence of leukaemia.

X

X chromosome A sex chromosome. In mammals paired in females, in birds paired in males.

X-inactivation The inactivation of one or other of each pair of X chromosomes to form the Barr body in female mammalian somatic cells. Thus tissues whose original zygote carried heterozygous X-borne genes should have individual cells expressing one or other but not both of the X-borne gene products. The inactivation is thought to occur early in development and leads to mosaicism of expression of such genes in the body. See also *Lyon hypothesis*.

X-linked diseases Any inherited disease whose controlling gene is carried on an X chromosome, eg. haemophilia. Most known conditions are recessive and thus since males have only one X chromosome they will express any such recessive character. Female expression of such diseases is uncommon because of the relative rarity of homozygotes.

X-ray diffraction Basis of powerful technique for determining the three-dimensional structure of molecules, including complex biological macromolecules such as proteins and nucleic acids, which form crystals or regular fibres. Low-angle X-ray diffraction is also used to investigate higher levels of ordered structure, as found in muscle fibres.

X-ray microanalysis See *electron microprobe*.

xanthine (2,6-dihydroxypurine) A purine, the starting point for purine degradation. Its methylated derivatives (theophylline, theobromine, caffeine) are potent cAMP phosphodiesterase inhibitors.

xanthine oxidases Dehydrogenases involved in conversion of hypoxanthine to xanthine, and xanthine to uric acid, as the final catabolism of purines. Deficient in the human disease xanthinuria.

xanthophylls *Carotenoid* pigments involved in photosynthesis. Consist of oxygenated carotenes, eg. lutein, violaxanthin and neoxanthine.

xenogeneic Literally, of foreign genetic stock; usually applied to tissue or cells from another species, as in xenogeneic transplantation.

xenograft A graft between individuals of unlike species, genus or family.

Xenopus The genus of African clawed toads; *X. laevis* is widely used in developmental biology and was formerly used in pregnancy diagnosis because the female ovulates easily under influence of luteinising hormone.

xenotropic virus A virus which can be grown on cells of a species foreign to the normal host species.

xeroderma pigmentosum Inherited (autosomal recessive) disease in humans associated with increased sensitivity to ultraviolet-induced mutagenesis, and thus to skin cancer. Sensitivity can be demonstrated in cultured cells, and appears to be due to deficiency in DNA repair, specifically in excision of ultraviolet-induced *thymine dimers*.

xylan Plant cell wall polysaccharide containing a backbone of $\beta(1–4)$-linked xylose residues. Side-chains of 4-O-methylglucuronic acid and arabinose are present in varying amounts (see *glucuronoxylan* and *arabinoxylan*), together with acetyl groups. Found in the *hemicellulose* fraction of the wall matrix.

xylem Plant tissue responsible for the movement of water and inorganic solutes from the roots to the shoot and leaves. Contains *tracheids*, *vessels*, *fibre cells* and *parenchyma*. Also provides structural support for the plant, especially in wood.

xyloglucan Plant cell-wall polysaccharide containing a backbone of $\beta(1–4)$-linked glucose residues to most of which single xylose residues are attached as side chains. Galactose, fucose and arabinose may also be present in smaller amounts. It is the major hemicellulose of dicotyledonous primary walls, and acts as a food reserve in some seeds.

xylose Monosaccharide pentose which is found in xylans, very abundant components of hemicelluloses.

xylulose A 5-carbon ketose sugar, whose 5-phosphate is an intermediate in the *pentose phosphate pathway* and the *Calvin–Benson cycle*.

Y

Y chromosome Chromosome found only in the heterogametic sex. Thus in mammals the male has one Y chromosome and one X chromosome. One region of the Y chromosome, the pseudoautosomal region, is homologous to and pairs with the X chromosome. The primary determinant of male sexual development is found on the unpaired, differentiated segment of the Y chromosome.

yabavirus A poxvirus of African monkeys causing benign tumours.

yeast Yeast is the colloquial name for members of the fungal families, ascomycetes, basidiomycetes and imperfect fungi, which tend to be unicellular for the greater part of their life cycles. Commer-cially important yeasts include *Saccharomyces cerevisiae*; pathogenic yeasts include the genus *Candida*.

yellow fever virus A togavirus (Class IV) with an RNA genome responsible for the disease of this name whose symptoms include fever and haemorrhage. About 10% of victims die. Transmitted by the mosquitoes *Aedes aegypti* and *Haemagogus*. Only one antigenic type of the virus known.

Yersinia Genus of Gram negative bacteria; all are parasites or pathogens. *Y. pestis* (formerly *Pasteurella pestis*) was the cause of the Black Death plague.

yolk cells In those eggs in which the yolk is not distributed evenly (telolecithal eggs) the cells formed when cleavage reaches the yolk region can be termed yolk cells.

yolk sac One of the set of extra-embryonic membranes, growing out from the gut over the yolk surface, in birds formed from the splanchnopleure, an outer layer of splanchnic mesoderm and an inner layer of endoderm.

Z

Z scheme of photosynthesis A schematic representation of the *light dependent reactions* of *photosynthesis*, in which the photosynthetic reaction centres and electron carriers are arranged according to their electrochemical potential (free energy) in one dimension and their reaction sequence in the second dimension. This gives a Z-shape, the two reaction centres (of photosystems I and II) being linked by the photosynthetic electron transport chain.

Z-disc Region of the *sarcomere* into which *thin filaments* are inserted. Location of α-*actinin* in the sarcomere.

Z-DNA Form of DNA adopted by sequences of alternating *purines* and *pyrimidines*. It is a left-handed helix with the phosphate groups of the backbone zigzagged (hence Z) and a single deep groove. It is still not clear whether Z-DNA occurs in genomic DNA.

Z-line See *Z-disc*.

zeatin A naturally-occurring *cytokinin*, originally isolated from maize seeds. Its riboside is also a cytokinin.

Zigmond chamber See *orientation chamber*.

zippering Process suggested to occur in phagocytosis in which the membrane of the phagocyte covers the particle by a progressive adhesive interaction. The evidence for such a mechanism comes from experiments in which capped B-cells are only partially internalised, whereas those with a uniform opsonising coat of antibody are fully engulfed.

zona pellucida A translucent non-cellular layer surrounding the ovum of many mammals.

zone of polarising activity The small group of mesenchyme cells in avian limb buds that is located at the posterior margin of the developing bud and which produces a substance, possibly retinoic acid, that pro-vides positional information to the developing limb bud.

zonula adhaerens Specialised intercellular junction in which the membranes are separated by 15–25nm, and into which are inserted microfilaments. Similar in structure to two apposed *focal adhesions*, though this may be misleading. Microfilaments inserted into the zonula adhaerens may interact (*via* myosin) with other microfilaments to generate contraction. Constitute mechanical coupling between cells.

zonula occludens (tight junction) Specialised intercellular junction in which the two plasma membranes are separated by only 1–2nm. Found near the apical surface of cells in simple epithelia; forms a sealing "gasket" around the cell. Prevents fluid moving through the intercellular gap and the lateral diffusion of intrinsic membrane proteins between apical and baso-lateral domains of the plasma membrane.

zovirax (acyclovir) Nucleoside analogue (hydroxyethoxymethyl-guanine) with antiviral properties; active against both type 1 and 2 herpes virus. Inactive until phosphorylated by specific viral enzyme, *thymidine kinase*, and then blocks replication.

ZPA See *zone of polarising activity*.

zwitterion A molecule carrying a positive charge at one end and a negative charge at the other. Also known as ampholyte or dipolar ion.

zygonema *Zygotene*.

zygospore Fungal spore produced by the fusion of two similar *gametes* or *hyphae*.

zygote *Diploid cell* resulting from the fusion of male and female gametes at *fertilisation*.

zygotene Classical term for the second stage of the *prophase* of *meiosis* I, during which the homologous chromosomes start to pair.

zymogen Inactive precursor of an enzyme, particularly a proteolytic enzyme. Synthesised in the cell and secreted in this safe form, then converted to the active form by limited proteolytic cleavage.

zymogen granule *Secretory vesicle* containing an inactive precursor (*zymogen*). The material is often very condensed.

zymogenic cells Cells of the basal part of the gastric glands of the stomach. They contain extensive rough endoplasmic reticulum and zymogen granules and secrete pepsinogen, the inactive precursor of *pepsin*, and rennin.

zymosan Particulate yeast cell-wall polysaccharide (mannan-rich) from *Saccharomyces cerevisiae* which will activate *complement* in serum through the alternate pathway. Becomes coated with C3b/C3bi and is therefore a convenient opsonised particle; also leads to C5a production in the serum.

Appendix

Prefixes

kilo	10^3
hecto	10^2
deca	10^1
deci	10^{-1}
centi	10^{-2}
milli	10^{-3}
micro	10^{-6}
nano	10^{-9}
pico	10^{-12}
femto	10^{-15}
atto	10^{-18}

Greek Alphabet

A	α	alpha
B	β	beta
Γ	γ	gamma
Δ	δ	delta
E	ε	epsilon
Z	ζ	zeta
H	η	eta
Θ	θ	theta
I	ι	iota
K	κ	kappa
Λ	λ	lambda
M	μ	mu
N	ν	nu
Ξ	ξ	xi
O	o	omicron
Π	π	pi
P	ρ	rho
Σ	σ	sigma
T	τ	tau
Y	υ	upsilon
Φ	φ	phi
X	χ	chi
Ψ	ψ	psi
Ω	ω	omega

Appendix

Useful constants

Avagadro's number (N) 6.022×10^{23} mol^{-1}
Boltzmann's constant (k) 1.381×10^{-23} J deg^{-1}
3.298×10^{-24} cal deg^{-1}
Faraday constant (F) 9.649×10^4 Coulomb mol^{-1}
Curie (Ci) 3.7×10^{10} disintegrations sec^{-1}
Gas constant (R)   8.314 J mol^{-1} deg^{-1}
π 3.14159
e   2.71828
$\log^e x = 2.303 \log^{10} x$

Single letter code for amino acids

A	Ala	Alanine
R	Arg	Arginine
N	Asn	Asparagine
D	Asp	Aspartic acid
B		Asparagine or aspartic acid
C	Cys	Cysteine
Q	Glu	Glutamine
E	Glu	Glutamic acid
Z		Glutamine or Glutamic acid
G	Gly	Glycine
H	His	Histidine
I	Ileu	Isoleucine
L	Leu	Leucine
K	Lys	Lysine
M	Met	Methionine
F	Phe	Phenylalanine
P	Pro	Proline
S	Ser	Serine
T	Thr	Threonine
W	Trp	Tryptophan
Y	Tyr	Tyrosine
V	Val	Valine

Revision form

The compilers invite your comments for incorporation into future, revised editions of this dictionary.

Please send completed forms or other communications regarding dictionary entries to:

Dr J M Lackie,
(Dictionary of Cell Biology)
Department of Cell Biology,
The University,
Glasgow, G12 8QQ
Scotland

It would be helpful if new entries followed the current brief format, but brief additional information would also be appreciated. The information must be legible and, most importantly, it must be possible to cross-check in published sources.

New Entry [   ] / Revised Entry [   ]

Entry word:_____   (Synonyms_____)

_____

_____

_____

_____

Cross-references to:_____, _____, _____ .

Add to Table   Y / N   (if Y, which?_____)

Example of usage:

_____

_____

Reference:

_____

Name_____

Full address _____

_____

_____

Postcode_____